Molecular Evolution and Population Genetics for Marine Biologists

Molecular Evolution and Population Genetics for Marine Biologists

Yuri Ph. Kartavtsev
A.V. Zhirmunsky Institute of Marine Biology
Far Eastern Branch of Russian Academy of Sciences; Far Eastern Federal University
Vladivostok
Russia

Edited by

M.S. Johnson
The University of Western Australia
Crawley
Perth

CRC Press is an imprint of the
Taylor & Francis Group, an **informa** business
A SCIENCE PUBLISHERS BOOK

Cover illustration reproduced by kind courtesy of Dr. A. Chichvarkhin

CRC Press
Taylor & Francis Group
6000 Broken Sound Parkway NW, Suite 300
Boca Raton, FL 33487-2742

First issued in paperback 2020

ISBN-13: 978-1-4987-0160-0 (hbk)
ISBN-13: 978-0-367-73788-7 (pbk)

Library of Congress Cataloging-in-Publication Data

Kartavtsev, Yuri Ph.
Molecular evolution and population genetics for marine biologists / Yuri Ph. Kartavtsev.
pages cm
"A CRC title."
Includes bibliographical references and index.
ISBN 978-1-4987-0160-0 (hardcover : alk. paper) 1. Population genetics. 2. Marine biology. I. Title.

QH455.K37 2016
576.5'8--dc23 2015021993

Visit the Taylor & Francis Web site at
http://www.taylorandfrancis.com

and the CRC Press Web site at
http://www.crcpress.com

PREFACE

This course originated from more than 40 years of personal research experience in the Population, Evolutionary, and Ecological Genetics of marine fish and shellfish along with teaching of upper division students at the Far Eastern Federal University in Russia (Vladivostok) and at two national universities in Republic of Korea: Korea Maritime University (Busan) and Chungbuk National University (Cheongju), i.e., more than 30 years in Russia and two years in Korea. In recent years the course was adapted for students' convenience to modern computer and web-net technologies, including MS Power-Point mode for lecture presentations, usage of different software, such as POPULUS, MEGA, BYOSIS, NTSYS, GENEPOP, ARLEQUIN, STATISTICA etc. for training courses (with numerical simulations and statistical analysis), and the website support for both. So, the lectures and training courses have a modern shape, and are easily updated each term. The lectures are normally given in a fluent verbal mode, and they are accompanied with questions to the audience, giving the students exact tasks to solve examples, and find solutions through special classroom sessions, seminars and preparing the abstracts. Each lecture is accompanied by a training section, where new terms are summarized and explained. In the book all these sections are combined in a special appendix. There is also an appendix in which methodical instruction for learning of the course, literature sources, and exam questions are given.

Originally each lecture presentation together with the training task was accompanied with the written text. These materials now form the foundation of the current textbook. The book contains 16 chapters, which are the foundation of the course. The material is given in accordance to its complexity, giving readers the opportunity to remember the basic genetic knowledge and later develop these skills within specialized fields and specific problems in immunogenetics, biochemical genetics, molecular genetics, general biology, ecology, evolution and speciation.

It is assumed that a student has basic comprehension of general genetics. That is why only two themes are devoted to basics of genetics: Material basis of genetics (Chapter 2) and Translation of genetic information (Chapter 3). Basic knowledge in genetics is partly covered in the other chapters. In particular, details of the genetic code and mutations, which have direct connection to the course, are given. Also, such questions as gene transcription, splicing, gene origin and several other notions are considered to a certain extent. The book is oriented not only towards students but may be of interest to a wider audience of specialists who are interested in genetics, population ecology, taxonomy and speciation.

The author is cordially thankful to those who inspired the book, Olga Koren and Alla Kholina, suggested publication as an integrated material after attending the course. My special warm thanks to the editors of the first, and second editions in Russian, Drs. Irina Kartavtseva and Olga Koren. I am also pleased to cordially thanks the first readers of the English version of the book, Dr. Cynthia Riginos, who proofread several beginning chapters, and Elena Cogan, who helped with editing of some chapters' summaries, table headings and figure captions.

ACKNOWLEDGEMENTS

The preparation of this book has been possible through the funding provided to the author by the A.V. Zhirmunsky Institute of Marine Biology Far Eastern Branch of Russian Academy of Sciences and Far Eastern Federal University (Vladivostok) where the author works. That is the reason the author shares part of intellectual property with the listed institutions. The author also gratefully acknowledges the financial support received from the Russian Science Foundation (grant #14-50-00034 *Molecular Phylogenetics and Biodiversity*), and in part from the Russian Foundation for Basic Research (grant #15-29-02456 *Application to DNA Barcoding, Population Genetics and Gene Introgression between Species*). Editorial help provided by Dr. M. Johnson is gratefully acknowledged.

INTRODUCTION

Population genetics and molecular evolutions may seem too dissimilar themes to be united in a single book. However, that is not so. Modern experimental and theoretical population genetics got great impulse from the development of molecular methods for the analysis of genetic variability. On the other hand, investigation of divergence of organisms over time or evolution at the molecular level is unimaginable without understanding the fundamental genetic properties of organisms themselves and their groups, which comprise in nature such reproductive units as local populations and biological species.

Population dynamics in time cannot be separated from spatial dynamics, from understanding the basics of intraspecific genetic variability and differentiation. Carried away by the vast abilities of molecular phylogenetic analysis of sequences of deoxy ribonucleic acid (DNA), some authors even completely refuse the necessity of spatial diversity analysis, contra-positioning the phylogenetic species concept to the biological species concept. However, many geneticists do not hold such extreme views, understanding the unity of mechanisms of intra- and interspecific divergence (Altukhov 1983; 1989; Ayala 1984; Nei 1987; Avise and Wollenberg 1997). These questions as well as many others are considered in the current book in quite a popular form, presenting information to the student audience and a wide range of non-expert readers. At the same time, the majority of the book chapters are short assignments that may be interesting to the experts in certain fields. The author of this textbook is a marine biologist. This peculiarity places its own perspective on the content, relating many examples to the marine realm, particularly by using fish and shellfish data; the latter led to the title of the book.

Chapter 1 is devoted to the history of the Earth, organic evolution and the origin of life, while ending the book, Chapter 16 is about the evolution of the genome. The basics of genetic knowledge are repeated in Chapters 2 and 3. Intense research on the molecular basis of evolution is introduced by developing a broad introduction to the relevant biological methods, firstly immunogenetic, then biochemical genetics, and finally molecular genetic methods of analysis of different macromolecules. Immunogenetic methods developed in the mid 20th century, and entered evolutionary genetics by quantifying genetic divergence by microcomplement fixation of purified protein. The protein used for this reaction is serum albumin. In connection with this, the main concepts of immunity and immunogenetics are given; also, some materials on population genetic and molecular phylogenetic analysis of certain groups of organisms are presented (Chapter 4). Biochemical genetics (Chapter 5) in the mid

1960s became an experimental base for modern population genetics of natural stocks. Proteins as a rule are enzymes, and are used as gene markers even today. They are competitive to more modern DNA markers by their ability to be used in vast population analysis and low cost. Databases that gathered in this field, which first of all included heterozygosity and distance/similarity records, have reasonable significance even now for comparative analysis of genetic variability and divergence that is very important for evolutionary, ecological and population genetics. In particular, in their relationship to protein variability many issues are considered regarding mutations (Chapter 6), genetic aspects of speciation (Chapter 7), population genetic structure of the species (Chapter 8), analysis of genetic diversity within the species (Chapter 9), hybridization and genetic introgression in nature (Chapter 10), heterosis and its connection with heterozygosity, as well as interrelationships of heterozygosity vs. quantitative traits in general (Chapter 12). Quantitative traits and their inheritance are also considered in a separate chapter (Chapter 11).

With deciphering of the molecular essence of genes, it appeared self-evident that evolutionary relationships of organisms could be investigated by means of comparison of nucleotide sequences of DNA or amino acids in protein chains, which are encoded by DNA. DNA polymorphism within and between populations is considered in two consecutive chapters (Chapters 13–14). The boom that was created by vast opportunities of analysis of divergence in different phyletic lines and taxa of organisms on the base of nucleotide sequences in genes gave rise to a new frontier in evolutionary biology, molecular phylogenetics. Its birth and development gave not only solutions for many hot questions of evolution of the organic world, but have demanded as well the development of new methods of mathematical analysis and software production. This book is not specially addressed to molecular phylogenetics; nevertheless a significant portion of the text is devoted to this field. To address it, the rate of mutation, DNA diversity and tree building methodology are considered. There is special chapter devoted to practical analysis of sequences of nucleotides (also applicable to proteins) and software needed for their editing, submission to gene banks, estimation of models of substitutions and finally gene tree building (Chapter 15). The textbook is equipped with several supplements, which allow the student to get themes and questions for seminars, find a list of questions for the exam, learn terms and solve tasks during special training courses developed for each chapter.

CONTENTS

Sun, and the latter's energy supply. Another source of energy are chemical reactions on the planet, which were originally geo-chemical and later biochemical reactions start their action as well. These later reactions and a biogenic-related circulation of molecules became in Earth's history an essential component of its chemistry in general.

The planet's chemical reactions developed over time, and all these events may be named the evolution of the Earth. During the first steps of Earth's history only inorganic chemical evolution was present, which starts with the transformation of the original inter planet pro-matter into the diversity of minerals, metals, and other associations of atoms and molecules. Later, when water condensation occurred, life originated. Fox and Doze (1972) in their book considered chemical evolution and origin of life. According to these authors, it is believed that chemical evolution started with the creation of the Earth; inorganic molecular evolution predominated from 4.5 up to 4.0 BY ago because most kinds of atoms were already present. When life originated, near 4.0 BY, the organic chemical or biochemical evolution appeared, and it is developing up to present time. Geologists and paleontologists divide all time scale into the certain eras and periods. Details of this division will be given in Section 1.2.

The geological structure of the Earth is also evolving. The properties of the surface, core and the Earth nucleus change in time, and these events cause the mobility of continental platforms that form the surface core of land and World Ocean. Such events determine the changes in the shape of the surface of land mass or continents. This is a geological evolution. Certainly it must be considered when the biological evolution, the evolution of living organisms, is studied in such a changing environment.

It is now generally accepted among geologists that continents on the Earth continually drifted, splitting and merging several times (Calder 1983) in the past. At the time of the Cambrian era, about 550 million years (MY) ago, there were four continents, but by the time of the Permian period, at about 200 MY ago, they had merged and formed a single supercontinent named Pangaea. About 20 MY later, Pangaea began to break into two supercontinents: Laurasia and Gondwanaland (Fig. 1.1.1; The image is from free source, URL: http://pubs.usgs.gov/publications/text/historical.html). Laurasia consisted of the present North America, Greenland, and Eurasia north of the Alps and the Himalayas, whereas Gondwanaland included the present South America, Africa, India, Australia and Antarctica. The super-continents later split into smaller land masses but they stayed relatively close; thus, the rifts between them did not become effective barriers to the movement of land animals until the Cretaceous period. However, by about 75 MY ago, North America was completely separated from Asia, and Gondwanaland split into South America, Australia, Antarctica, etc. Several important aspects of the present geographical distribution of animals and plants can be explained by this continental drift. Continental drift continues to operate today, and even in the Cenozoic era, about 60 MY ago, the land surface was not the same as it is at present time. Information on continental drift has been used by many authors to date the times of divergence among organisms (Table 1.1.1). Maxson and Wilson (1975) constructed the phylogeny of tree frogs from several continents by using the albumin clock (see below on molecular clocks). They showed that South American and Australian species were separated about 75 MY ago. This agrees with the time estimated from geological information on the separation of the two continents. Sibley and Ahlquist (1984) used the time of separation between the ostrich in Africa and the

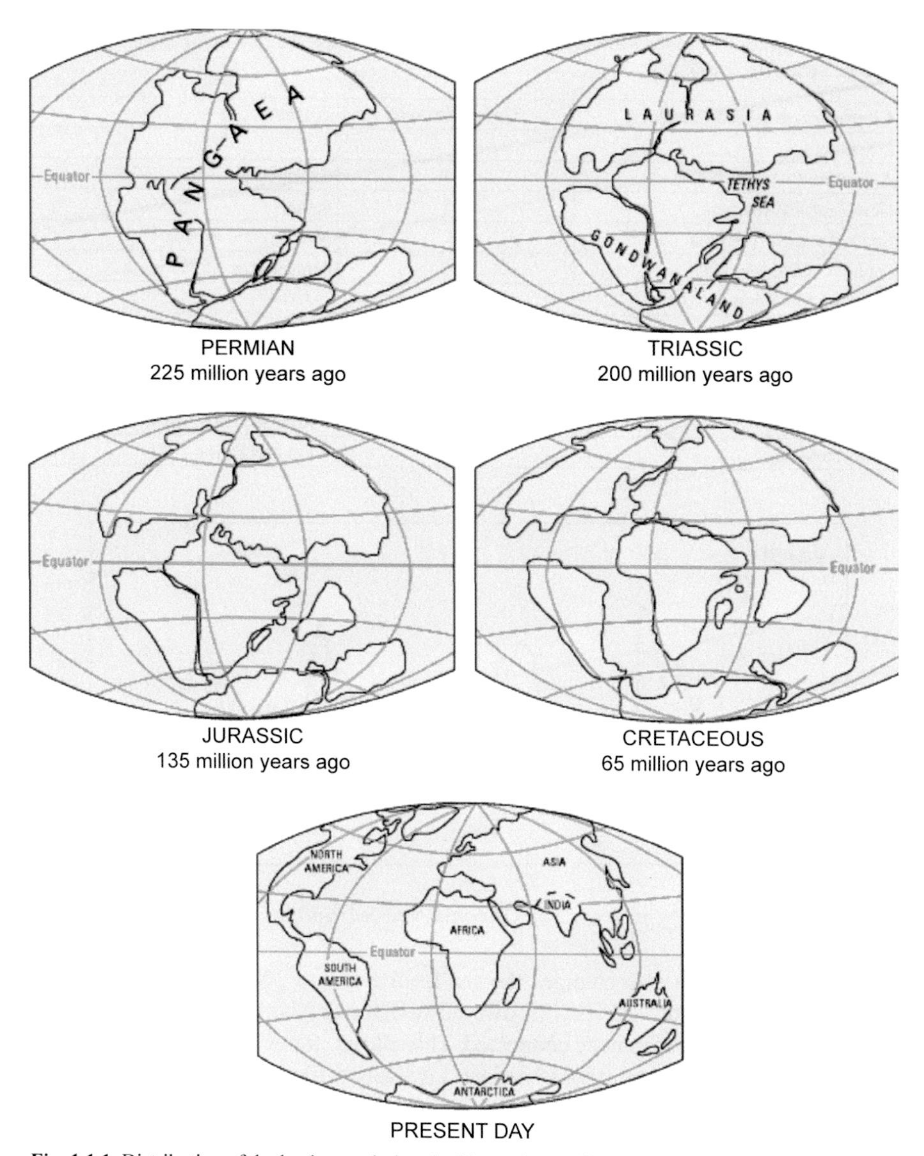

Fig. 1.1.1. Distribution of the land mass during the Mesozoic era (from An On-Line Biology Book 2002; with permission from USGS).

common rhea in South America, which happened about 80 MY ago, to calibrate their molecular clock for DNA hybridization data. Vawter et al. (1980) used the time of the formation of a land bridge between North and South America for calibrating the electrophoretic clock.

Another geological event that is also important for the study of biological evolution is glaciation. In the Earth's history there have been warm and cold periods (Fig. 1.1.2). Glaciation during cold epochs and deglaciation in a warmer time cause drastic changes in planetary climate, which in turn affect the geographical distributions of animals

Table 1.1.1. Times of various geological events that have been used for calibrating molecular clocks (Adopted from Nei 1987).

Continents or islands involved	Time (MY ago)	Authors
I. Separation		
Africa/South America	80	Sibley and Ahlquist 1984
New Zealand/Australia	80	same
South America/Australia	75	Maxson and Wilson 1975
Pacific/Gulf of Mexico	2–5	Vawter et al. 1980
II. Island formation		
Hawaii	0.8	Hunt et al. 1981
Oahu	4	same
Kauai	6	same
Lesser Antilles	3–5	Kirn et al. 1976
Galapagos	0.5–4	Yang and Patton 1981

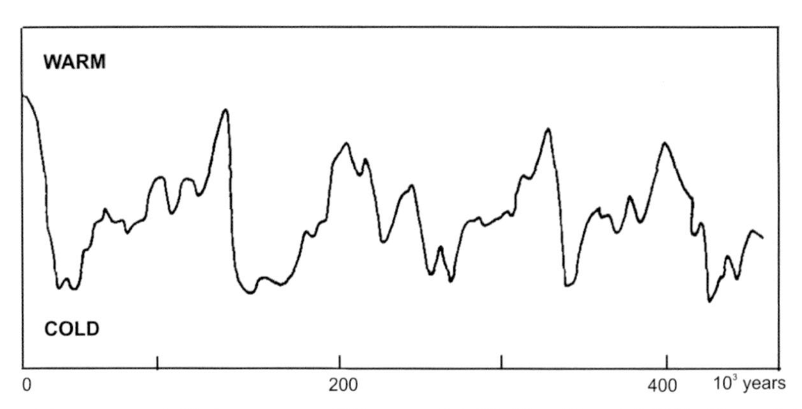

Fig. 1.1.2. Ice ages recorded by heavy oxygen (^{18}O) content (Modified from Hays et al. 1976 with permission).

and plants. Glaciation also changes the sea level down to 200 m (Lindberg 1948). During an ice period, the sea level is lowered so that two islands or continents which were previously isolated may be connected. This allows, for example, land plants and animals to move through previous barriers; inversely different river basins may join due to this low sea level and create hybrid zones or simply extend inhabiting area of a species and give new opportunities for a species evolution. Conversely sea basins may be separated by ace bodies during cold time thus providing barriers to gene flow among the populations, such events may start the formation of new species after a long set of independent existence of descendant generations. If during a glacial period migration between two geographical areas is terminated, that may lead to the formation of new species.

1.2 PALEONTOLOGY DATING OF EVOLUTION

Some conventional dating of appearances of living organism are given in Fig. 1.2.1 (The image is from free source, URL: http://www.clearlight.com/~mhieb/WVFossils/

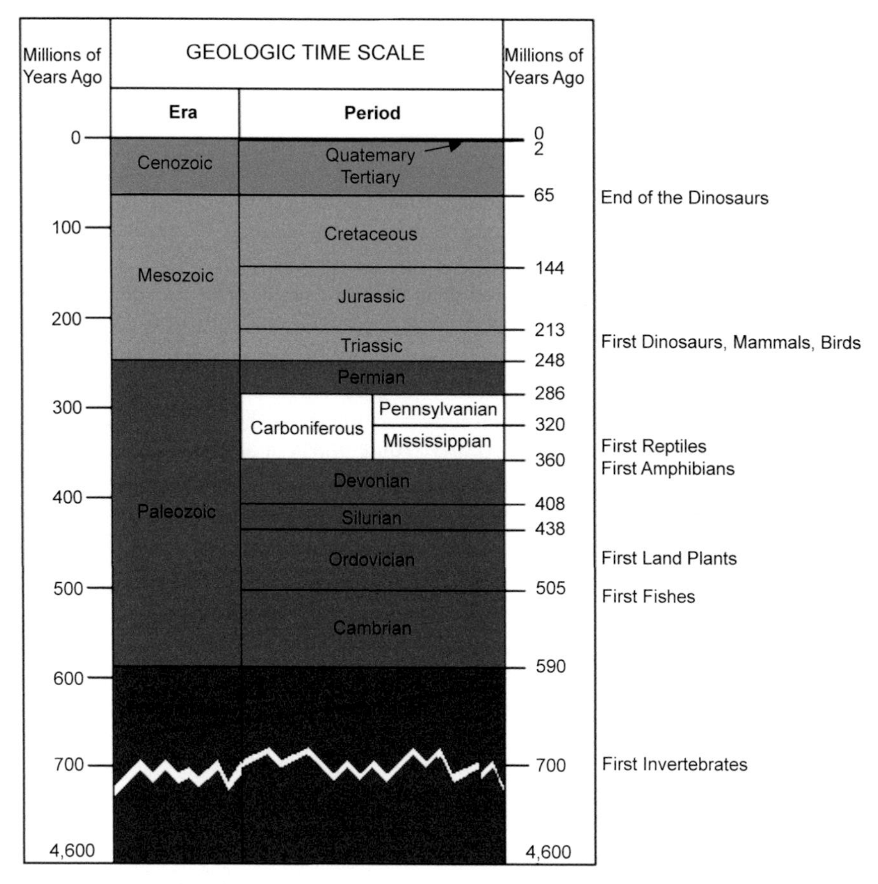

Fig. 1.2.1. Geological time and the early history of life (from An On-Line Biology Book, 2002; with permission from USGS).

GeolTimeScale.html). It is not known exactly when the first life or self-replicating living form was formed. Since Barghoorn and Schopf (1966) reported the discovery of presumed fossilized bacteria, numerous claims of microfossils from the Precambrian period (more than 570 million years ago) have been reported. Although most of them have not been sustained by careful reexamination (Schopf and Walter 1983; Hoffmann and Schopf 1983), the bacteria-like microfossils reported by Awramik et al. (1983), Walsh and Lowe (1985) and later (Science Daily 2011) seem to be authentic; they have been dated 3.5, 3.5 and 3.4 BY old, respectively.

Considering these microfossils and other fossilized organic matter, Schopf et al. (1983) suggested that life probably arose around 3.8 BY ago. By 3.5 BY ago, both anaerobic and photosynthetic bacteria seem to have originated. The next two billion years were the age of prokaryotes. According to Schopf et al. (1983), unicellular mitotic eukaryotes originated around 1.5 BY ago, and the divergence between plants and animals occurred somewhere between 600 MY and 1 BY ago, probably close to the latter time. More dates for recent geological epochs are in Table 1.2.1. Also, some

Table 1.2.1. Scale of a geological time for the Neozoic (Cenozoic) Era.

Period	Epoch	Time for lowest border, Million Years
Quarter, Pleistocene, or Anthropogenic		1.75
Neogene	Pliocene	5.32
	Miocene	23.8
	Oligocene	33.7
Palaeogene	Eocene	54.5 (54.8)
	Paleocene	65.0

phylogenetic relationships among organisms and dating are presented in the next figure (Fig. 1.2.2). Fossil records of Paleontology give rise to many dating points (Fig. 1.2.2). Extensive information on paleontology is available on the website given in the reference list for this course (On-Line Biology Book 2002). A common problem with the phylogenies is that nodes in the tree that denote the phyla origin are inexact (Fig. 1.2.2). Most of them are not known, and are products of different experts' suggestions and reconstructions. These diagrams are largely the products of morphological, developmental and paleontological reconstructions, and are far from precise at many points; in such situations the molecular phylogenetic approach can play an important role both in bringing more precision in dating points and in phylogenetics, giving more knowledge of the genetic kinships of lineages in question. Molecular phylogenetics is a new and world-wide developing science field, incorporating molecular biological knowledge and hardware-and-software achievements. Naturally, since publication

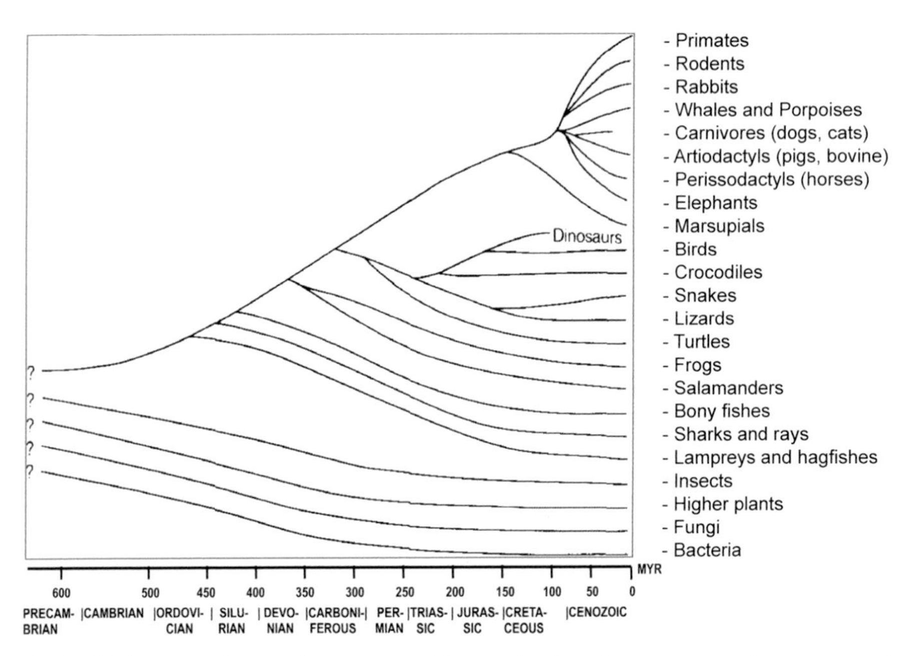

Fig. 1.2.2. Divergence of the vertebrate groups and lineages for a few other groups, according to geological and morphological evidence (MYR, Million of Years; Modified from McLaughlin and Dayhoff 1972).

by McLaughlin and Dayhoff (1972) many nodes of the tree presented (Fig. 1.2.2) have been specified or changed demonstrating the vast utility of molecular genetics and comparative genomics for improving phylogeny of higher taxa (Aleshin 2013; Koonin 2014).

1.3 MOLECULAR BIOLOGICAL DATING

When the molecular basis of genes was elucidated, it became obvious that the evolutionary relationships of organisms can be studied by comparing nucleotide sequences in DNA or amino acid sequences in proteins (Crick 1958). Zuckerkandl and Pauling (1962; 1965) and Margoliash and Smith (1965) later showed that the rate of amino acid substitution in proteins is approximately constant, when time is measured in years. This finding has provided a new method of constructing phylogenetic trees. The principle of constant rate of gene substitution was then extended to RNAs and DNAs, and many authors have used such methods to clarify phylogenetic relationships of many different groups of organisms (Dayhoff 1969; 1972). The phylogenetic trees constructed by these methods are subject to large topological and sampling errors, as well as to gene-dependent systematic biases. Nevertheless the results obtained are frequently consistent with former knowledge. Recent studies indicate that the molecular clocks are not as accurate as was originally expected, but this does not principally affect the suitability of molecular data for constructing phylogenetic trees (Chapter 14).

There is an advantage of molecular methods for phylogenetics, i.e., that the rate of nucleotide or amino acid substitution varies considerably among different genes, and thus it is possible to study both short-term and long-term evolution if a researcher will choose appropriate genes. This is similar to the procedure applied in a radioactive isotope dating, using different radioactive elements with diverse decay rates. The genes for bacterial ribosomal RNA (rRNAs), rRNAs in nuclear genomes, and transfer RNAs (tRNAs) evolve very slowly, and were used widely for clarifying the early evolution and radiation on the Earth (McLaughlin and Dayhoff 1970; Kimura and Ohta 1973; Fox et al. 1977; Hori and Osawa 1979). For example, Hori and Osawa (1979) studied the nucleotide differences of 5S RNAs from various eukaryotic and prokaryotic species and constructed the phylogenetic tree given in Fig. 1.3.1.

The time scale given in this figure was derived under the assumption of a constant rate of nucleotide substitution. According to this tree, eukaryotes diverged from prokaryotes about 1.5 BY ago, whereas plants and animals separated about 1.0 BY ago. These estimates are dependent on a number of assumptions but still, they are quite reasonable in view of the available fossil records. It is notable, that various groups of prokaryotes diverged a long time ago, and that some of them diverged before the prokaryote—eukaryote divergence (Nei 1987, p. 13). Thus, we can see that the deepest divergence is between the three bottom lineages and the other taxa and that Halobacteria first clustered with eukaryotes and then this complex group joins group of three other prokaryotes (Fig. 1.3.1). As stated above and as we will see in Chapter 14, such trees have big topological and sampling errors, so it is necessary be very careful with such wide-scale generalizations.

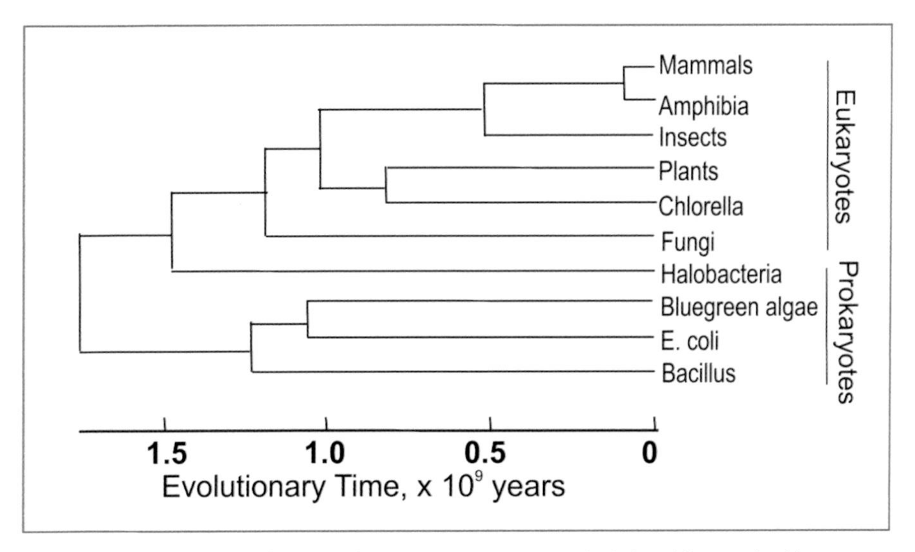

Fig. 1.3.1. Phylogenetic tree of various eukaryote and prokaryote species inferred from nucleotide sequence differences of 5S RNAs (from Hori and Osawa 1979).

Let us remember that in first half of the 20th century it was common to divide life forms into two kingdoms: plants and animals. This classification was later found to be overly simplified, and new kingdoms, i.e., fungi, algae, and bacteria, were proposed (Whittaker 1969). Vorontsov (1989) continued this logic, and introduced in megasystematics the Empire Celulata, which includes Subempire Mesokaryota, Subempire Prokaryota, with kingdoms Archeobacteria and Eubacteria, as well as Subempire Eukaryota. The organic system was further developed into a multikingdom system that incorporated the knowledge of molecular genetics and cell biology (Kusakin and Drozdov 1998). Now up to 26 kingdoms are accepted in biology for the cellular organisms without viruses (Kusakin and Drozdov 1998; Fig. 1.3.2).

For higher organisms, numerous evolutionary trees have been constructed by using various proteins and genes (e.g., Fitch and Margoliash 1967a; Dayhoff 1969; 1972; Goodman et al. 1982; Sibley and Ahlquist 1981; 1984; Sarich and Wilson 1967; Maxson and Wilson 1975; Lakovaara et al. 1972; Fenis et al. 1981; Brown et al. 1982; Avise et al. 1983; Kartavtsev et al. 1983; 1984; 2003). Because these trees are subject to large sampling errors none of them may be considered as final. Yet they have already provided new positive insight into the systematics and evolutionary history of many organisms (see examples in Chapter 7). In particular for apes and hominids these data give much information for understanding common evolutionary history (Fig. 1.3.3).

Until recent time, humans and chimpanzees were considered to have diverged about 30 MY ago by anthropological fossil records. When Sarich and Wilson (1967) estimated by means of immunological study, that the divergence time was only near 5 MY ago, most anthropologists criticized the work. However, with more careful paleontological study of the time of divergence between humans and apes made, the

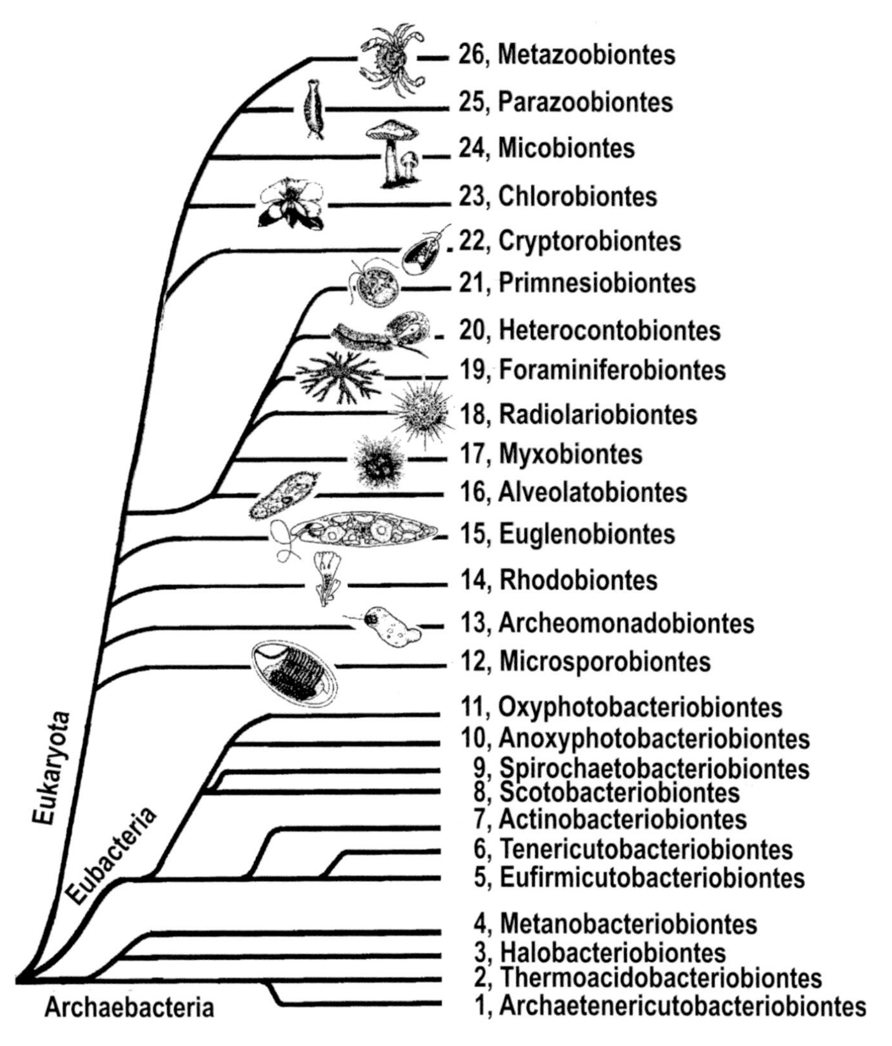

Fig. 1.3.2. Phylogenetic scheme of the multikingdom system of the cellular living beings (from Kusakin and Drozdov 1998 with permission).

conclusion was changed and it was accepted that molecular data and fossil records did not contradict each other and the divergence time between *Homo sapiens* and chimpanzees could be some 5 MY ago (Pilbeam 1984). The results presented in Fig. 1.3.3 support this view more precisely, with a divergence time of 5.2–6.9 MY ago (Arnason et al. 1996). The evolutionary tree in this figure is also in rough agreement with the trees obtained from data on protein sequences (Goodman et al. 1982), chromosome banding pattern (Yunis and Prakash 1982) and other mitochondrial DNA markers (Chapter 14).

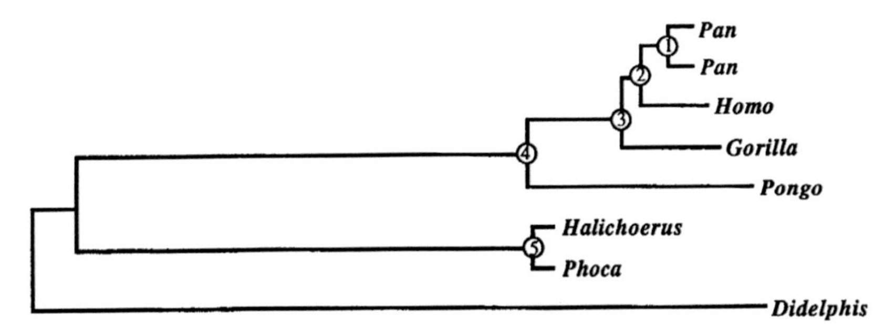

Fig. 1.3.3. Phylogenetic tree for hominoids, two phocids (true seals) and an outgroup (opossum) based on a concatenated sequence of all the 13 protein-coding mitochondrial genes.

The inferred mean divergence times (and 95% confidence limits) for the marked nodes are: (1) 3.0 MY (2.3–3.6), (2) 6.1 MY (5.2–6.9), (3) 8.4 MY (7.3–9.4), and (4) 18.1 MY (16.5–19.6) (from Arnason et al. 1996).

1.4 TRAINING COURSE, #1

- Chose the abstract theme (Take it from the list in the Methodological Instruction; see Appendix I).
- Tell the teacher what theme and when the abstract will be ready to make presentation (Date of defending the theme as oral presentation).
- Bring calculators for the next classroom tasks.
- Genetic vocabulary. Repeat some terms.
- Questions on the statistic software knowledge. General statistics and genetic software acquaintance.

2
THE MATERIAL BASIS OF HEREDITY

MAIN GOALS

SUMMARY

1. There are four main conclusions by Mendel or **Mendel's Laws**.
   - Genetic characters are controlled by unit factors that exist in pairs in individual organisms.
   - In many cases, one of the unit factors appears to be dominant to another unit factor, which results in phenotypic uniformity of the first generation of cross, F_1 hybrids.
   - In the second generation of hybrids, in F_2, segregation of unit factors occurs in certain and predictable proportions.
   - Unit factors that control contrasting pairs of traits assort independently from one to another.
2. Notions of a **gene**, **allele**, **locus**, and **cistron** are the most important for genetics as a science. Students should precisely understand both the similarity and difference between these terms.
3. **Pleiotropy** is a gene action mode when one gene has an impact on two or more phenotypic traits.

Epistasis is a case of intergene action, when alleles of one locus inhibited the action of alleles at another locus.
Complementation is an inter-locus interaction, when one or more genes add to the expression of the original gene a new quality, resulting in a new phenotype.
A polygene is one of the series of genes that control one phenotypic character.

4. In accordance with the modern concept a gene structure comprises the following parts: I, a site of initiation of transcription, R, **Regulator** (**CAAT**) and O, **Operator** sites of a **Promoter**, **Exon**, coding part of the gene, **Intron**, non-coding part of the gene, T, a site of termination of transcription. Other elements may also be pointed out such as an enhancing site of the Promoter (in some viruses) and complex pathways both for regulation of transcription and translation.

2.1 CHROMOSOME STRUCTURE

A **chromosome** is a cell particle which can be seen during metaphase in an optical microscope (Fig. 2.1.1; The image is from freely available web resource, A service of the U.S. National Library of Medicine®, URL: http://ghr.nlm.nih.gov/handbook/basics/chromosome). There are three main types of chromosome morphology, according to the position of the centromere: acrocentric, metacentric, and sub-metacentric, which have the centromere on one end, in the middle of the chromosome, or located intermediately, respectively (Fig. 2.1.1; A sub-metacentric with short, p arm, and long, q arm, are shown). In each chromosome, two chromatid strands are seen, representing DNA molecules. The chromosome consists of chromatin, which is of two types: The Euchromatin, where most structural genes are located, and Heterochromatin, where there are few such genes but a lot of a repetitive DNA. Differential staining of the chromatin is presented in two photos (Fig. 2.1.2), which gives ideas on the individuality and morphology of chromosomes.

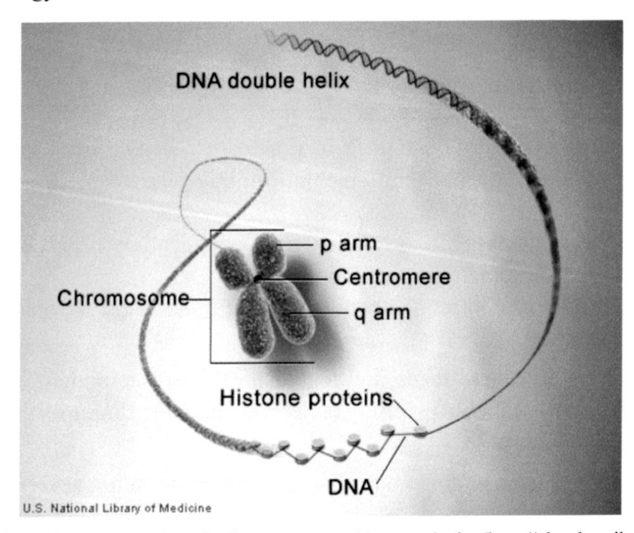

Fig. 2.1.1. A schematic representation of a chromosome and its complexity (http://ghr.nlm.nih.gov/handbook/illustrations/chromosomestructure; free use allowed).

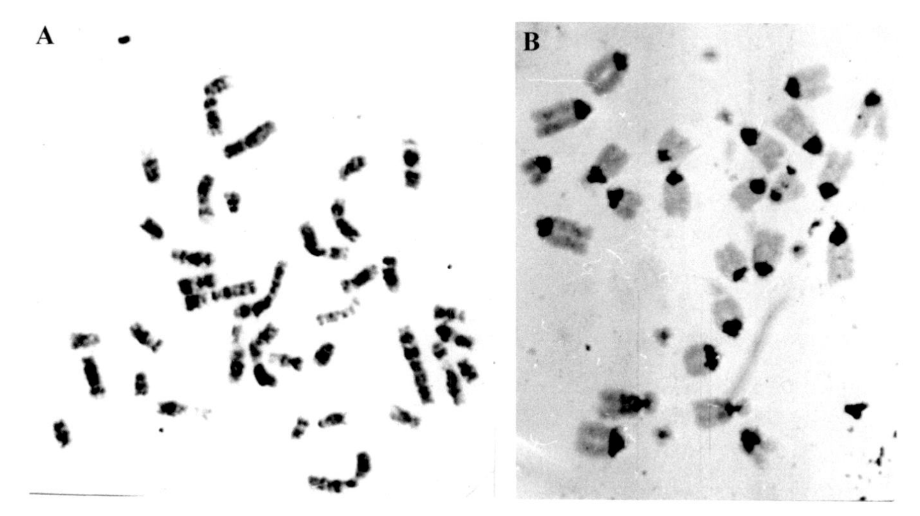

Fig. 2.1.2. (A) Chromosomes of a gerble (Meriones vinogradovi) in metaphase, 2n = 44, Gimza G-banding. (B) Chromosomes of a hamster (Tscherskia triton) in metaphase, 2n = 28. C-banding.
Three morphology types of chromosomes seen for both species (A, B) and heterochromatin location is observed in precentromeric regions (B, black zones) (Photographs presented and permitted for distribution by Dr. I.V. Kartavtseva).

2.2 MEIOSIS AND GAMETE FORMATION

Meiosis is a mode of cell division; under this process cells are created that contain only one member of each pair of chromosomes present in the premeiotic diploid cell. When a diploid cell with two sets of chromosomes undergoes meiosis, the result is four daughter cells, each genetically different and each containing one haploid set of chromosomes (**n**).

Meiosis consists of two successive nuclear divisions. The essentials of chromosome behavior during meiosis are outlined in Figs. 2.2.1–2.2.2. This outline affords an overview of meiosis as well as an introduction to the process as it takes place in a cellular context of males and females.

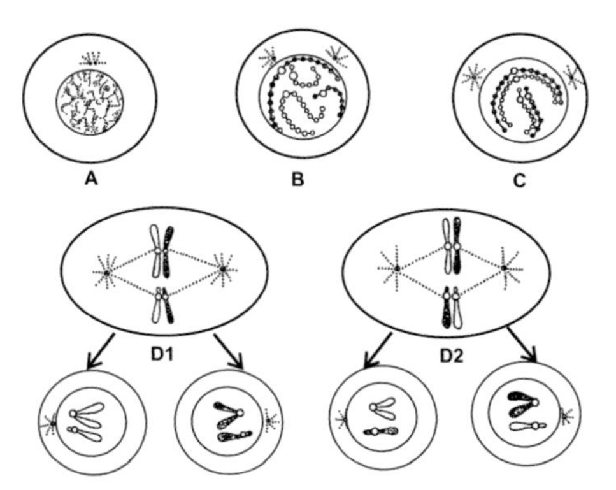

Fig. 2.2.1. Overview of major events and outcomes of mitosis and meiosis; for two chromosome pairs.

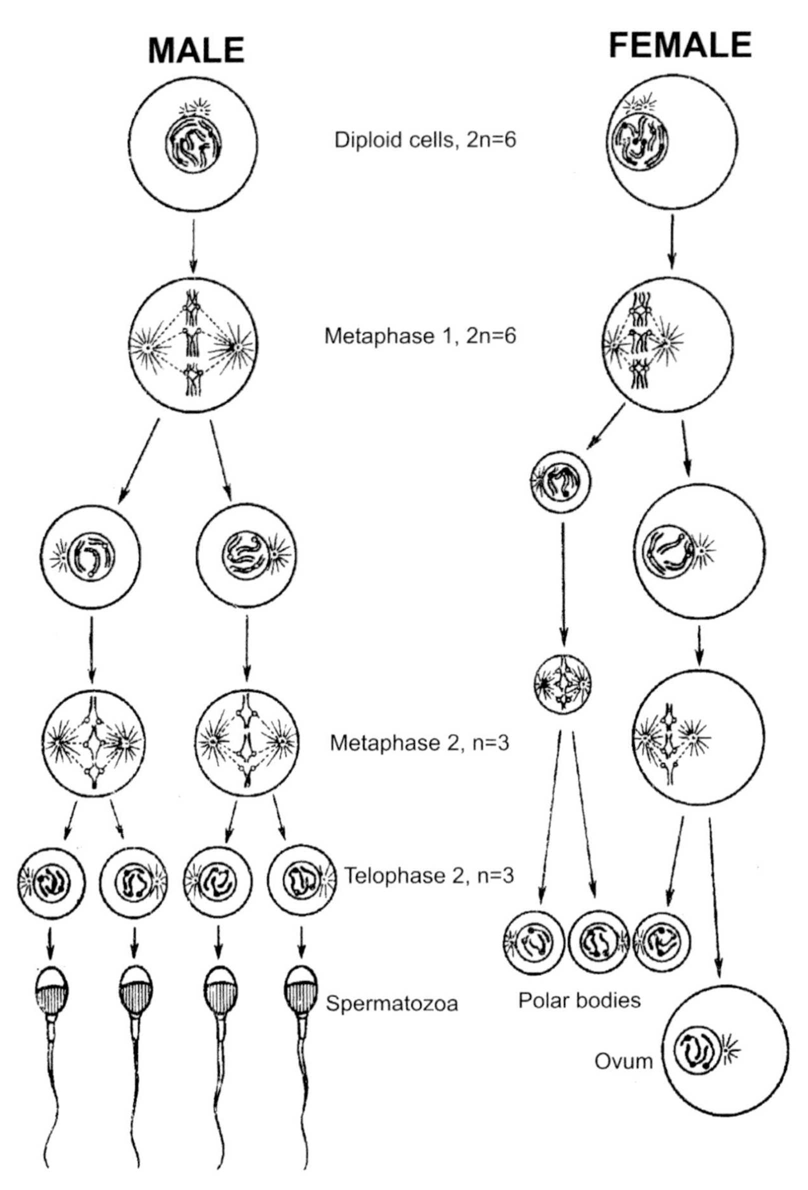

Fig. 2.2.2. The sub-stages of meiotic behavior in two sexes.

Although events that occur during the meiotic divisions are similar in all cells that participate in gametogenesis, in most animal species, there are certain differences between the production of a male gamete (spermatogenesis) and a female gamete (oogenesis) (Fig. 2.2.2).

Spermatogenesis takes place in the testes, the male reproductive organs. The process begins with the expanded growth of an undifferentiated diploid germ cell called a **spermatogonium**. This cell enlarges to become a **primary spermatocyte**, which undergoes the first meiotic division. The new cells produced by this division, **secondary spermatocytes**, have a haploid number of chromosomes (dyads), **n**.

The secondary spermatocytes then go to the second meiotic division; each of these cells produces two haploid **spermatids**. Spermatids go through developmental changes and become specialized, labile **spermatozoa (sperm)**. All sperm cells produced during spermatogenesis receive equal amounts of genetic material and of cytoplasm.

In contrast, in animal **oogenesis**, the formation of **ova (ovum)**, or eggs, takes place in the ovaries, the female reproductive organs. The daughter cells resulting from the two meiotic divisions receive equal amounts of genetic material; however they have unequal amounts of cytoplasm. During each division, the cytoplasm of the **primary oocyte**, derived from the **oogonium**, is unequally distributed and concentrated in one of the two daughter cells. This mechanism of ovum formation is essential for the function of the developing embryo after fertilization and creation of a new organism. During the first meiotic division in anaphase, the tetrads of the primary oocyte separate, and the dyads move to opposite poles. In the first telophase, the dyads present at one pole are pinched off with very little cytoplasm to form a non-functional **first polar body**. The process ends with the formation of an ovum that contains a haploid number of **n** chromosomes.

2.3 MENDEL'S OR TRANSMISSION GENETICS. MONO-, DI-, AND TRIHYBRID CROSSES AND MENDEL'S LAWS

Heritable transmission of simple discrete traits, or qualitative traits, such as color, shape or other phenotypic traits, is realized by certain laws. The first of these laws were founded by G. Mendel in the mid 19th century. In the simplest case, the genetic control of a trait (phenotype) occurs when only a single gene acts. This is a case of mono-genic or monohybrid cross. With our cytological knowledge and modern genetic designations we may explain the transmission of single trait or its gene control by a monohybrid cross (Fig. 2.3.1). A Punnet square gives a simple table representation of these results (Fig. 2.3.2).

Mendel considered several traits of a garden pea morphology and obtained certain rules. Let us consider a scheme (Fig. 2.3.1). The symbol *d* and *D* designate dwarf and tall unit factors (Mendel's term for our modern notion of gene) in the genotypes of mature plants and gametes. Individuals are shown in rectangles and gametes in circles. Segregation proportion of phenotypes for the case of dominance of one gene upon other is 3:1 (Fig. 2.3.2, bottom). The test-crosses are also used by geneticists to prove the results. Usually the recessive homozygote crossed in back direction to hybrid line, $\mathbf{F}_b$ or $\mathbf{F}_a$ (Last called **analyzing cross**). In last case *dwarf* × *tall:* $\mathbf{F}_b$ – *Dd* × *DD,* $\mathbf{F}_a$ – *Dd* × *dd.* To understand in another simple mode these results a Punnet table is convenient (Fig. 2.3.2).

More complex is the case of inheritance when two genes control the traits (Fig. 2.3.3).

Combination of genes present in gametes into genotypes gives peas with different phenotypes (cells of the table), their proportions are 9:3:3:1. Even more complex cases of inheritance are frequent. When three genes control the traits we deal with the

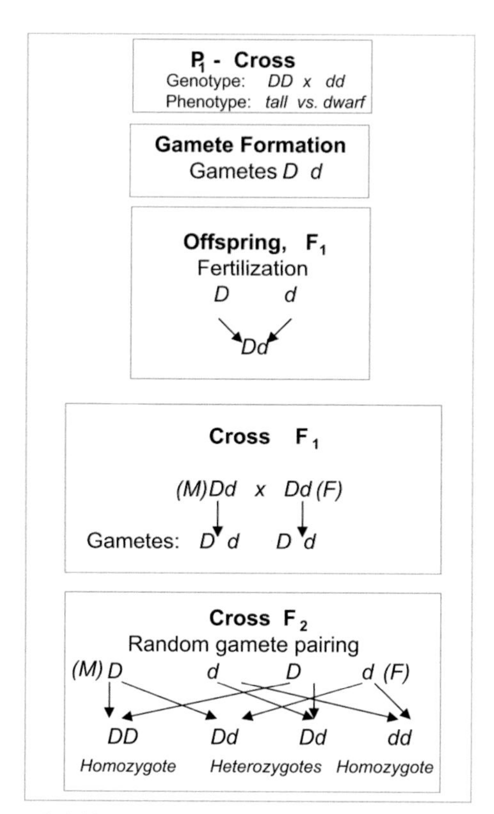

Fig. 2.3.1. The monohybrid cross between tall and dwarf pea plants (M, males; F, females).

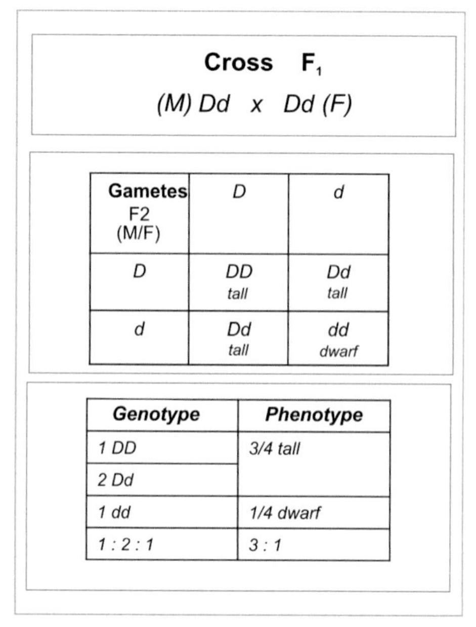

Gametes F2 (M/F)	*D*	*d*
D	*DD* *tall*	*Dd* *tall*
d	*Dd* *tall*	*dd* *dwarf*

Genotype	*Phenotype*
1 DD	*3/4 tall*
2 Dd	
1 dd	*1/4 dwarf*
1 : 2 : 1	*3 : 1*

Fig. 2.3.2. A Punnet square generates the F_2 of the $F_1 \times F_1$ cross, that shown in Fig. 2.3.1.

Trihybrid cross (Fig. 2.3.4). It is much less difficult to consider each contrasting pair of traits separately and then to combine these results using the **forked-line method**. This method, also called a **branch diagram**, relies on the simple application of the laws of probability established for the monohybrid cross. Each gene pair is assumed to segregate independently during gamete formation.

Cross F_1: (♂) CcWw x (♀) CcWw

Gametes	CW	Cw	cW	cw
CW	CCWW Yellow, Round	CCWw Yellow, Round	CcWW Yellow, Round	CcWw Yellow, Round
Cw	CCWw Yellow, Round	CCww Yellow, Wrinkled	CcWw Yellow, Round	Ccww Yellow, Wrinkled
cW	CcWW Yellow, Round	CcWw Yellow, Round	ccWW Green, Round	ccWw Green, Round
Cw	CcWw Yellow, Round	Ccww Yellow, Wrinkled	ccWw Green, Round	ccww Green, Wrinkled

Fig. 2.3.3. The dihybrid cross between yellow-green and round-wrinkle pea plants. The Punnet square is shown to explain the results.

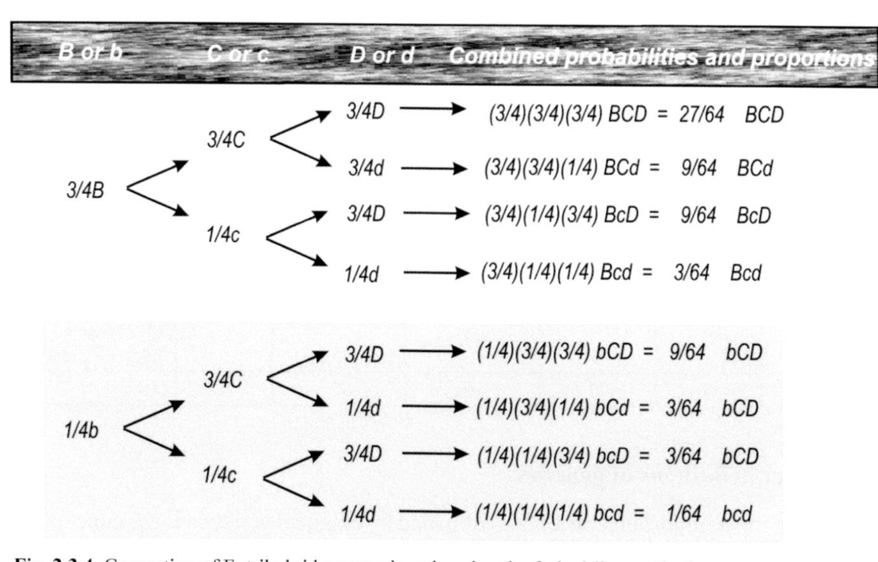

Fig. 2.3.4. Generation of F_2 trihybrid genotypic ratio using the forked-line method.

When the monohybrid cross $AA \times aa$ is made, we know that

1. All F_1 individuals have the genotype *Aa* and express the phenotype represented by the *A* allele, which is called the **A** phenotype in the following discussion.
2. The F_2 generation consist of individuals with either the **A** phenotype or the **a** phenotype in ratio 3:1.
3. The same generalization can be made for the $BB \times bb$ and $CC \times cc$ crosses.

Closing this section, let us reformulate Mendel's main postulates or laws, keeping in mind that unit factors are genes in modern genetics understanding.

Mendel's Laws:

- Genetic characters are controlled by **unit factors** that **exist in pairs** in individual organisms.
- One of the unit factors appeared to be **dominant** to another unit factor, which resulted in uniformity of first generation of cross, F_1 hybrids.
- In the second generation of hybrids, the F_2, the **segregation** of unit factors occurred in certain and stable proportions.
- Unit factors that control contrasting pairs of traits give an **independent assortment** of random combination of each to other.

Finally let us note that in general the independent assortment of gametes and chromosomes give rise to a huge variety of genes and genotypes (Table 2.3.1). Recombination due to a crossover that was discovered later increases this variation even more.

Table 2.3.1. Number of classes in hybrid offspring and the mode of segregation in F_2 under different traits combination.

CROSS	Number of traits in cross	Number of possible gametes	Number of gamete combinations	Number of classes		Phenotype ratio
				phenotype	genotype	
Monohybrid	1	$2^1 = 2$	$4^1 = 4$	$2^1 = 2$	$3^1 = 3$	3:1
Dihybrid	2	$2^2 = 4$	$4^2 = 16$	$2^2 = 4$	$3^2 = 9$	9:3:3:1
Trihybrid	3	$2^3 = 8$	$4^3 = 64$	$2^3 = 8$	$3^3 = 27$	27:9:9:9:3:3:3:1
Tetrahybrid	4	$2^4 = 16$	$4^4 = 256$	$2^4 = 16$	$3^4 = 81$	81:27:27: 27: 27:9:9:9: 9:9:9:3:3:3:3:1
Polyhybrid	n	2^n	4^n	2^n	3^n	$(3:1)^n$

Some other definitions of genetics

Among the most important concepts established by classical genetics is the concept of **linkage**. What is linkage? **Linkage** is the joint inheritance of two or more genes that code for independent traits. Basic researches on linkage in the beginning of the 20th century were made by Morgan, Sturtevant and Bridges. Linkage occurs because of the linear gene location in DNA of a chromosome; one linkage group is determined by location of genes on a certain chromosome. Linkage was discovered as deviation from independent assortment under investigation of a crossover results. That is why linkage is measured as crossover rate; the less is the crossover rate, the closer are genes located. When the crossover rate is near 50%, no linkage is observed via breeding experiments.

Another important concept of genetics is genetic interaction. There are intralocus interactions like dominance, codominance or incomplete dominance, overdominance, additivity and neutrality. Another kind of interaction is between loci, like epistasis and complementation; also the important notions are pleiotropy and polygenic inheritance, which are special cases of complex gene action. All intralocus cases of gene interactions and their quantitative effects will be considered separately in Chapter 12. However, dominance is explained above by one of Mendel's laws, and the codominance mechanism will be met later in Chapter 5. On this step it is principally important to remember that alleles of one gene can act in different modes, with dominant, codominant, additive, multiplying or neutral effects to each other. **Epistasis** is an inter-locus interaction, such as when one gene inhibits the expression of another gene. **Complementation** is an inter-locus interaction, when one and another gene add to the expression of the original genes a new quality, resulting in the new phenotype. Both epistasis and complementation change segregation proportions from those expected under Mendel's random assortment. **Pleiotropy** is a case of gene action, when it controls more than one trait. Many genes that control morphology, for example, have pleiotropic effects on viability. A **Polygene** is one in the series of genes that control one phenotypic character (most quantitative traits are controlled by polygenes). We will consider polygenic traits and their inheritance in Chapter 11.

2.4 EUKARYOTE GENE STRUCTURE. MUTATIONS AND MODIFICATIONS

The main latest understanding on gene structure in a Eukaryote organism is outlined in Fig. 2.4.1.

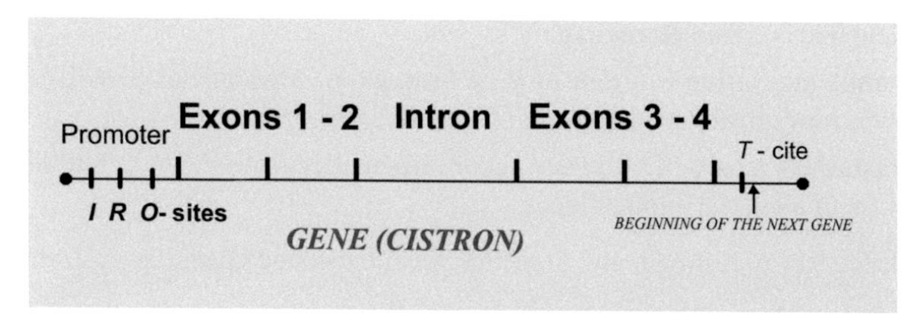

Fig. 2.4.1. A scheme of a gene. I, is a site of initiation of transcription, R, is a regulator (**CAAT**) and O, operator sites of a Promoter. Exon, is coding part of the gene, Intron, is non-coding part of the gene, T, is a site for termination of transcription. Other elements may also be pointed out, including enhancing site of the Promoter (in some viruses) and complex pathways both for the regulation of transcription and translation.

Mutation. Genes provide a stable function for an organism's needs. However, quite rarely, the gene is suddenly changed, or undergoes a mutation. The mutation rate (μ) can vary widely as for specific genes, as for different organisms: $\mu = 10^{-3}–10^{-7}$, roughly. Gene or site mutations are simple base changes in the DNA. There are different types of chromosome ***rearrangements*** too: ***inversions***, ***translocations***, ***deletions***, and ***duplications***. These changes are sometimes incorrectly also referred to as mutations. We will learn more about mutation in Chapter 6.

Modification is another kind of a genome change, an epigenetic change, when different phenotypes occur for the same genotype in different environments. In the marine realm modifications are best known in fish, like salmon for example, where they may represent different ecotypes or biotypes: anadromous or landlocked, lake or river forms. They have major roles in adaptation. However, some fishery researchers (e.g., B. Mednikov, Russia) think that they may play big role in fish speciation too.

2.5 TRAINING COURSE, #2

2.1. Genetic Vocabulary: 1. Give definitions for the following notations—*AA*, *Aa*, *BbCc*, ×, *P*, *F1*, *Fa*. 2. Explain Terms: *locus*, *gene*, *allele*, *cistron*, *inversion*, *translocation*, *mutation*, *deletion*, and *duplication*.
LOCUS ≈ GENE ≈ ALLELE ≈ CISTRON: these terms are among the most important, and we should comprehend both their meanings and their differences (Jointly with the audience to recall and explain these terms).

- **Locus** is a place on a chromosome, where the gene is located.
- **Gene** is a coding sequence of DNA, which determines an elementary function of an organism (like polypeptide synthesis).
- **Allele** is a variant of a gene with the certain peculiarities of a function (e.g., differences in a charge of proteins).
- **Cistron** is a unit of genetic function, which includes DNA coding sequence and regulatory elements for macromolecular synthesis.
- **Dominant** allele controls a trait, which is expressed in the presence of any other allele that is called **Recessive**.
- **Complementation** is a case of gene interaction, when alleles of different loci give a new phenotype that is not present in both parents.
- **Epistasis** is a case of intergene action, when alleles of one locus inhibited the action of alleles at another locus.

2.2. Independent Assortment and Statistics. Qualitative and Quantitative Traits.

2.3. Chi-Square Test of Independence in Frequency Distribution at Qualitative Traits. $\chi^2 = \Sigma[(f-f')^2/f'], f'$ and f are expected and observed numbers, d.f. = n – 1, where n is number of phenotypic classes.

2.4. Check Mendel's results in Table 2.5.1 (Data from Klug and Cummings 2002, p. 49).

Table 2.5.1. Two examples of segregation in Mendel's crosses.

Monohybrid Cross			
Observed frequencies (f)	Expected frequencies (f′)	Deviation, $d^2 = (f-f')^2$	$(d^2)/f'$
740	3/4 (1000) = 750	$-10^2 = 100$	100/750 = 0.13
260	1/4 (1000) = 250	$+10^2 = 100$	100/250 = 0.40
Total = 1000			$\chi^2 = 0.53$
			P = 0.48
Dihybrid Cross			
587	567	+20 = 400	0.71
197	189	+8 = 64	0.34
168	189	–21 = 441	2.33
56	63	–7 = 49	0.78
Total = 1008			$\chi^2 = 4.16$
			P = 0.26

SELF-EDUCATION TASK

WEB-support: Solve 3 Exercises on any one of the three Mendel's Laws. Take a printout (or write it down) and show it to the teacher in the classroom.

3

TRANSLATION OF GENETIC INFORMATION. INTRODUCTION TO PROTEOMICS

MAIN GOALS

SUMMARY

1. **Translation** describes synthesis of polypeptide chains in cells, under the direction of mRNA and in association with ribosomes. This process ultimately converts the information stored in the genetic code of the DNA forming a gene into the corresponding sequence of amino acids making up the polypeptide.
2. Translation is a complex energy-requiring process that also depends on charged tRNA molecules and numerous protein factors. Transfer RNA (tRNA) serves as an adaptor molecule between an mRNA triplet and the appropriate amino acid.
3. Investigation of nutritional requirements in *Neurospora* by Beadle with colleagues made it clear that mutations cause the loss of enzyme activity. Their work led to the concept of **one-gene: one-enzyme**. The one-gene: one-enzyme hypothesis was later revised. Pauling and Ingram's investigations of hemoglobin from patients with sickle-cell anemia led to the discovery of the fact that one gene directs the synthesis of only one polypeptide chain.
4. The proposal suggesting that a gene nucleotide sequence specifies the sequence of amino acids in a polypeptide chain in a collinear manner was confirmed by experiments involving mutations in the tryptophan synthetase gene in *E. coli.*

5. Proteins, the end products of genes, demonstrate four levels of structural organization that collectively provide the chemical basis for their three-dimensional conformation, which is a basis for function of a molecule.
6. Of the myriad functions performed by proteins, the most influential role is assumed by enzymes. These highly specific, cellular catalysts play the central role in the production of all classes of molecules in living systems. Proteins consist of one or more functional domains, which are shared by different molecules. The origin of these domains may be the result of exon shuffling during evolution.
7. **Proteome** is a set of proteins expressed and modified during a cell's entire lifetime. In a narrower sense, the term also describes a set of proteins expressed in a cell at a given time. **Proteomics** is the study of a proteome, and it uses technologies ranging from genetic analysis to mass spectrometry.

3.1 COMPONENTS THAT ARE CRITICAL FOR PROTEIN SYNTHESIS

Five major components are essential for protein synthesis, which are described below in paragraphs 1–5. Then we will consider more specifically two essential components in the translation of genetic information (Sections 3.2 and 3.3).

1. ***Messenger RNA.*** Messenger RNA (mRNA) is needed to bring the ribosomal subunits together (described below) and to provide the triplet coding sequence of bases that determines the amino acid sequence in the polypeptide chain. Original mRNAs in higher organisms undergo the process of maturation, the main step of which is processing of a pro-mRNA by cutting from its full sequence the introns and regulatory sites. We will learn more of mRNA in the following sections.
2. ***Transfer RNA.*** The **transfer RNAs** (tRNA) are specific ribonucleic molecules. This class of molecules adapts certain triplet codons in mRNA to their target amino acids. This adaptor or transfer role of tRNA was postulated by F. Crick in 1957.

   tRNA Structure. Because of their small size and stability in the cell, tRNAs have been investigated extensively and are the best-characterized RNA molecules. They are composed of only 75–90 nucleotides, and have nearly identical structure in bacteria and eukaryotes. In all organisms, tRNAs are transcribed as larger precursors, which are processed into mature *4S* tRNA molecules by a special enzymatic process. In *E. coli,* for example, tRNA is composed of 77 nucleotides, while its precursor contains 126 nucleotides. Holley and his colleagues (1965; after Klug and Cummings 2002) reported the complete sequence of $tRNA^{ala}$ (the superscript identifies the specific tRNA that binds to an amino acid) isolated from yeast. A very important original finding was that a number of nucleotides are unique to tRNA. Each tRNA consists of four nitrogenous bases common to RNA (G—guanine, C—cytosine, A—adenine, and U—uridine). These also include inosinic acid, which contains the purine hypoxanthine, ribothymidylic acid, and pseudouridine, among others.

   Holley's et al. sequence analysis led them to propose the two-dimensional **cloverleaf model of tRNA**. It was known that tRNA demonstrates a secondary structure due to base pairing. Holley et al. discovered that they could arrange the linear model in such a way that three, four or more complementing base pairing would result. This arrangement created paired stems and unpaired loops resembling the shape of a cloverleaf. Loops consistently contained modified bases that do not generally form base pairs. Holley's et al. model is shown in Fig. 3.1.1. Because the triplets GCU, GCC, and GCA specify alanine, Holley and colleagues looked for an anticodon sequence complementary to one of these codons in his $tRNA^{ala}$ molecule. They found it in the form of CGI (the 3'-to-5' direction), in one loop of the cloverleaf. The nitrogenous base I (inosinic acid) can form hydrogen bonds with U, C or A, the third members of the triplets. The **anticodon loop** was recognized namely by this approach.

3. ***Ribosomes.*** The ribosomes are cell particles on which protein synthesis takes place. They move along an mRNA molecule, and align successive tRNA

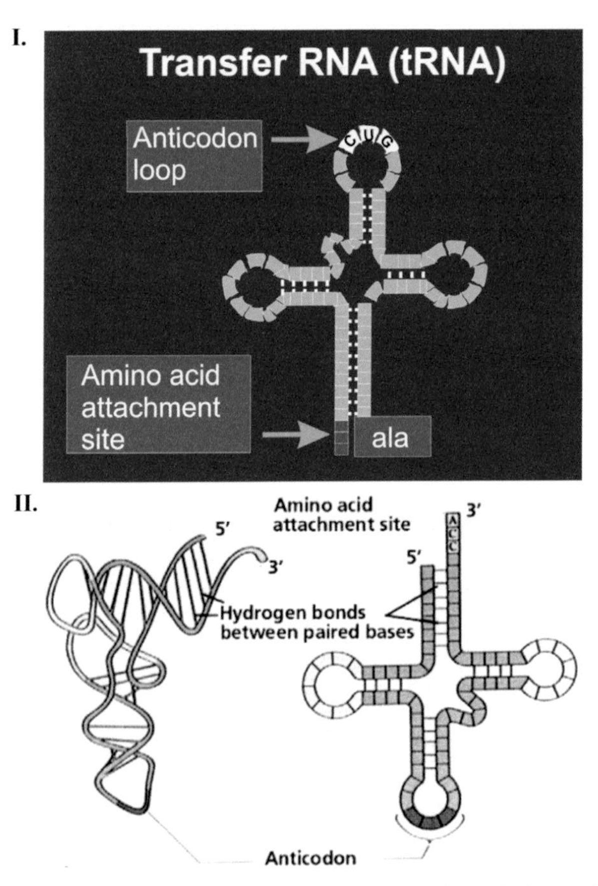

Fig. 3.1.1. Two-dimensional cloverleaf model of transfer RNA. (I) General scheme with attachment site for alanine. Blocks represent nucleotides that paired in loops by hydrogen bonds (white squares). (II) Two- and Three-dimensional models of tRNA, based on X-ray research and their comparison. The anticodon recognize consequent codon in mRNA by base pairing. In part II the amino-acid is bonded on top at 3'-end which is currently free (The image in II is from free WEB-site, URL: www2.estrellamountain.edu/faculty/farabee/biobk/BioBookPROTSYn.html).

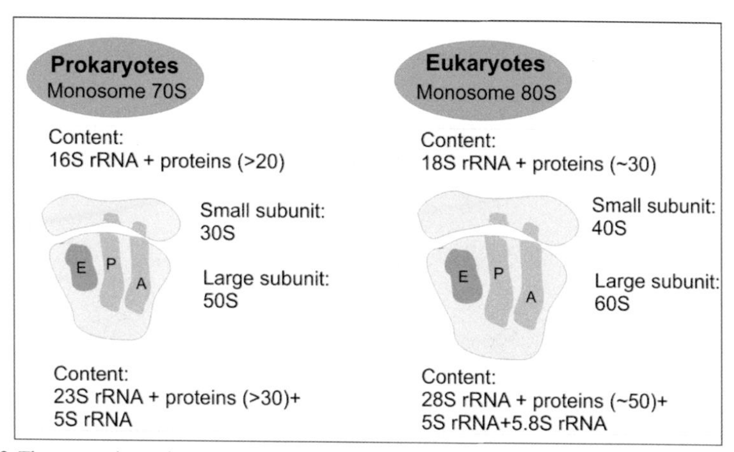

Fig. 3.1.2. The comparison of components in prokaryotic and eukaryotic ribosome. S, sedimentation weight; A, P, and E are aminoacyle, peptidyle and exit sites.

molecules; the amino acids are attached one by one to the growing polypeptide chain by means of peptide bonds. Ribosomes consist of two separate ribonucleic-and-protein particles (the small subunit and the large subunit), which come together in polypeptide synthesis to form a mature ribosome or monosome (Fig 3.1.2). Their molecular weights (MW) could be measured as sedimentation rate. The sedimentation (S) rates of ribosome subunits are shown in the figure and below for different subunits (50S, 30S, 60S, and 40S).

Different ribosomal RNAs (rRNA) that jointly with proteins comprise subunits are also depicted.

4. ***Aminoacyl-tRNA synthetases.*** Each enzyme in the biosynthetic set of molecules catalyzes the attachment of a particular amino acid to its corresponding tRNA molecule. A tRNA attached to its amino acid is called an **aminoacylated** tRNA or a **charged tRNA**.

5. ***Initiation, elongation and termination factors.*** Polypeptide synthesis can be divided into three stages: initiation, elongation and termination. Each stage requires specialized molecules. Let us consider them more attentively.

3.2 THE PROCESS OF TRANSLATION: FROM RNA TO POLYPEPTIDE

Initiation. In eukaryotes, it takes place by scanning the mRNA for an initiation codon. The process of translation begins with an mRNA^{met} molecule binding to a ribosome by an enzymatic reaction with the participation of several initiation factors, the **IF**. The charged tRNAs are brought along sequentially, one by one, to the ribosome that is translating the mRNA molecule. The initiation complex forms at the 5' end of the mRNA. In Eukaryotes this consists of one 40S ribosomal subunit, the initiator tRNA^{met} and initiation factors. The initiation complex recruits a 60S ribosomal subunit in which the tRNA occupies the P (peptidyl) site of the ribosome. This complex travels along the mRNA until first AUG, the start codon is found and translation begins. Initiation factors (**IF**) are realized (Fig. 3.2.1, arrows) and the elongation starts.

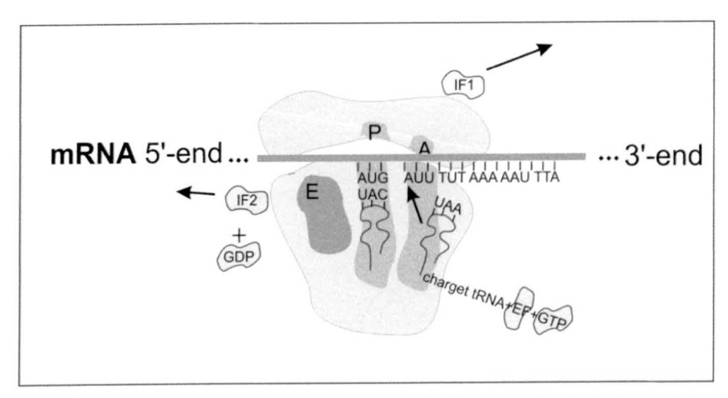

Fig. 3.2.1. Elongation cycle in protein synthesis. An uncharged tRNA occupies the E site, and the polypeptide is attached to the tRNA in the **P** site. The function of **EF1** is to release the uncharged tRNA and bring the next charged tRNA into the A site. A peptide bond is formed between polypeptide and the amino acid held in the A site. This time, the 60S subunit is shifted relative to the 40S subunit, re-creating the pretranslocation state. The function of **EF2** is to translocate the 40S ribosome to the next codon, once again generating the posttranslocation state.

Elongation. It takes place codon by codon through a ratchet mechanism. Recruitment of the elongation factor **EF + GTP** into the initiation complex begins the elongation phase of polypeptide synthesis (Fig. 3.2.1). Elongation consists of three processes executed iteratively: 1. Bringing each new charged tRNA into line. 2. Forming the new peptide bond to elongate the polypeptide. 3. Moving the ribosome to the next codon along the mRNA. The first and the next steps in elongation are illustrated in Fig. 3.2.1. Several components are omitted for simplicity in the scheme.

Termination. A termination of codon signals release of the finished polypeptide chain. Compared to initiation and elongation, the termination of polypeptide synthesis or the **release** phase is simpler. When a stop codon is met, the tRNA holding the polypeptide remains in the **P** site, and a release factor (**RF**) binds with the ribosome. GTP hydrolysis provides the energy to cleave the polypeptide from the tRNA, to which it is attached, as well as to eject the release factor and dissociate the 80S ribosome from the mRNA. Peptide bonds are made between successively aligned amino acids, each time joining the amino group of the incoming amino acid to the carboxyl group of the amino acid at the growing end. In the end, the chemical bond between the last tRNA and its attached amino acid is broken, and newly synthesized polypeptide is removed from ribosome.

Polyribosomes. As elongation proceeds and the initial portion of the mRNA has passed through the ribosome, this mRNA can again associate with another small subunit to form a second initiation complex. This process is repeated several times with a single mRNA, forming so called **polyribosomes (polysomes)**. Polysomes can be isolated and analyzed following soft lysis of cells. In Fig. 3.2.2, these complexes are seen under the electron microscope and you can see mRNA (the thin line) between the individual ribosomes. The polypeptide chains emerging from the ribosomes during translation are seen too. The formation of polysome complexes represents an efficient use of the components available for protein synthesis during a unit of time.

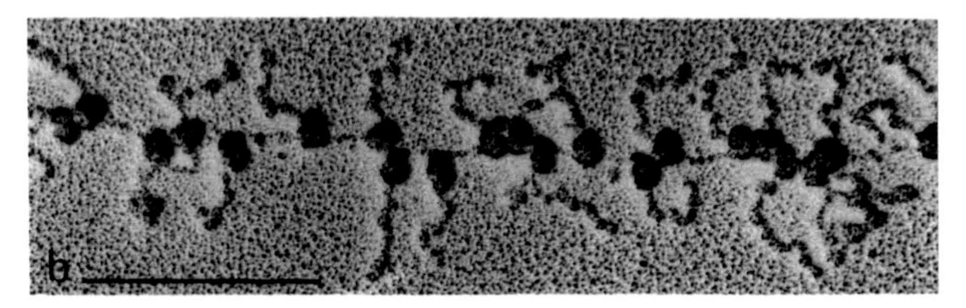

Fig. 3.2.2. Polyribosomes from giant salivary gland cells of the midgefly, Chironomus thummi, visualized under the electron microscope. The polypeptide chain is apparent as it emerges from each ribosome. Its length increases as translation proceeds from left (5') to right (3') along the mRNA (Image from E.V. Kiseleva 1989; p. 252, Fig. 1b).

Prokaryotes often encode multiple polypeptide chains in a single mRNA, a fact known long ago, that is the operon structure. In prokaryotes, mRNA molecules have no cap, and there is no scanning mechanism to locate the first AUG. In *E. coli,* for example, translation is initiated when two initiation factors (**IF-1** and **IF-3**) interact with the 30S subunit at the same time that another initiation factor (**IF-2**) binds with a special initiator tRNA charged with formylmethionine ($tRNA^{fMet}$). These components come together and combine with an mRNA, but not at the end (Fig. 3.2.5). The attachment occurs by hydrogen bonding between the 3' end of the 16S RNA present in the 30S subunit and a special sequence, the **ribosome-binding site**, in the mRNA. Together, the 30S + $tRNA^{fMet}$ + mRNA complex recruits a 50S subunit, in which the tRNA is positioned in the **P** site and aligned with the AUG initiation codon. In the assembly of the completed ribosome, the initiation factors dissociate from the complex.

3.3 HEREDITY, PROTEINS AND FUNCTION

The One Gene—One Enzyme Hypothesis

In two investigations beginning in 1933, George Beadle provided the first precise experimental evidence that genes are directly responsible for the synthesis of enzymes. In a collaborative investigation, conducted with Boris Ephrussi, *Drosophila* eye pigments were analyzed. They confirmed that mutant genes that alter the eye color of fruit flies can be linked to biochemical errors that likely involve the loss of enzyme function. Beadle then joined with Edward Tatum to investigate nutritional mutations in the pink bread mould Neurospora crassa. This investigation led to the **one gene—one enzyme hypothesis.** ***Neurospora Mutants Role in a Concept***. Beadle and Tatum in the early 1940s, fixed their attention on and started to work with *Neurospora*. This object was well-characterized for its biochemical requirements, and so nutritional mutations could be induced and isolated quite easily. By inducing a number of mutations, these authors produced several strains that had genetic blocks of reactions essential to the growth of the organism. It was known that *Neurospora* could manufacture nearly all the essential components for its normal development. Using inorganic carbon and nitrogen sources, this organism can synthesize water-soluble vitamins, amino acids, carotene pigments and necessary nucleotides. Beadle and Tatum supplied X-rays to conidia or spores to increase the frequency of mutations and then allowed them to be grown on "minimal" medium containing all the necessary growth factors: purines and pyrimidines, vitamins, amino acids, etc. Under such conditions, a mutant strain that was unable to grow on the minimal medium was able to develop with additional supplements included in new, enriched medium. After that all the colonies were moved to the minimal medium. If growth was detected even on the minimal medium, then the organisms were able to synthesize all the necessary growth factors themselves; thus the conclusion was made that the colony did not contain a mutation. In the cases where no growth occurred, the researchers had concluded that the culture contained a nutritional mutation. On the final step they are able to determine the mutation type. Their principal experimental design is shown in Fig. 3.3.1.

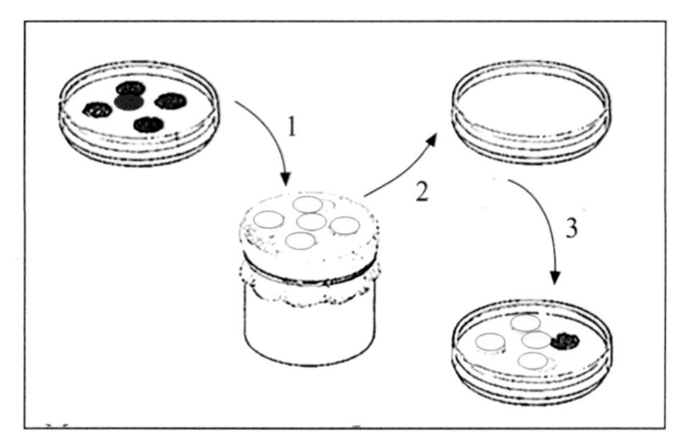

Fig. 3.3.1. Mutation isolation and characterization. Stamp method applied for detection of mutations in bacteria that are resistant against T1 Phage. 1, getting "stamp" replicas from bacterial colonies. 2, stamps' transfer to medium. 3, incubation of a stamp. Bacterial colony with a mutation that produced T1-resistence is shown in a bottom Petri plate with dark color.

The findings derived from testing over 80,000 spores led Beadle and Tatum to conclude that biochemical mutations are caused genetically. It seemed likely that each nutritional mutate on leads to the loss of the enzymatic activity that determines an essential reaction in wild-type organisms. It also became evident that a mutation could be found for nearly any enzymatically controlled reaction. Beadle and Tatum had thus provided sound experimental evidence for the hypothesis that one gene specifies one enzyme. With modifications this concept became a major principle of genetics.

Genes and Enzymes: Analysis of Biochemical Pathways

The one gene—one enzyme concept and its attendant methods have been used over the years to work out many details of metabolism in *Neurospora, Escherichia coli* and many other microorganisms. One of the first metabolic pathways to be investigated in detail was that leading to the synthesis of the amino acid arginine in *Neurospora.* By studying seven mutant strains, each requiring arginine for growth (arg^-), Adrian Srb and Norman Horowitz (*opt. cit.* Klug and Cummings 2002) ascertained a partial biochemical pathway that leads to the synthesis of this molecule. Their work demonstrates how genetic analysis can be used to establish biochemical information.

Srb and Horowitz tested the ability of each mutant strain to reestablish growth if either citrulline or ornithine (two compounds with close chemical similarity to arginine) was used as a supplement to the minimal medium. If either was able to substitute for arginine, they reasoned that it must be involved in the biosynthetic pathway of arginine. They found that both molecules could be substituted in one or more strains.

Of the seven mutant strains, four of them (*arg* 4–7) grew if supplied with citrulline, ornithine or arginine. Two of them (*arg* 2 and *arg* 3) grew if supplied with citrulline or arginine. One strain (*arg* 1) would grow only if arginine was supplied; neither citrulline nor ornithine could be a substitute for it. From these experimental observations, the following pathway and metabolic blocks for each mutation were deduced (Fig. 3.3.2):

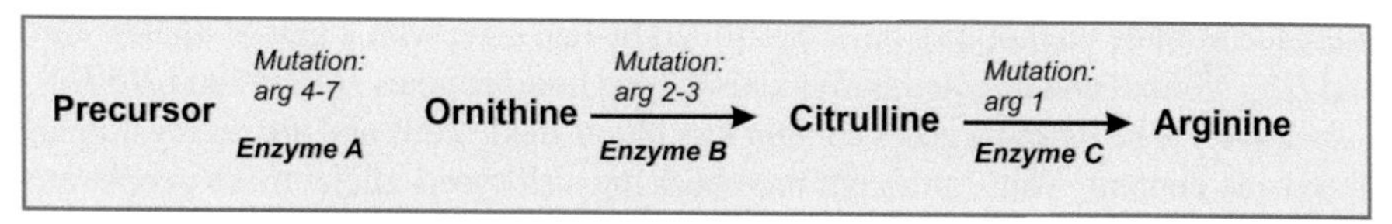

Fig. 3.3.2. Chain of chemical reactions in *Neurospora*.

One Gene—One Polypeptide Chain

The concept of one gene—one enzyme, which was mentioned above, was not accepted immediately by geneticists. This is not surprising because it was not yet clear how mutant enzymes could cause variation in phenotypic traits like in *Drosophila* mutants, which demonstrate altered eye size, wing shape, wing vein pattern, etc. Similar mutations were known in other animals, and plants also exhibit varieties of qualities. Nevertheless, many things soon become clear. First of all, scientists understood that while all enzymes are proteins, not all proteins are enzymes. When biochemical genetics developed, it became obvious that each protein is coded by a certain gene. Consequently, more accurate terminology, that is, **one gene—one protein** coined. Secondly, proteins often show a subunit structure, consisting of conformational aggregation of two or more polypeptide chains. Their combination is the basis of the quaternary structure of proteins, which will be discussed later in this chapter and in Chapter 5. Because each type of polypeptide chain is encoded by a separate gene, a more accurate statement of **one gene—one polypeptide chain** was established. These modifications of the original hypothesis became apparent after research by Jacob and Monod (1965) and during the analysis of hemoglobin structure in individuals afflicted with sickle-cell anemia.

Sickle-Cell Anemia. Serious evidence that genes can specify protein synthesis other than enzymes obtained from the work on mutant hemoglobin molecules. Hemoglobin (Hb) derived from humans subjected to the sickle-cell anemia disease and analyzed in a laboratory. Humans with this disorder have erythrocytes that, under low oxygen tension, become curve-and-sickle shaped during polymerization of hemoglobin (Fig. 3.3.3, right). Normal erythrocytes in contrast have a disc-shaped surface (Fig. 3.3.3, left). Their deficiency leads to heavy oxygen anemia.

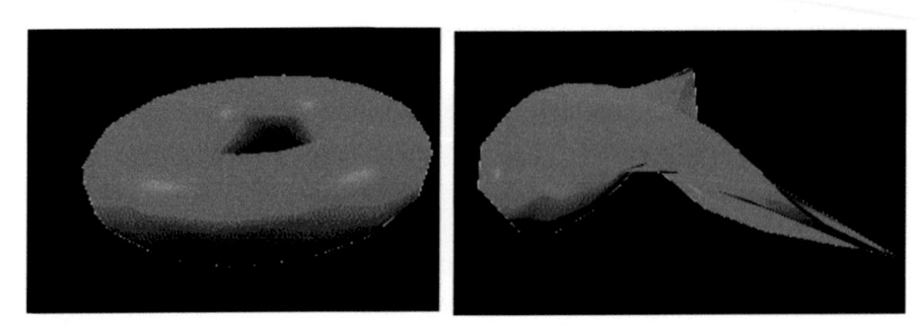

Fig. 3.3.3. A comparison of erythrocytes from a normal individual (left) and from individual with sickle-cell anemia (right).

Neel (1949) and Beet (1949) demonstrated that this disorder is inherited according to Mendelian laws. In a pedigree analysis the knowledge was obtained that the presence of three phenotypes have controlled by one gene with a pair of alleles, Hb^A and Hb^S. Normal and affected individuals are the homozygotes Hb^AHb^A and Hb^SHb^S, respectively. The heterozygotes contain less blood sickle cells and are nearly normal in oxygen content. Thus, although they bear the sickle-cell allele, these people are resistant to the disorder because the remaining hemoglobin alleles produce normal molecules. Moreover, later it was discovered that these heterozygotes are less sensitive to malaria. Heterozygous individuals being bearers of the defective gene transmit it to 50% of their offspring: segregation in Fb = 1:1 (see Chapter 2). Also in 1949, Pauling and colleagues (1949) provided some insight on the molecular basis of the disorder. These authors showed that hemoglobin molecules isolated from diseased and normal individuals differ in the speed of electrophoretic migration of the relevant protein bands. In this technique (see Chapter 5 also), charged molecules migrate in an electric field. If the cumulative charge of two molecules is different, the speed of migration is different and could be detected in the gel after staining the corresponding bands. The data obtained allowed Pauling and his coauthors to conclude that there is a chemical difference between normal and sickle-cell hemoglobin. These two hemoglobin molecules are designated **Hb-S** and **Hb-A**, respectively, because of their mobility in electric flow (Fig. 3.3.4A). The migration pattern of Hb obtained from individuals of all three possible genotypes, "slow" and "quick" homozygotes as well as corresponding heterozygote, when subjected to starch-gel electrophoresis is depicted in the upper part of the image. The amino acid substitutions that cause these mobility changes are given in bottom of the image (Fig. 3.3.4B).

Pauling's findings may be interpreted in two ways: either a heme (iron-containing groups) or a globin containing four polypeptide chains are the causes of Hb-A and Hb-S difference. Later in 1954–1957 Ingram (1963) resolved this question. He was able to demonstrate that the chemical change occurs in the primary structure of the globin portion of the hemoglobin molecule as visualized above (Fig. 3.3.4B).

Using the fingerprinting technique, Ingram showed that Hb-S differs both in peptide content and in amino acid composition compared to Hb-A. Human adult hemoglobin contains two identical $\boldsymbol{\alpha}$-chains of 141 amino acids and two identical $\boldsymbol{\beta}$-chains of 146 amino acids in its quaternary structure. Further analysis revealed just one amino acid change in the $\boldsymbol{\beta}$-chain: Valine was substituted for glutamic acid at the sixth position, defining these notable polypeptide differences (Fig. 3.3.4b).

A. Electrophoresis of human hemoglobins

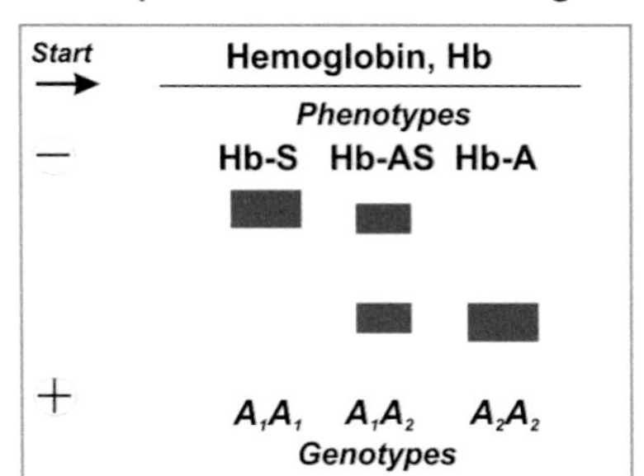

B. Partial polypeptides of normal, Hb-A and sickle-cell hemoglobin, Hb-S (upper chain)

NH_2 - val - his - leu - thr - pro - glu - glu - - - COOH

NH_2 - val - his - leu - thr - pro - val - glu - - - COOH

Fig. 3.3.4. Investigation of hemoglobin derived from Hb^SHb^S (A_1A_1) and Hb^AHb^A (A_2A_2) individuals using gel electrophoresis and amino acid analysis. (A) Hemoglobin from humans with sickle-cell anemia (Hb-S) migrates slowly in the electrophoretic field as shown. (B) Bottom part demonstrates an altered amino acid, valine (val), at the sixth position from amino-group in the polypeptide chain. During electrophoresis, heterozygotes Hb^AHb^S (A_1A_2) reveal both forms of hemoglobin, Hb-AS.

Protein Structure and Biological Diversity. Now let us discuss in brief a protein structure. How can these molecules take a central place in determining the complex of a tremendous variety of cellular activities? As you will see, the fundamental aspects of the structure of proteins provide the basis for this vast complexity and diversity. First, we should define a difference between the terms **polypeptides** and **proteins**.

Polypeptides are the precursors of proteins. Several steps of aggregation, transportation and modification occur before a fully functional molecule will act. The polypeptide chains of proteins, like the strands of nucleic acids, are linear nonbranched polymers. There are 20 amino acids that serve as monomers of proteins. Each amino acid has a **carboxyl group**, an **amino group**, and side **radicals**, **R-groups** bound covalently to a **central carbon atom** (Fig. 3.3.5). The R group gives each amino acid its chemical identity. Set of Rs and the length of the polypeptide define the uniqueness of a primary structure.

The formation of a peptide bond occurs between carbon and nitrogen atoms in the main chain (Fig. 3.3.5). Polypeptide components that lead to a variety of configurations can be classified into four main categories: (1) **nonpolar (hydrophobic)**, (2) **polar (hydrophilic)**, (3) **negatively charged** and (4) **positively charged**. Polypeptides are long polymers, and because each position in a chain may be occupied by any one of the 20 amino acids, an enormous number of combinations in the polypeptides is possible. This leads to huge variation in chemical conformation and activity, with unique chemical properties of such polymers. For instance, for a polypeptide that is composed of 250–300 amino acids (an average polypeptide chain), upto $\mathbf{20^{250}–20^{300}}$ different molecules can be created using at random combinations of 20 amino acids.

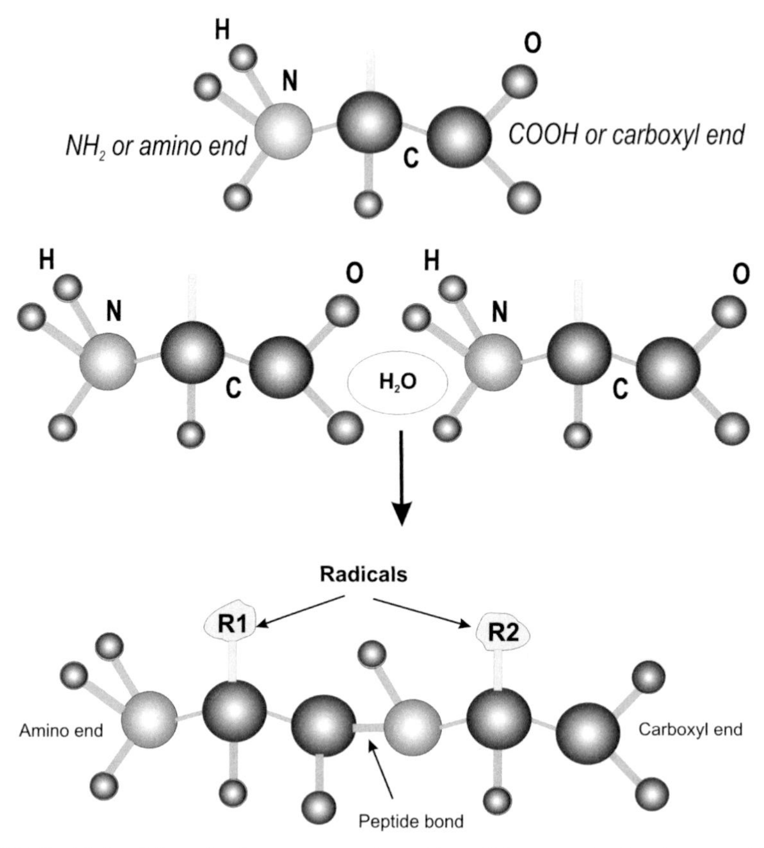

Fig. 3.3.5. Peptide bond formation between two amino acids. In light gray color the nitrogen and in dark gray carbon (large balls for both) are depicted. In small light gray balls the hydrogen or oxygen (slightly larger balls) atoms are shown.

In a protein four different levels of molecular structure are recognized: primary, secondary, tertiary and quaternary. The **primary structure** is determined by the sequence of amino acids in the linear set of the polypeptide (e.g., two of them combined in the dipeptide depicted in Fig. 3.3.5, bottom). Primary structure is specified by the sequence of nucleotides in DNA via an mRNA during translation as described above. Exact correspondence between DNA and amino acid sequences is called **colinearity**; the colinearity is an important quality for gene markers to be used as scientific tools. The primary structure of a polypeptide determines the specific properties of the higher orders of organization (conformation) when a protein is formed; i.e., it defines the space configuration. The **secondary structure** is defined by a regular or repeating conformation in space assumed by amino acids that closely align in the polypeptide chain, forming an **α-helix** as one type of secondary structure. Secondary structure was predicted by Pauling and Corey (1951; after Klug and Cummings 2002), on theoretical grounds. Another variant of this structure is the so-called **β-pleated-sheet** configuration. Most proteins exhibit a mixture of α- and β-structures (see 3d images at web-net, https://www.google.ru/search?q=molecules+structure+3d&newwindow=1&

biw=1430&bih=778&tbm=isch&tbo=u&source=univ&sa=X&ei=CnhdVMfWBoHU OdrCgOAL&ved=0CDIQsAQ&dpr=0.8, etc.). In the β-pleated-sheet model, a single polypeptide chain folds back on itself, or several chains run in parallel or antiparallel fashion beside each other. A combined structure then stabilized by hydrogen bonds formed among atoms present on the nearest chains. **Tertiary protein structure** (see web site above and many others) occurs due to the three-dimensional conformation of the whole chain in space. During formation, a protein aggregates its polypeptides, the sections of which turn and loop around themselves in a very specific fashion, forming the specific molecular shape of the protein. The **quaternary level of organization** is attributed only to oligomeric proteins that are composed of more than one polypeptide chain (subunit); thus this level is the conformation of the various subunits of one to the other(s). The individual subunits (protomers) have conformations that fit together with other subunits in a specific complementary way; for most enzymes the activity (catalytic) center becomes functional only when the quaternary structure is established. The forming of an active protein molecule is frequently accompanied by bonding a metal or co-factor, which binds, for example, oxygen, as in a case of hemoglobin and myoglobin (Fig. 3.3.6). Myoglobin, an oligomeric protein consisting of four polypeptide chains, was studied in more detail than many others. Its quaternary structure consists of four subunits (Fig. 3.3.6). Most enzymes of any their classes, like oxidoreductases, isomerases, ligases, and transferases, including DNA and RNA polymerases, demonstrate the quaternary structure.

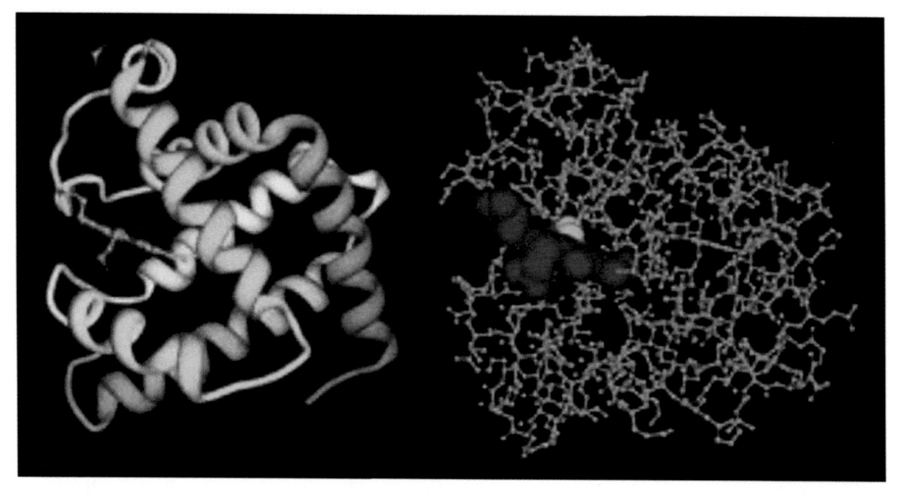

Fig. 3.3.6. The quaternary level of protein structure as seen in myoglobin. The left image shows 3d myoglobin molecule. The right image shows the detailed structure with surrounding atoms. (From http://www.rcsb.org/pdb/101/static101.do?p=education_discussion/Looking-at-Structures/coordinates.html; Phillips S.E. Structure and refinement of oxymyoglobin at 1.4 A resolution. Journal Mol. Biol. 1980. 142. 531–554. PUBMED: 7463482.)

Posttranslational Modification of Proteins

An important achievement of modern biochemistry is the understanding that after synthesis of macromolecules most of them are subjected to modification. Both the polypeptide chain and the RNA transcript are often modified once they have been synthesized. For proteins this additional processing is named as **posttranslational modification**. Although details of these modifications are beyond the scope of our course, it is necessary to be aware that they occur and that they are very important to the functioning of the protein as the final gene product. A few examples of posttranslational modification are given below. Posttranslational modifications are obviously important in achieving the functional status that is specific to any protein. Because the final three-and-four-dimensional structure of the molecule is intimately related to its specific function, how polypeptide chains ultimately fold into their final conformations is also an important topic. For many years it was accepted that protein folding was a spontaneous process, whereby the molecule achieved maximum thermodynamic stability, based largely on the combined chemical properties defined by the original amino acid sequence in the polypeptide chain(s), which compose this protein. Contrary to this, numerous recent investigations have shown that folding depends upon a group of other, ubiquitous proteins named **chaperones** (called also chaperonins or molecular chaperones). A chaperone's function is to facilitate the folding of other proteins.

Protein Function

For cellular organisms, which are most of the developed life forms on the Earth, one of the main functions is protein synthesis and protein functioning. It may even be said that *life itself in its main peculiarity is the self-replicating transformation of DNA (RNA) information into a variety of functioning proteins, realized in a regulated unity like a cell and an organism.* On the other hand, the diversity of tissue and organ functions in higher organisms is to a great extent dependent upon the diversity of protein functions and their qualities.

Proteins are the most abundant macromolecules of cells, where they play diverse roles. Thus, the respiratory function and oxygen transport occur because of **hemoglobin** and **myoglobin**. **Actin** and **myosin** are contractile proteins, found in abundance in muscle tissue, but present also in all dividing cells, making possible intracellular contraction like chromosome movement to the poles in mitosis and meiosis. **Collagen** and **keratin** are structural proteins of skin, connective tissue and hair of organisms such as mammals. Other examples are the **immunoglobulins**, which function in the immune system. **Histones**, which bind to DNA in eukaryotic organisms, are very conservative proteins giving stability to the chromosome during numerous packing-and-unpacking of chromosomes in cell circles. The list may be very vast, perhaps as great as the diversity of functions and structures in living matter is.

The largest group of proteins, that supports numerous biochemical catalytic reactions, is the **enzymes**, four groups of which were mentioned above. We will pay special attention to these proteins in Chapter 5, specifically from the viewpoint of biochemical genetics as to gene markers. Here it may be useful to extend our discussion and include more general information on their biological role. Enzymes increase the rate at which a chemical reaction reaches its equilibrium asymptotic state, but they do not alter the end point of the chemical equilibrium. **Biological catalysis** is a process by which the energy of activation for a given reaction is lowered. Because enzyme reactions are asymptotic, the achievement of an optimum condition is very important; it is created by factors including the correct temperature, pH, substrate concentration, etc. Conversion of a substrate into a product is a reversible reaction and depends frequently upon the concentration of two components. Numerous macromolecules like peptides, amino acids, vitamins, sugars, and alcohols are used as the substrates in many enzyme reactions. That is why, in many cases, the auto-regulation of metabolic reactions within cells occurs due to enzymes.

The catalytic properties and specificity of an enzyme's efficacy are determined by the chemical configuration of the molecule's **active site**. This site is associated with a crevice (a pit) on the surface of the enzyme, which binds the substrates, facilitating their interaction with other reagents. Enzymatically catalyzed reactions control most metabolic activities in the cell. There are two kinds of reactions **anabolic** and **catabolic**. The synthesis from simple components occurred through anabolic reactions, yielding the various components such as proteins, nucleic acids and lipids. Catabolic reactions are the decomposition reactions of large molecules into smaller ones with the release of chemical energy.

Protein Domains and Exon Shuffling

We concluded above that the active center is formed as some pit, which depends mostly on the structure of specific amino acid sequences in protein molecules. Not all such sequences are equally important. Usually only 50–300 amino acids constitute what are called **protein domains**, and are represented by modular portions of the protein that fold into stable, unique conformations. Different domains serve for different functional capabilities of a protein. There are one, two or several domains in a molecule. The domains and the tertiary-and-quaternary structure of proteins are interrelated. Each domain is a mixture of secondary structures, including both α-helixes and β-pleated sheets. The specific, unique conformation that is realized in a domain defines a specific function of the protein. For example, a domain may serve as the basis of an enzyme's active center, or it may provide the capability for specific binding with a membrane or another molecule. That is why the terms catalytic domains, DNA-binding domains, etc. are commonly used. The essence of the matter is that a protein must be considered as a set of structural and functional units or modules. Obviously, the presence of multiple domains within a single protein increases the ability of each molecule to take part in diverse functions and minimizes the energy requests. A genetic viewpoint for origin of protein domains was suggested by Gilbert (1978), who guessed **exon shuffling** as a source for a variety of domains.

Relationships among exons may be quite complex (Fig. 3.3.7). For the case of low-density lipoprotein (LDL), the first exon controls a signal sequence that is removed from the protein before the LDL receptor joins with the membrane. The next five exons comprise the domain, which specify the binding site for cholesterol. The next domain, that have large homology to the peptide hormone epidermal growth factor of a mouse, consists of a sequence of 400 amino acids. Eight exons encode this region, which contains three repeated sequences of 40 amino acids. The fifteenth exon defines

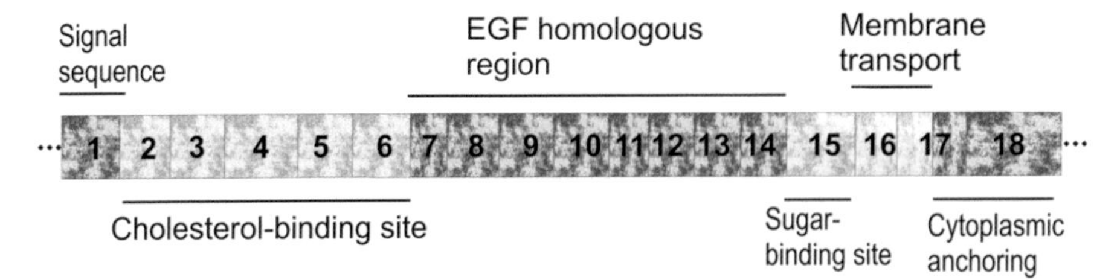

Fig. 3.3.7. A comparison of 18 exons making up the gene sequence encoding the LDL receptor protein. The exons are organized into five functional domains and one signal sequence.

the domain for the posttranslational addition of the carbohydrate. Lastly, the remaining two exons specify parts of the protein that belong to the membrane, attaching the LDL to sites named coated pits on the cell surface.

These observations on the LDL exons are compelling support for the theory of exon shuffling during evolution, as initially developed by Gilbert (1978). Also, there is agreement that protein domains are responsible for specific molecular interactions.

3.4 ESSENTIAL PROTEOMICS

A **proteome** is a set of proteins expressed and modified during the life circle of a cell. In a more strict sense, this term also describes the set of proteins expressed in a cell at a certain time. Correspondingly, **proteomics** is a science that focuses on the investigation of the proteome. It uses a range of techniques, beginning from genetic analysis and to mass spectrometry.

After realizing the genome sequencing projects, main attention of researchers is attracted to the question of protein structure and function. Genome projects make it clear that many newly discovered genes have unknown functions and others have only presumed functions known by analogy with similar genes. For example, in such a well-studied organism as the yeast, *S. cerevisiae*, more than half the genes in the genome studied have no known function. The Human Genome Project discovered that more than 40% of the predicted proteins have no assigned functions (see Chapter 16, Fig. 16.1.1).

From the above text it is clear that gene function involves more than identifying the gene products. Polypeptides, the gene products, are modified by the addition of chemical groups (methyl-, acetyl-, etc.) or linkage to sugars, lipids and other polypeptides. Polypeptides may be cleaved by the removal of end groups such as signal sequences, pro-peptides or initiator methionine residues. Proteins are internally and externally cross-linked, and even processed by removing internal amino acid sequences (called **inteins**). Over a hundred mechanisms of posttranslational modification are known. Thus, the human genome, which has 35,000–40,000 protein-coding genes, may produce about half a million different gene products. The goal of proteomics is to provide for each protein encoded by a genome, the information that characterizes function, structure, posttranslational modifications, cellular localization, variants and relationships to other proteins.

Proteomics Technology. Proteomics is using the very many methods of molecular biology that developed when genome projects became realized, including PCR (Polymerase Chain Reaction), different vectors, automated DNA sequence analysis and specialized software packages. Proteomic methods are quickly developing now, comprising one of the largest genetic frontiers. The basic techniques in proteomics are separation and identification of proteins that are isolated from cells. Most frequently, proteins are separated by two-dimensional gel electrophoresis (Fig. 3.4.1). In brief, by this technique, proteins extracted from cells of different tissues are placed in a polyacrylamide gel (PAG), and an electric power is applied to the gel. This separates the proteins according to the molecular charge. Then the gel is rotated 90 degrees, and new electrophoresis is performed, and proteins this time are separated in a second dimension according to their molecular size (weight) using the effect of PAG as a molecular sieve.

After the staining of the gels with the coomassie blue or other stain, the proteins that are highlighted by this stain are revealed as spots; typical gels may show 200–10,000 spots or fractions (Fig. 3.4.1). To identify individual proteins, spots are then cut out of the gel, and the proteins are partly digested with proteases such as trypsin. After that a set of characteristic fragments occur. These fragments are investigated by mass spectrometry using a peptide mass fingerprinting method. In the end, to identify the protein, the peptide mass is compared with the predicted masses using genetic-and-protein databases. To increase the speed and accuracy of proteome analysis, researchers widely used DNA software technology and other statistical tools.

The Bacterial Proteome. Simple organisms like, *Mycoplasma genitalium,* with a reduced genome of 480 genes, are the most convenient for analysis. Several groups are analyzing the proteome of *M. genitalium* in an attempt to define the minimum set of

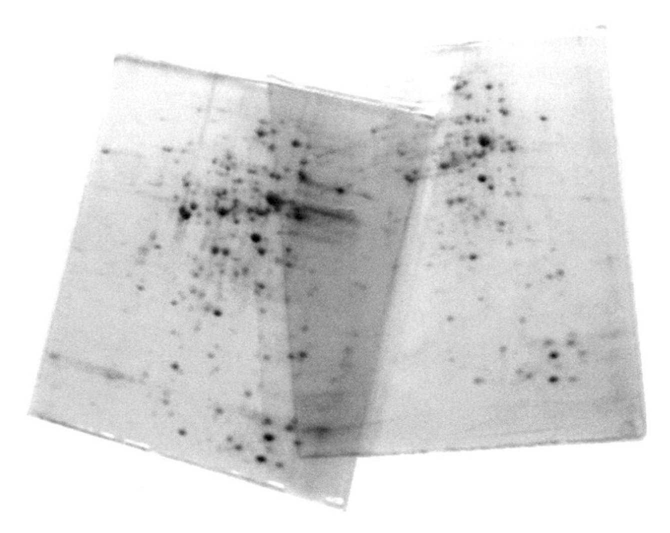

Fig. 3.4.1. A two-dimensional protein gel-electrophoresis results, showing the separated proteins as spots. Up to several thousand proteins can be displayed on such gels (from: http://de.wikipedia.org/wiki/Bild:Coomassie-2D-Gels.jpg; posted for free access by Jörg Bernhardt).

biochemical and metabolic reactions needed for a living system. In a study, Wasinger and colleagues (1995) used proteomics to provide an outcome of gene expression during the exponential and post-exponential growth phases in *M. genitalium.* Using two-dimensional electrophoresis, they identified a total of 427 protein spots-fractions in exponentially growing cultures.

Of these, 201 proteins were analyzed and identified by peptide digestion, mass spectrometry and comparison to known proteins. The analysis revealed 158 known proteins (33% of the proteome) and 17 unknown proteins. The remaining spots include a mixture of different fragments: derived from larger proteins, different isoforms of the same protein and posttranslationally modified polypeptide products. The identified proteins appeared to be major enzymes, including those that are involved in energy metabolism, transport of molecules within the cell membrane, DNA replication, transcription and translation. When exponential growth ends and the stationary phase starts, some 42% reduction occurred in the number of proteins that synthesized in cells. During the phase change, some new proteins are expressed, while other proteins decrease seriously their abundance. Such changes are apparently a consequence of nutrient change, increased acidity of the growth medium and other adaptations to changes in the environment.

The analysis performed helps to establish the minimum number of expressed genes that are required for independent existence, and to understand changes in gene expression associated with transition in growth conditions. Wasinger's study also shows that it is unlikely that conditions can be found under which a cell will express its entire proteome. In *M. genitalium,* only 33 percent of the proteome was expressed under conditions optimized for maximal growth. By technical limitations, these represent the most abundantly expressed proteins. The remaining 67 percent of the proteome are proteins likely to be expressed under a variety of environmental conditions, those present in too low an abundance to be detected by two-dimensional gel-electrophoresis or proteins that cannot be solubilized and extracted by experimental methods used (e.g., some rare enzymes). In any case, proteomic analysis provides new information that cannot be obtained by genome sequencing alone.

3.5 TRAINING COURSE, #3

3.1. Internet Support. Consider one example for protein synthesis on the University of Illinois web-site: http://www.gene.com.ac/AB/GG/protein_synthesis.html

3.2. Let us consider a few new terms.

- **Translation** is the biological polymerization of amino acids into polypeptide chain on mRNA matrix.
- **Protein** is linear nonbranched polymer of amino acids.
- **Primary structure** of protein is defined by the sequence order of amino acids in a molecule. Secondary, tertiary and quaternary structures of protein molecule are also present as specified above.

4

IMMUNOGENETIC ANALYSIS AND EVOLUTIONARY DISTANCES

MAIN GOALS

4.1 Introduction to Immunogenetics
4.2 Inheritance of Immunogenetic Traits and Population Variability
4.3 Immunogenetic Dating of Evolution
4.4 Dating of Evolution by Amino Acid Sequence Analysis
4.5 TRAINING COURSE, #4

SUMMARY

1. The best technique for immunogenetic polymorphism detection is probing with antisera.
2. Inheritance of immunogenetic traits is simply Mendelian. Hardy-Weinberg (H-W) law holds as a rule in natural populations.
3. Polymorphism of immunogenetic traits was found in more than 100 fish species. Thus, introduction of the immunogenetic approach made it possible to conduct experimental population genetic research in the wild.
4. More than two alleles were found in many blood group loci, proving the existence of multiple alleles and wide genetic variation for natural populations. Fish populations differ in allele frequencies, and sometimes even have unique alleles.
5. Immunogenetic dating provides reliable results for many groups of animals, both vertebrate and invertebrate, and helps to improve phylogeny in these groups.
6. Estimation of the number of amino acid substitutions through Poisson theory gives quite accurate empirical results when the considered evolutionary time period is relatively short. In the case of long history the Amino Acid Substitution Matrix method is used. The obtained data provide the evidence on the reliability of the Molecular Clocks.

4.1 INTRODUCTION TO IMMUNOGENETICS

Nowadays immunogenetics has moved mostly to the fields of medicine and genetic engineering. Nevertheless, I feel that students should know this technique despite the fact that it is not widely used now in phylogenetics and population genetics. Antibodies are used for molecular biology, and so students should have some ideas on this subject. Furthermore, the ideology introduced in the chapter is well related with main content of the book.

All vertebrate animals have developed an immune system. It is not so complicated and perfect in fish as in mammals but it still exists, and permits researchers to detect antigen-antibody reactions, which are the key to immunogenetics.

Immune reaction may occur in a tissue or in a blood system. In the last case, of the so-called humoral response, the antibodies (immunoglobulins: the labile proteins, e.g., α, β, γ-globulins) are secreted and come into sera where immune reactions occur throughout the blood system. The antibodies develop against different alien or "foreign" sources and their parts that are called antigens. The antigens are any macromolecules (peptides, lipoproteins and glycoproteins), viruses or bacteria. Sometimes even an organism's own cells can serve as antigens, which creates the autoimmune deseases, like rheumatic arthritis (6.5 million people in USA suffer from this desease), juvenile diabetes (0.5% of human population is impacted) and the disperse sclerosis (brain autoimmune reaction). In this chapter we will consider two figures on the basics of immunogenetics; their content must remind you about the antigen-antibody action. For a more detailed and general view of immunology and immunogenetics see the monograph by Kuby (1994).

Antibody-mediated (humoral immunity). Humoral immunity is regulated by B cells and the antibodies they produce (Fig. 4.1.1). Cell-mediated immunity is controlled by T cells. Antibody-mediated reactions defend against invading viruses and bacteria. Cell-mediated immunity concerns cells in the body that have been infected by viruses and bacteria; that protect against parasites, fungi, and protozoans; and also kill cancerous body cells.

The antibody binds with the antigen by a variable part of its immunoglobulin molecule chain (Fig. 4.1.1, 1; This and images 2–4 are from Purves et al. 1994; URL: www2.estrellamountain.edu/faculty/farabee/biobk/BioBookIMMUN.html), and its further action against antigens may continue in two major immune responses: Humoral (a) and Cellular (b) (Fig. 4.1.1, 2–4). At the end of both responses, antigen-bearing agents are inactivated (see Effector phase, Fig. 4.1.1, 4). If antigens are located on the surface of red blood cells, then the antibodies react with them, bind them in the dense aggregations and coagulate the blood in general. The next slide explains such a reaction for sharks (Fig. 4.1.2).The next step of the immunogenetic analysis is to find a source (serum) which reacts most specifically to the blood antigens. The left part in Fig. 4.1.2 (bottom) shows the iso-agglutination (a serum used from the same species) and the two phenotypes in immune reaction (S+ or S– cells), and the right part in this figure presents the sera reaction of a rabbit after his immunization. Here already three

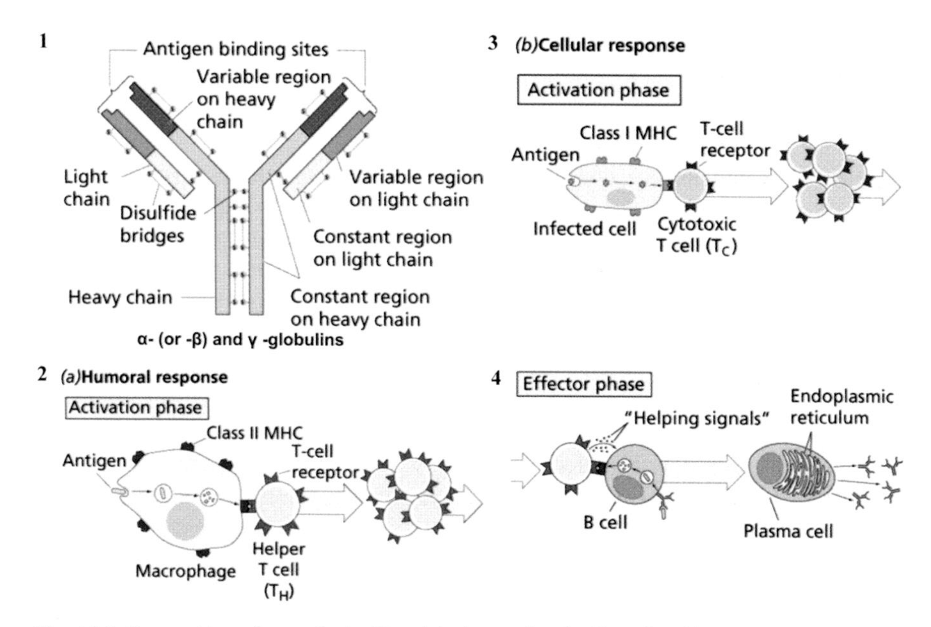

Fig. 4.1.1. Composition of an antibody (1) and the humoral and cell-mediated immune responses (2–4). MHC is Main Histo-compatibility Complex (from On-Line Biology Book 2002).

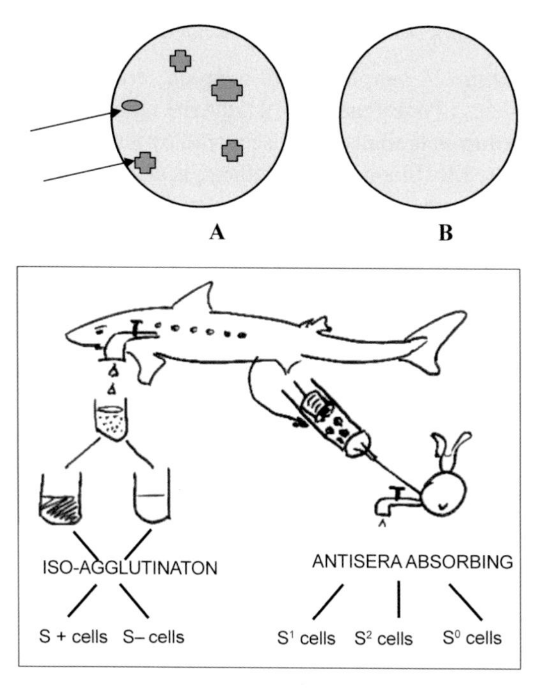

Fig. 4.1.2. A detection of the S-system of the blood group in the shark Squalus acanthias. (A) There is an immune reaction of the agglutination (Blood cell aggregations are shown by arrows, left) and no reaction (right), (B) There is no such reaction. Below are shown different steps of the immunogenetic analysis (Adopted from Altukhov 1974, with modifications).

phenotypes can be detected (S^1-, S^2-, S^0-cells). According to typical genetic approach, we can now suggest (hypothesize) a gene control of the S-system and find the evidence in its support. The first step in this direction is to arrange the repetition test, and then the second step is get crosses and perform segregation analysis. For wild populations, the most frequent approach consists in the analysis of naturally occurring offspring of the crosses (because most species are absent in captivity).

To summarize the Introduction let us define what a **normal** and **immune serum (pl. sera)** is. **Normal serum** is the usual serum of blood. **Immune serum** is a serum that developed in an organism after entering into the blood system of an antigen source. In Fig. 4.1.2 two reactions with normal and immune sera were considered (on the left and right side respectively). The immune sera are more sensitive, and in the case of the shark, three phenotypes were detected with the cells S^1, S^2 and S^0. Such immune sera are frequently called the **immune antisera**. To increase the power of the antisera they may be absorbed (or depleted) repeatedly with the same or different organism's blood. This gives the possibility of detecting antigens with smaller immune effects. Agglutination and watching of glued erythrocytes in a microscope is not the only way to detect the immune reaction. Precipitation is another technique to detect immune reaction. Antibodies that developed in an organism's serum against an antigen can be subjected to sedimentation or precipitation, and then investigated, for example by the immuno-electrophoresis. Note that sometimes the lectins, substances isolated from legumes and later found in other plants (they are now known to be widely present in many other organisms), served like sera in detecting an immunogenetic variation.

The ABO Blood Group. A simple case of a strong serum reaction is that in which three alternative alleles of one gene control the ABO human blood group (Fig. 4.1.3). The **ABO blood group** in humans was discovered by Karl Landsteiner in 1900. The ABO system, like the MN blood types or others, is characterized by the presence of specific antigens on the surface of red blood cells. The A and B antigens are distinct from the MN antigens, and are under the control of a different gene, located on chromosome 9. As in the MN system, one combination of alleles in the ABO system exhibits a codominant mode of inheritance. When individuals are tested using antisera that contain antibodies against the A or B antigen, four phenotypes are revealed. Each individual has either the A antigen (A phenotype), the B antigen (B phenotype), the A and B antigens (AB phenotype) or neither antigen (0 phenotype). In 1924, based on the family analysis of the blood types, it was hypothesized that these phenotypes were inherited as the result of three alleles of a single gene. Although different designations can be used, we use the symbols I^A, I^B and I^0 to distinguish these three alleles. The "*I*" designation stands for isoagglutinogen (another term for antigens). We may assume that I^A and I^B are alleles responsible for the production of their respective A and B antigens and that I^0 is an allele that does not produce any detectable A or B antigens. Our knowledge of human blood types has several practical applications, the most important of which is compatible blood transfusions. The human blood groups or blood phenotypes can be explained in the following schematic fashion (Fig. 4.1.3):

Phenotype	**Antigen**	**Genotype**
O (I)	No	I^0I^0,
A (II)	A	I^AI^A, I^AI^0,
B (III)	B	I^BI^B, I^BI^0,
AB (IV)	A and B	I^AI^B.

Fig. 4.1.3. The ABO human blood system. The I^A and I^B alleles expressed dominantly to the I^0 allele, but codominantly to each other.

Similar to the ABO system, distinct blood groups were detected in many fish species. The best studied are S- and A-systems of antigens in the shark (see above) and herring.

The herring A-system may be designated as follows:

Phenotype	**Antigen**	**Genotype**
A1 (I)	A1	A_1A_1, A_1A_2, A_1A_0,
A2 (II)	A2	A_2A_2, A_2A_0
A0 (III)	No	A_0A_0

$A_1 > A_2 > A_0$.

As we can see not all the herring genotypes can be distinguished by this reaction, and it is necessary to apply the techniques of depletion of antisera if we want more precision in the analysis. Before we finish the Introduction section, let us define what a blood group is. *The **blood group** is a phenotype of immune reaction, which is expressed by single or several genotypes, either in the homozygote or in the heterozygote.*

4.2 INHERITANCE OF IMMUNOGENETIC TRAITS AND POPULATION VARIABILITY

Immunogenetic traits are inherited usually in a simple Mendelian fashion. Alleles of the same locus may exhibit either codominance or dominance. In the last case the null alleles with no gene product occur as a counterpart of the active alleles. The examples of the shark S-system and herring A-system can be cited here. Similar inheritance modes were found for some blood groups of skipjack tuna.

Breeding experiments that are necessary for proving Mendelian inheritance are not common in fish or shellfish. One example was reported for the rainbow trout *S. gairdneri* (Sanders and Wright 1962). There are two antigens in this species the R-1 and R-2 that produce three phenotypes or the blood groups: R1 (Genotype R_1R_1), R2 (R_2R_2) and R1-2 (R_1R_2). When crosses were made, the cited authors found: R1 × R1-2, Fa = 1:1; R1-2 × R1-2, F2 = 49 R1 + 91 R1-2 + 42 R2 or 1:2:1, as expected in the monohybrid crosses with codominance.

In carp, the lipoprotein *Lpt-1* is inherited as a dominant-recessive trait. In the trout *S. trutta,* an ABO-like blood system was found but it was determined by two alleles. In the latter case, it was difficult to perform the analysis due to an ontogenetic variation.

Quite frequently, the immunogenetic traits are coded by a series of multiple alleles: the herring has 3 alleles of the A-system, the shark 3 alleles of the S-system and the skipjack tuna upto 12 alleles of one of the four blood systems obtained.

The following are the general peculiarities of the inheritance of the immunogenetic traits:

1. Dominance, in which in the heterozygote or in the hybrid, only one antigen is available;
2. Codominance, where both alleles are active in the heterozygote;
3. Epistasis, where one locus product inhibits the product of another locus;
4. Newborns, in which the heterozygote phenotype does not represent simple addition of two homozygote features (We will better understand why it is so when we learn more about biochemical genetics);
5. Multiple alleles, where more than two alleles are found in many blood group loci.

The Hardy-Weinberg Law. Because of the difficult rearing and artificial breeding for most commercial fish and shellfish species, the inheritance at the immunogenetic traits is usually proved via a special approach, by using offspring of the crosses that routinely occurred in nature. If matings occur at random, certain proportions among gametic and genotypic frequencies are expected. These relations were established by Hardy and Weinberg, and after them the law or ratio **(H-W)** originated. Before we define this law, let us consider what the expected conditions for this law or model are.

First, we will consider the large (infinite in a theory) population of a diploid bisexual organism, mating randomly with single diallelic, autosomal locus taking part in segregation. So, we can write, for example:

Genotypes: R_1R_1 R_1R_2 R_2R_2

Numbers: D H R,

D+H+R = N individuals sampled (2N genes). Allelic (or equal gametic) frequency is than:

$p_{(R1)} = (2D + H)/2N = (D + 1/2\ H)/N$ and $q_{(R2)} = (2R + H)/2N = (R + 1/2\ H)/N$. The frequency of the second allele can be defined simply: $q_{(R2)} = 1 - p_{(R1)}$, because $p_{(R1)} + q_{(R2)} = 1$. Frequencies in such conditions are probabilities. Now we can turn from allelic or gametic frequencies to their expectations. Under random mating, if *p* and *q* are gamete frequencies (or probabilities), we have the frequencies of zygotes as the products of these probabilities in Females (F) and Males (M) as expected in F_1:

Female	***Male (M)***	
(F)	*(R1) p*	*(R2) q*
(R1) p	p^2	*pq*
(R2) q	*pq*	q^2

In other words: $(p + q)^2 = p^2\ (R_1R_1) + 2pq\ (R_1R_2) + q^2\ (R_2R_2)$. These binomial or equilibrium proportions among gametic and genotypic frequencies hold as long as stochastic conditions exist for the model. They can be disturbed by the systematic evolutionary factors (natural selection, drift, migration), as well as other causes that can change condition of equilibrium, like assortative mating and inbreeding.

The proportion above is an analytical representation of the **H-W** law. In words the **H-W** law formulation is *"In a large, randomly mating population, starting at the first generation, there are equilibrium binomial proportions between gametic and genotypic frequencies, which exist infinitely long until the population is subjected to a systematic action of one or more evolutionary forces, like natural selection, random drift, migration or action of other disturbing causes"*.

The **H-W** ratio gives a very easy way to check the genetic control scheme in the wild populations. You may define a genetic hypothesis, calculate allele frequencies, compare them with expected ones under H-W ratio and then make a decision on correctness of the hypothesis suggested (We are going to do such a task in the Training Course #4).

Comparison of Fish Species

When immunochemical markers were introduced to genetics in 1950s, a wide variety of natural populations had been investigated. Beyond those already mentioned, that is, the shark, herring and tuna, the analysis of the European anchovy, cod, trouts and others also performed. In the cod *Gadus morhua,* the A- and E-, in the skipjack tuna *Katsuvonus pelamis* the C-, Tg, G, Keyvamfar, in the sea bream *Sebastes mentella* and in the European anchovy *Engraulis encrasicholus* the A-blood groups were discovered (Altukhov 1974; Kirpitchnikov 1987).

The allele frequencies were calculated, and different populations compared. Polymorphism of immunogenetic traits was obtained among more than 100 fish species. The general conclusion was that separate populations of the species have, as a rule, different gene pools. The H-W ratio holds in most cases, but at the same time the populations in many different species in nature are not alike. In the tuna, for example, allele differences were available, but only in a vast area, like different parts of the Pacific Ocean. In contrast, genetically distinct populations of the anchovy exist in the comparatively small Black Sea and Azov Sea. In fact, spatial genetic variation was recorded even in the Azov Sea (Fig. 4.2.1). A unique allele was found for the Azov Sea population (Altukhov 1974).

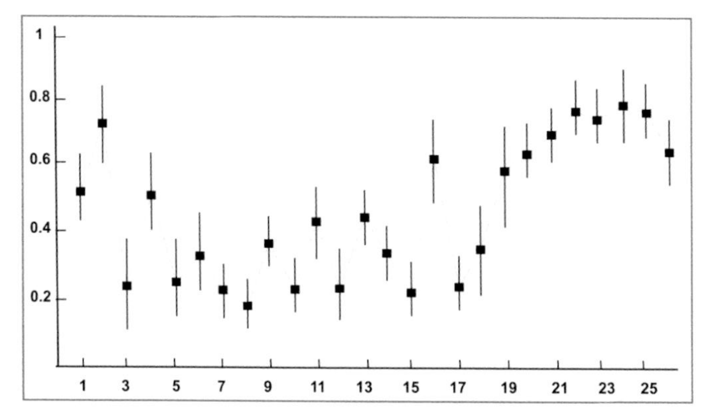

Fig. 4.2.1. Space variation of blood group A_0 in spanning gatherings of anchovy *E. encrasicholus* in the Azov Sea. On Y-axis is allele frequency, on X-axis are samples (from Altukhov 1974).

4.3 IMMUNOGENETIC DATING OF EVOLUTION

Amino acid sequence data are useful, beyond other things, for estimating evolutionary time and clarifying the genetic relationships of organisms, although acquisition of sequence data is time-consuming. Nucleotide sequencing is much easier than amino acid sequencing but, takes a lot of time too. For taxonomic purposes, we may need a simpler method (Avise 1994). One such method is to use the intensity of the immunological reaction between antigens and antisera prepared from different species. There are several methods for measuring the intensity of the immunologic reaction, but the simplest and most useful method seems to be that of the quantitative microcomplement fixation of a purified protein, introduced by Sarich and Wilson (1966). The protein often used for this purpose is a serum albumin (Champion et al. 1974). Briefly, the method is as follows (Fig. 4.3.1).

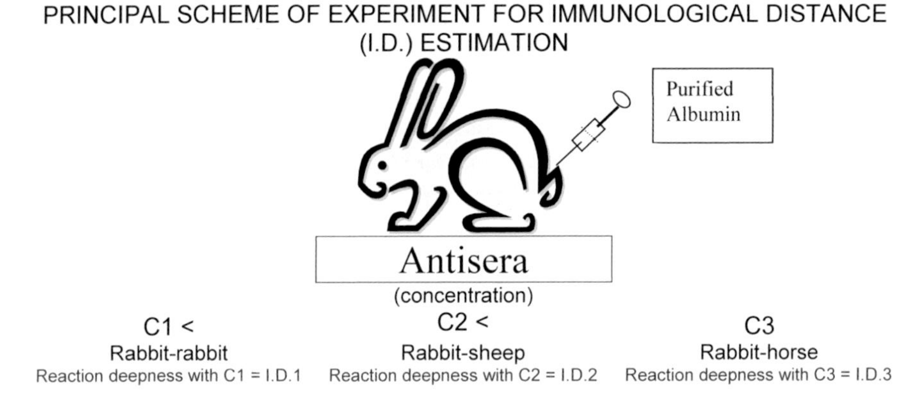

Fig. 4.3.1. Obtaining and using antisera for I.D. estimation. The explanation is in the text.

The antisera to be used are produced by immunization of the rabbit with a purified serum albumin from an organism of the group to be tested. The antisera strongly react with the albumin from the same organism or species (homologous antigen) but less strongly with that from another species (heterologous antigen) for a given concentration of antisera. If the serum concentration is raised, however, the level of the reaction with heterologous antigen may become the same as with homologous antigen. So, we may measure the antigenic difference between two albumins by the degree to which the antiserum concentration must be increased to produce the same reaction as that with a homologous albumin. This degree was called the index of dissimilarity (*I.D.*). The antigen-antibody reaction is measured by a method called quantitative complement fixation. Sarich and Wilson (1967) showed that the logarithm of *I.D.* is approximately linearly related to the time of divergence between the organisms investigated. They called $d_i = 100 \times \log_{10} (I.D.)$ the immunological distance. In many proteins (albumin, lysozyme, ribonuclease, etc.) this d_i is linearly related to the proportion of different amino acids between the two sequences (Prager and Wilson 1971; Benjam et al. 1984). It is unclear, why d_i is linearly connected to the proportion of different amino acids but this is an experimental fact. So, the empirical property of d_i may be used for the estimation of genetic distance among species relatively easily because this technique is much more feasible than amino acid sequencing. According to Maxson and Wilson (1974), one divergence unit corresponds to roughly one amino acid difference in albumin. The relationship between d_i and the time (t) since divergence between two species may be written as

$$T = cd_i \qquad \textbf{(4.1)}$$

In equation 4.1 c is the proportionality constant, which varies with the protein used and also may change depending on taxa, e.g., for mammals, reptiles, and frogs $c = 5.5 - 6.0 \times 10^5$, and for birds $c = 1.9 \times 10^6$ (Prager et al. 1974; Wilson et al. 1977; Collier and O'Brien 1985). As we see, the rate of evolution of albumin is nearly three times slower in birds than in other organisms compared. That difference possibly may be connected with the relatively recent origin of most bird orders (Wyles et al. 1983). Other reasons for these differences will be discussed when we consider biochemical genetic distances and DNA sequence divergence database.

Albumin cannot be used as an immunogenetic marker in invertebrates. Beverley and Wilson (1984; 1985), therefore, used a larval hemolymph protein in their study of the phylogenetic relationship of various species of Drosophilidae and higher Diptera. Using information on fossils in amber, continental drift, island formation, etc., they estimated that the proportionality constant for this protein is $c = 8 \times 10^5$. Obviously, this value is similar to that for albumin. The estimate of t obtained by equation (4.1) is subjected to four different types of errors. The first type is experimental error. According to Sarich and Wilson (1966), this error is generally less than 2 percent of the estimate, even if the estimate is relatively small. The second type of error arises when the antigen (protein) used is polymorphic in the species examined. If distantly related species are compared, however, this kind of error is small. The third type of error is generated by the fact that amino acid substitution in antigenic proteins occurs stochastically rather than deterministically. As mentioned earlier, this kind of error will make the variance of d_i larger than the mean. Thus, the third mistake is significantly larger than the former two. The fourth kind of error occurs because d_i is not strictly proportional to the number of amino acid substitutions (Champion et al. 1974). Probably, this kind of error is not less significant than the third kind. Application of immunogenetic methods has had a big impact on the study of evolution. Thus, Sarich and Wilson (1966; 1967) used this method for reconstruction of phylogenetic relationships among primates. They obtained surprising results. In disagreement with the prevailing view, they found that chimpanzees and gorillas are more closely related to humans, while latter belong to a different family (Hominidae), compared to orangutans and gibbons, with which they share same family (Pongidae). These observations shook anthropology at that time, indicating that humans, chimpanzees, and gorillas diverged about 5 MY ago, as mentioned earlier (Chapter 1).

One of the interesting observations found in modern data is that different frog species belonging to the same genus often have large immunological distances, comparable to values for different families or orders of mammals (Post and Uzzell 1981; Maxson 1984; Maxson et al. 1975). Later a similar pattern will be shown for protein markers (Chapters 5 and 7). Such evidence suggests that frog species may be separated for long times but have a slow evolution of morphological characters. Similarly, some pairs of *Drosophila* species may diverge over a long time. Beverley and Wilson (1984; 1985) found that the subgenus *Drosophila* (e.g., *D. virilis, D. mulleri*) diverged from the *D. melanogaster* species group about 60 MY ago, while Hawaiian *Drosophila* species diverged from the subgenus *Drosophila* some 40 MY ago (Fig. 4.3.2).

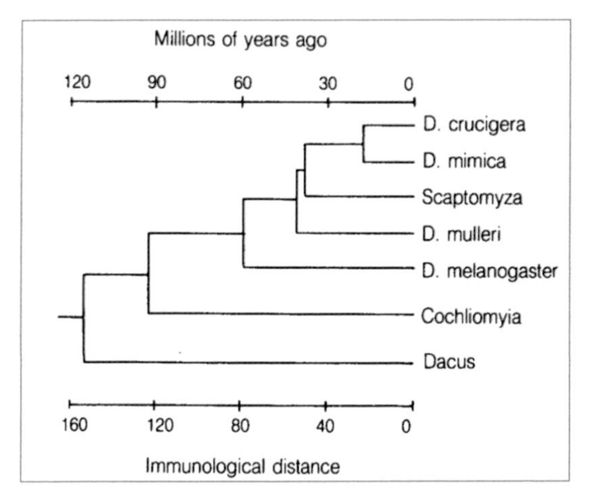

Fig. 4.3.2. Phylogenetic tree reconstructed from immunological distance data for seven fly lineages. *D. crucigera* and *D. mimica* are Hawaiian drosophilids, whereas *D. mulleri* and *D. melanogaster* are continental drosophilids. Scaptomyza is a fly genus closely related to Drosophila. Drosophila, Cochliomyia, and Dacus belong to different families (from Beverley and Wilson 1984).

Geological records on the Hawaiian archipelago give dating of the order of 5–6 MY (Table 1.1.1, Chapter 1), so one might expect that all Hawaiian *Drosophila* species have originated relatively recently. However, immunogenetic data suggest that such species as *D. crucigera* and *D. mimica* of Hawaiian species diverged about 20 MY ago. Some close groups of fly originated even earlier, some 40 MY ago. Nevertheless, these observations agree with the history of the Koko Seamount – Midway – Hawaii Archipelago occurring during the past 70 MY (Beverley and Wilson 1985).

4.4 DATING OF EVOLUTION BY AMINO ACID SEQUENCE ANALYSIS

In the previous section we dealt with the immunogenetic distances, which depend on the number of amino acid differences in polypeptides of antibodies. This technique has its advantages due to its relative simplicity as noted earlier. However, there is a necessity to detect amino acid differences in all coding genes not only immunogenetic. Data on a sequence of amino acids in a peptide can provide such information. Sequencing of proteins is a laborious task, and is now substituted by DNA sequencing methods. Nevertheless, we have to consider the subject of the approach in brief to be better prepared for future material on DNA techniques.

Proportion of Different Amino Acids and the Number of Amino Acid Substitutions

The study of the evolutionary change of proteins or polypeptides starts from the comparison of two or more amino acid sequences of a given polypeptide from different organisms. A simple quantity to measure the evolutionary divergence between a pair of amino acid sequences is the proportion (p) of different amino acids between the two sequences. This proportion is estimated by

$$\hat{p} = n_d/n, \quad (4.2)$$

where n is the total number of amino acids compared, and n_d is the number of different amino acids. If all amino acid sites are subject to substitution with an equal probability, n_d follows the binomial distribution. Therefore, the variance of $\hat{p}$ is given by

$$V(\hat{p}) = p(1 - p)/n. \quad (4.3)$$

When p is small, it is approximately equal to the number of amino acid substitutions per average site. When p is large, however, it is no longer a good measure of this number, because there might have been two or more amino acid substitutions at sites where amino acids are different between the two sequences. To estimate the number of amino acid substitutions from p, we need a mathematical model.

Poisson Process. A simple mathematical model that can be used for relating p to the expected number of amino acid substitutions per site is the Poisson process in probability theory (Nei 1987). Let λ be the rate of amino acid substitution per year at a particular amino acid site, and assume for simplicity that it is the same for all sites. This assumption does not necessarily hold in reality, but the error introduced by this assumption is small until a very long evolutionary time is involved. The mean number of amino acid substitutions per site during a period of t years is then λt, and the probability of occurrence of r amino acid substitutions at a certain site is derived from the following Poisson distribution:

$$Pr(t) = e^{-\lambda t} (\lambda t)^r/r! \quad (4.4)$$

Therefore, the probability that no change has occurred at a given site is $P_0(t) = e^{-\lambda t}$. Thus, if the number of amino acids in a polypeptide is n, the expected number of unchanged amino acids is $ne^{-\lambda t}$. In reality, we generally do not know the amino acid sequence for an ancestral species, so that (4.4) is not applicable. The number of amino acid substitutions is usually computed by comparing homologous polypeptides from two different organisms that diverged t years ago. Assuming the probability of no amino acid substitution occurrence during t years as $e^{-\lambda t}$, the probability (q) that neither of the homologous sites of the two polypeptides undergoes substitution is

$$q = e^{-2\lambda t}. \quad (4.5)$$

This probability can be estimated by $\hat{q} \equiv 1 - \hat{p} = n_i/n$, where n_i is the number of identical amino acids between the two polypeptides. The equation $q = e^{-2\lambda t}$ is approximate because backward mutations and parallel mutations (the same mutations occurring at the homologous amino acid sets in two different sequence lineages) are not taken into account. The effects of these mutations are generally very small, however, unless a long evolutionary time is considered.

If we use (4.5), the total number of amino acid substitutions per site for the two polypeptides ($d = 2\lambda t$) can be estimated by

$$\hat{d} = -log_e \hat{q}. \qquad \textbf{(4.6)}$$

Therefore, if we know t, λ is estimated by $\hat{\lambda} = \hat{d}/(2t)$. On the other hand, if we know λ, t is estimated by $\hat{t} = \hat{d}/(2\lambda)$. The large-sample variance of d is

$$V(\hat{d}) = [\mathrm{d}d/\mathrm{d}q]^2 * V(q) = (1 - q)/qn, \qquad \textbf{(4.7)}$$

since $V(\hat{q}) = q(1 - q)/n$ (Elandt-Johnson 1970). Obviously, the variances of $\hat{\lambda}$ and $\hat{t}$ are given by $V\hat{d}/(2t)^2$ and $V(\hat{d})/(2\lambda)^2$, respectively.

It should be noted that if we knew the numbers of amino acid substitutions for all amino acid sites, the variance of the number of amino acid substitutions per site would have been $2\lambda t/n$ under the Poisson process (The variance of a Poisson variable is equal to the mean). In practice, it is impossible to know these numbers, so we must estimate d by equation (4.6). The equation (4.6) is based on incomplete information on amino acid substitutions, so (4.7) gives a variance larger than $2\lambda t/n$.

We assumed above that the rate of amino acid substitution is the same for all amino acid sites. However, this assumption usually does not hold, because the rate is higher at functionally less important sites than at functionally more important sites (King and Jukes 1969; Dickerson 1971; Dayhoff 1978; Li et al. 1984; Nei 1987; Li 1997). Indeed, Fitch and Margoliash (1967b) and Uzzell and Corbin (1971) have shown that the distribution of the number of amino acid substitutions has a larger variance than the Poisson variance. However, equation (4.6) is quite robust, and approximately holds even if the rate varies considerably from site to site. This can be seen by considering the extreme case, where proportion a of amino acid sites is invariant and proportion $1 - a$ is subjected to the Poisson change. The expected proportion of identical amino acids for this case is given by $q = a + (1 - a)\ e^{-2\lambda t}$. Strictly speaking, therefore, $d \equiv -log_e q$ is not linear-dependent with evolutionary, time. However, d is approximately linear with time when $2\lambda t <= 1$ (Fig. 4.4.1). It should be noted that when $2\lambda t << 1$, $e^{-2\lambda t} \approx 1 - 2\lambda t$. Therefore, $d \approx -log_e[1 - 2(1 - a)\ \lambda t]$. Because $(1 - a)\ \lambda$ is the average rate ($\bar{\lambda}$) of amino acid substitution for all amino acid sites, d can be written as $2\ \bar{\lambda}\ t$. Namely, when λ varies with amino acid, $d = 2\ \bar{\lambda}\ t$ still holds unless $2\ \bar{\lambda}\ t$ is large.

In the formulations above, we assumed that the two amino acid sequences in the comparison have the same number of amino acids and that the divergence between them occurred solely by amino acid substitutions. When the two sequences are distantly related, however, insertions and deletions are often involved. In this case, the locations of insertions and deletions must first be identified. When the number of insertions or deletions involved is small, as in the case of the example that will be considered in Training Course task #4, this can be done relatively easily. When the number of insertions and deletions is relatively large, however, the alignment of amino acid sequences is quite troublesome. Because this problem is usually more serious with DNA sequences, and the problem is nearly identical for the two types of data, we will discuss it in the subsequent chapter on DNA sequence analysis.

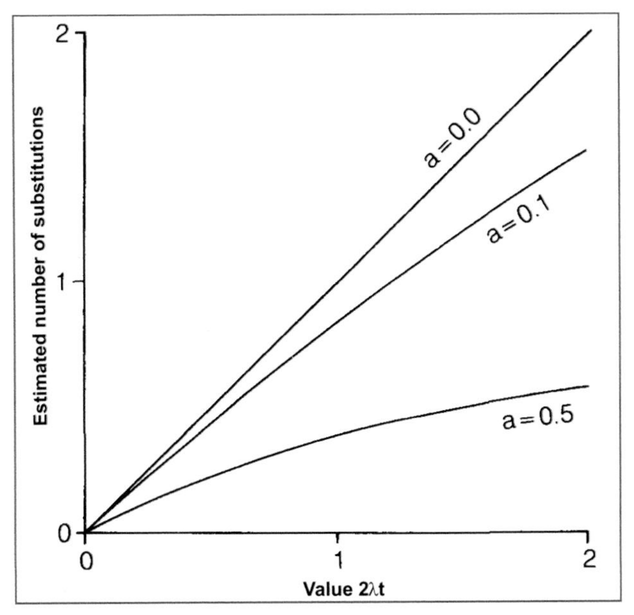

Fig. 4.4.1. Effects of invariable amino acids on the estimate of amino acid substitutions (d); a = proportion of invariable amino acids in a polypeptide (from Nei 1987 with permission).

Amino Acid Substitution Matrix

The above Poisson motion method for estimating the number of amino acid substitutions gives quite an accurate estimate when the evolutionary time considered is relatively short. When the amino acid sequences from distantly related organisms are compared, however, the effects of backward and parallel mutations cannot be excluded, and the Poisson process method is expected to give an underestimation. Unequal rates of substitution at different amino acid sites will also contribute to the inaccuracy of the estimate obtained. To minimize these problems, Dayhoff (1978) proposed another approach, the so called **Amino Acid Substitution Matrix** method, based on the accepted point mutations (PAM) number. Principles of this approach are explained in the original paper and summarized by Nei (1987, pp. 45–46).

4.5 TRAINING COURSE, #4

We have to be acquainted with the following main topics.

1. Hardy-Weinberg Ratio: Repeat formulation and solve the numerical example for testing deviation between observed and expected frequencies by chi-square formula.
2. Estimates of standard errors (SE) of allele frequency basing on expression (4.3); manual example with calculators. Chi-Square Test of Heterogeneity.

3. Software: SPECSTAT. Solve Numerical Examples on H-W Ratio and allele frequency heterogeneity in Pacific Herring.
4. Make recalculation of *d*^ for sample of sequence data on α chains of hemoglobins of 4 organisms.

   Figure 4.4.2 shows the amino acid sequences of hemoglobin α chains from the human, horse, bovine, and carp. The three mammalian hemoglobins consist of 141 amino acids, whereas the carp hemoglobin has 142 amino acids. Comparison of these sequences suggests that deletions or insertions occurred at three different positions after the divergence between fish and mammals. If we ignore these deletions/insertions, the proportion of different amino acids (p^) and the estimate of the number of amino acid substitutions for each pair of organisms becomes as given in Table 4.4.1 (Nei 1987, Table 4.1). For example, in the case of the human and carp p^ = 68/140 = 0.486, and *d*^ = $-\log_e(1 - \text{p}^\wedge) = 0.666$. On the other hand, the variance of *d*^ becomes *V*(*d*^) = 0.486/(0.514*140) = 0.006754, the standard error being 0.082.

```
Human    VLSPADKTNVKAAWGKVGAHAGEYGAEALERMFLSFPTTKTYFPHF-DLSHGSAQVKGHG
Horse    VLSAADKTNVKAAWSKVGGHAGEYGAEALERMFLGFPTTKTYFPHF-DLSHGSAQVKAHG
Bovine   VLSAADKGNVKAAWGKVGGHAAEYGAEALERMFLSFPTTKTYFPHF-DLSHGSAQVKGHG
Carp     SLSDKDKAAVKIAWAKISPKADDIGAEALGRMLTVYPQTKTYFAHWADLSPGSGPVK-HG
Human    KKVA-DALTNAVAHVDDMPNALSALSDLHAHKLRVDPVNFKLLSHCLLVTLAAHLPAEFT
Horse    KKVA-DGLTLAVGHLDDLPGALSDLSNLHAHKLRVDPVNFKLLSHCLLSTLAVHLPNDFT
Bovine   AKVA-AALTKAVEHLDDLPGALSELSDLHAHKLRVDPVNFKLLSHSLLVTLASHLPSDFT
Carp     KKVIMGAVGDAVSKIDDLVGGLASLSELHASKLRVDPANFKILANHIVVGIMFYLPGDFP

Human    PAVHASLDKFLASVSTVLTSKYR
Horse    PAVHASLDKFLSSVSTVLTSKYR
Bovine   PAVHASLDKFLANVSTVLTSKYR
Carp     PEVHMSVDKFFQNLALALSEKYR
```

Fig. 4.4.2. Amino acid sequences in the α chains of hemoglobins from four vertebrate species. Amino acids are expressed in terms of one-letter codes. The hyphens indicate the positions of deletions or insertions (from Nei 1987 with permission).

Table 4.4.1. Numbers (frequencies) of amino acid differences (above the diagonal) between hemoglobin $\boldsymbol{\alpha}$, chains from the human, horse, bovine and carp. Deletions and insertions were excluded from the computation, the total number of amino acids used being 140. The figures in parentheses are the proportions of different amino acids. The values given below the diagonal are estimates of the average number of amino acid substitutions per site between two species (*d*^) (from Nei 1987).

Organism	***Human***	***Horse***	***Bovine***	***Carp***
Human		18(0.129)	17(0.121)	68(0.486)
Horse	0.138 ± 0.032		18(0.129)	66(0.471)
Bovine	0.129 ± 0.031	0.138 ± 0.032		65(0.464)
Carp	0.666 ± 0.082	0.637 ± 0.080	0.624 ± 0.079	

Note. Table 4.4.1 indicates that *d*^ is nearly the same for the three pairs of mammalian species, whereas the *d*^ values between the carp and the mammalian species are considerably larger. This observation is in agreement with the view that the number of amino acid substitutions is roughly proportional to evolutionary time, since the human, horse and bovine diverged about 75 MY ago, whereas the carp (bony fish) and mammals diverged about 400 MY ago (Fig. 1.2.2, Chapter 1). The average d^ for the three pairs of mammalian species is 0.135, whereas the average for the pairs of the carp and the three mammalian species is 0.642, the latter being about five times the former. This ratio is close to the ratio of the corresponding divergence times.

5. Consider and learn the following terms.
   Normal/Immune Sera are the blood sera that developed in an organism without/with the immunization and introduction of an antigen respectively.
   Multiple Alleles are several alleles that segregate at the same genetic locus in a population.
   Null Allele is an allele of the locus that gives no active gene product (Produces no antigen).
   Hardy-Weinberg Law. In a random mating population there is an equilibrium binomial ratio between gametic and genotypic frequencies, which stay infinitely stable in space and in time in the absence of action of evolutionary forces.

5

BIOCHEMICAL GENETICS

MAIN GOALS

5.1 Introduction to the Methods of Protein Variability Detection
5.2 Interpreting Protein Variability
5.3 Genetics of Isozymes
5.4 Evolution of a Genome Regulation
5.5 TRAINING COURSE, #5

SUMMARY

1. Isozymes or rather allozymes are very good markers of intra-locus or allelic variability.
2. Expression of enzyme loci is usually codominant.
3. Multiple alleles are common for allozyme loci.
4. Inheritance of allozymes is simply Mendelian. H-W law is normally held in nature for these traits.
5. Isozymes are the best markers to detect tissue activity of genes.
6. Isozymes permit investigation of the time of gene activation and expression of genes in ontogenesis.

5.1 INTRODUCTION TO THE METHODS OF PROTEIN VARIABILITY DETECTION

There are a lot of methods for detection of protein variability: chromatography, gel filtration, ultra-centrifuge sedimentation and others. The main and most sensitive, however, remains gel electrophoresis of proteins with subsequent histochemical detection of their activity.

The Modes of Electrophoresis

Five of 20 common amino acids, which form proteins, have electric charges. The charges of arginine, histidine and lysine are positive, while those of aspartic acid and glutamic acid are negative. Proteins have different amino acid assemblages and hence, different net electrical charges. Electrophoresis uses this physical-and-chemical property of proteins to separate their mixtures on the basis of the charge. If there are allelic differences at a protein-coding locus, the net charge of the polypeptide or protein, defined by these alleles, often changes too. Gel electrophoresis makes it possible to identify such allelic differences. The principal techniques of gel electrophoresis are simple (Fig. 5.1.1). The electrophoretic procedure includes a gel (commonly starch or polyacrylamide, PAG) and a buffer, which gives a flow of ions when DC power is applied. The solutions of proteins are introduced in the gel and separated by the passage of the electrical current. Before that, the mixtures of proteins are extracted with water-based solutions from tissues such as skeletal muscle, kidney, liver, etc. Also protein solutions contained in body fluids like a blood serum, for instance, may be used directly. Frequently buffered aqueous solvents with low ionic concentration and neutral pH are used for the extraction; additional stabilizers of chemical bonds and nonionic detergents may also be added. The water-soluble protein mixtures are typically introduced to the gel on a piece of special filter (chromatography) paper that is saturated with the protein extract (solution). In PAG the solution is introduced by a pipette. Protein extracts (specimens) sampled from 25–50 or more individuals could be then introduced to a single gel for further electrophoresis, although only 8 individuals are pictured in Fig. 5.1.1.

Electrophoresis in general is based on the ionization of molecules in the special solutions (buffers) and movement of the charged ions (molecules) to the anode (+) or cathode (–). In biochemical genetics the electrophoresis is performed in a different

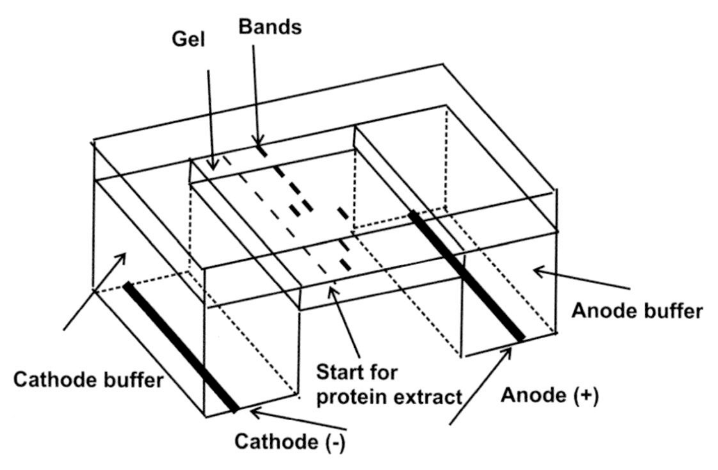

Fig. 5.1.1. General scheme of electrophoresis. Protein extract (specimen) is taken from a tissue such as muscle or liver. Specimens from different individual are introduced one by one to the gel by soaked filter paper inserted in the slots. Different forms of a particular protein often move different distances from the point of applications when electric current ("–" and "+") is applied because they have different charges. The bands are identified later by a specific staining.

medium, like the previously mentioned starch-gel or PAG-gel, as well agarose (or agar-agar), carboxy-methyl cellulose, desoxy-ethyl amino-ethyl cellulose and others.

The macromolecules are separated largely according to their electric charge but the medium may also act as a molecular sieve, i.e., separating molecules by molecular weight. Especially effective in this sense is PAG-electrophoresis; PAG-gel polymer is made from artificial monomers, and the size of pores may be exactly planned. There are two main modes of electrophoresis: (1) continuous and (2) discontinuous. In the first mode, the content of the buffer is the same in the gel, cathode and anode parts of the electrophoretic system. In the second, the content and concentration of a buffer may be different in these parts of the electrophoretic system. What is the buffer? The buffer is a special solution (electrolyte), whose content has two components: one is a strong acid and the other is the salt of a weak base, or vice versa. Frequently, both of the components, the acid and base, is not strong. I will give three examples that represent these cases: TRIS-HCL buffer, Boric-NaOH buffer and TRIS-citric buffer (TRIS is a popular base in biochemistry, and its whole name is Tris hydroxy-methyl amino-methane).

Histochemical Staining of Macromolecules

After electrophoresis we must find a way to detect the location of the proteins on a gel; they are not visible because the majority of the proteins are colorless and present in a small amount. Histochemical staining permits us to visualize the proteins. There are three types of such a staining: (i) Specific staining for a different macromolecule type, (ii) Specific staining for a certain group of the proteins and (iii) Oxidoreductase (Dehydrogenase) reaction with a specific substrate.

i. In the first type of reactions the specific stains of a wide spectrum are used. For example, all the proteins are stained with Amido Black-B or Comassie Blue. DNA is stained with the Cytoorsein, Methylene Blue and others.
ii. The second mode is a specific stain that permits us to detect a certain chemical group in the proteins. With this staining reaction it is possible to resolve 9 or more groups. Let us remember three: (1) amino groups of ending α-amino acids, (2) ending and side carboxyl groups, (3) Sulfhydryl (SH) groups of cysteine.
iii. The third is the dehydrogenase reaction. That is the most important reaction for the detection of enzymes. The typical reaction is as follows:

DH
Substrate Product

However, some other components participate in the reaction. Let us consider as an example the staining for the detection of lactate dehydrogenase (LDH):

REACTION: **Lactate ⟷ LDH ⟷ Pyruvate**

Reaction Components	Reaction Products
Lactate in a buffer with pH- and temperature optimum	Pyruvate
NAD+, NaCN+, $MgCl_2$+	NADH
PMS+, NBT+	PMS-, NBT- (Formazan in a zone of reaction: Blue stain occurs as an indication of the enzyme fraction)

Most of the proteins studied by the electrophoresis are enzymes. Because of this, it is easy to develop a histochemical staining procedure to visualize the activity of a specific enzyme (Hunter and Markert 1957). Several sources give detailed descriptions of hundreds of receipts for visualizing the enzymatic activities, following electrophoresis (e.g., Shaw and Prasad 1970; Harris and Hopkinson 1976; Manchenko 2002). Each procedure uses a product of the enzyme's specific activity to locate precisely that enzyme in the gel, and separate it as well from the mixture of proteins present in a source solution.

5.2 INTERPRETING PROTEIN VARIABILITY

Specific staining for an enzyme activity permits researchers to distinguish even isoforms of the same enzyme (isoenzymes or simply isozymes). This technique allows the detection of several isozymes at a time, if there are some in a certain species or population sample. The final result of the electrophoretic procedure is bands or fractions (Fig. 5.2.1), which identify the location of various forms of a single type of protein on a gel. The banding pattern of an individual may be interpreted as this individual's genotype with respect to the locus (loci) coding for that particular enzyme.

We may show the interpreting scheme in a somewhat different way, depending on the quaternary structure of the enzyme (Fig. 5.2.2).

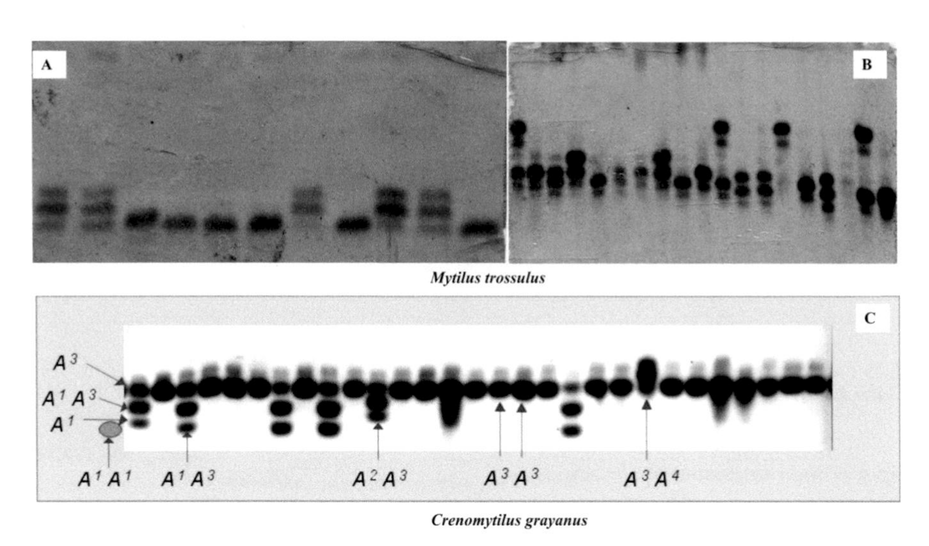

Fig. 5.2.1. Examples of starch-gel electrophoretic zimograms, enzymes stained in the gel. (A) aspartate aminotranspherase (**AAT-1***), (B) leucine aminopeptidase (**LAP-1***), C, formaldehyde dehydrogenase (**FDH***). With arrows the isozymes and their genetic interpreting are shown. Dotted arrows show non-genetically determined fractions (MMFE, see Section 5.3). On the bottom left (C) a fraction depicted that was actually absent in the sample but may be predicted.

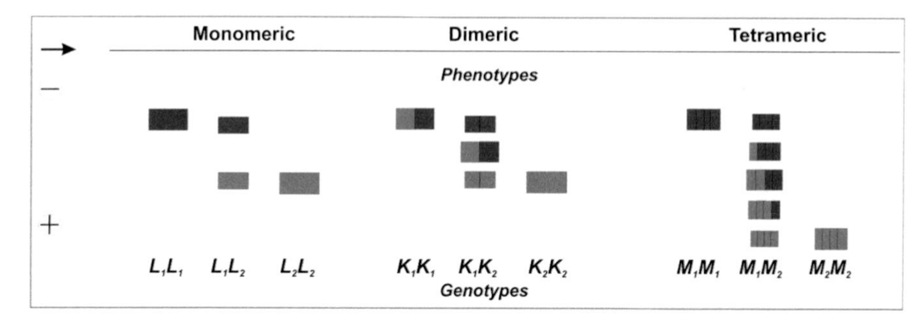

Fig. 5.2.2. Interpretation of allozyme pattern variability in the terms of genotype depending on their quaternary structure. A polymorphic locus codes for a monomeric, dimeric and tetrameric enzyme having two alleles designated L_1 and L_2 (or K_1 vs. K_2 and M_1 vs. M_2). These alleles produce two types of subunits (sections within rectangles). The protein encoded by the L_1 allele (or K_1 and M_1) migrates more slowly than the protein encoded by the L_2 allele (shown by the dark blue and blue color respectively). Combinations of the same subunits are called homomeric, while different subunits form heteromeric isozymes. Three different genotypes are possible for an individual at a locus: L_1L_1, L_1L_2, and L_2L_2.

In genetic interpretation of protein polymorphism it is most important to remember (see Chapter 4):

1. That one gene determines one polypeptide (subunit).
2. That there is a colinearity or strict correspondence between the sequence of nucleotides in DNA and amino acid sequence in the polypeptide.

Evidence in support of the first statement is from the results of the artificial synthesis of peptides on mRNA template. The second statement is supported by comparison of sequenced genes and polypeptides, which were synthesized on it (Fig. 5.2.3). Other results were introduced prior to the development of sequencing techniques (Stent 1971), by which the obvious connection defined between the distance of a mutation at gene map from its beginning and the length of the polypeptide (Fig. 5.2.4).

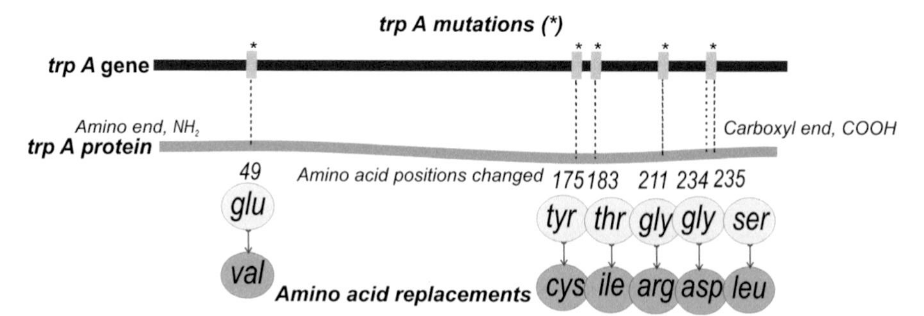

Fig. 5.2.3. Correlation of the positions of mutations in the genetic map of the *E. coli* triptophan gene, trpA with positions of amino add replacements in the TrpA protein.

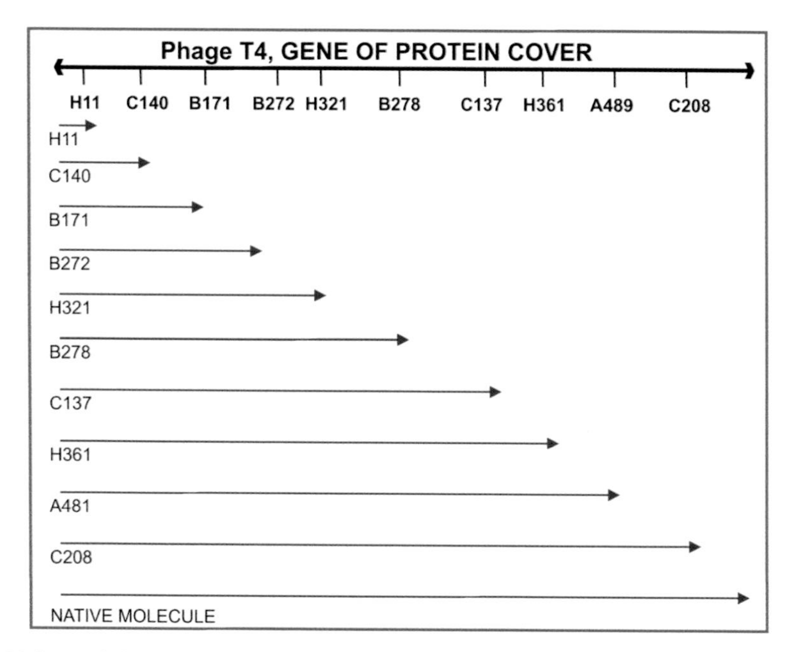

Fig. 5.2.4. Association of the positions of mutations in the genetic map of the Phage T4 at the gene of protein cover of the phage and the length of polypeptides. Polypeptides produced by certain mutation shown with an arrow. Location of each mutation noted at the gene map.

Thus, in general the scheme of the interpretation of an enzyme polymorphism, and its allelic variants the allozymes, is very simple, especially when only one locus is segregating and the tertiary (subunit) structure of the enzyme is known. When an enzyme is coded by two loci, the interpretation also may not be too complicated (Fig. 5.2.5).

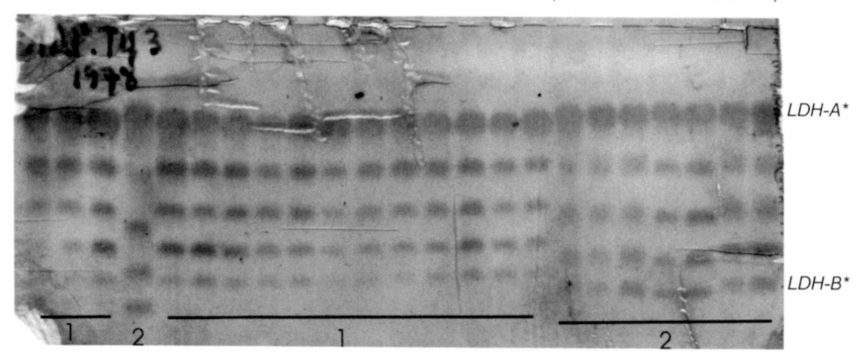

Fig. 5.2.5. Isozyme pattern and genetic interpreting for the case of two loci.

Locus LDH-A*: 1 – A_1A_1, 2 – A_1A_1; Locus LDH-B*: 1 – B_2B_2, 2 – B_1B_1.

5.3 GENETICS OF ISOZYMES

In the previous chapter we defined the difference between the terms polypeptide and protein, and noted many modifications that happen with proteins, including enzymes. Enzymes can form numerous isoforms of different natures. Enzymes and their isoforms, isoenzymes (isozymes) are the most important gene markers in biochemical genetics. Let us consider what are the sources of the origin of **Multiple Molecular Forms of Enzymes** (MMFE). MMFE (Korochkin et al. 1977) have complex natures and only some of them are isozymes. Proteins may be modified *in vivo* and *in vitro.* Many modifications are natural processes, which lead to the active molecule. For example, as we have seen in Chapter 4, in low density lipoprotein (LDL), the fifteenth exon specifies the domain for the posttranslational addition of the carbohydrate, while the sixteenth and seventeenth exons specify the regions of the protein that are part of the membrane, anchoring the receptor to specific sites on the cell surface. Only after modification does LDL become active and the membrane itself takes new functional qualities. We may classify MMFE into the two categories: (i) nongenetic MMFE and (ii) genetically caused MMFE, the isozymes. *Isozymes, or genetically caused MMFE, (i) originate through substitutions in amino acid sequence caused by base changes in an allele (allozymes) or/and (ii) are single-locus or multiple-locus products occurring through different combinations of subunits (homomeric and heteromeric isozymes; see Fig. 5.2.2).* Other sources for MMFE like, for example, alternative splicing are also possible (see below). In this sense the LDL modifications described previously are not isozymes because no structural gene change occurred here. In Section 5.2 we considered only genetic causes for MMFE and genetic interpreting for the only single-locus phenotype variability or **allozyme variability**. The real situation is more complex. Let us return to Fig. 5.2.1. In parts B and C of this figure several extra minor bands (sub-bands) are seen. These are not isozymes but conformers, post-translation modifications (either *in vivo* or *in vitro*). They must be excluded in our genetic interpretations. Another complication arises when two or more loci coding for an enzyme are expressed in the same tissue. In these cases subunits of different loci can interact, forming additional isozymes; in this situation they are the inter-locus products (Fig. 5.3.1).

Schematic representation of the expression pattern for polymorphic duplicated loci that encode enzymes with different quaternary structure are given in Utter et al. (1987, Fig. 2.5).

Exceptions in Codominant Expression Mode

The phenotypes of Figs. 5.2.1–5.2.2 and 5.3.1 are due to **codominant expressions** of the respective genes. The contribution of both parent alleles in heterozygotes can be identified in these cases; also for oligomeric enzymes new isozymes are formed, represented by heteromeric molecules or subunits. Due to the formation of heteromeric molecules, new qualities of heterozygous individuals occur, which are not simply equal to the sum of two homozygous parent phenotypes (see Fig. 5.2.2). The first evidence of such heterozygote qualities were obtained under immunogenetic

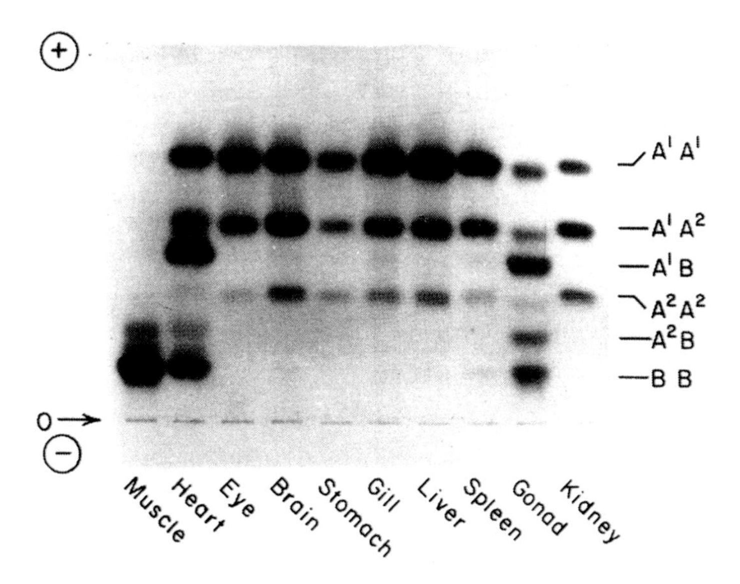

Fig. 5.3.1. Differential expression of glucosephosphate isomerase-A loci (**GPI***) in tissues of the fish *Hypentelium nigricans*. Extensive divergence in the expression of duplicate **GPI-A*** loci have occurred within and among tissues. The **GPI-B*** locus is singly expressed in tissues of this species due to gene diploidization (from Ferris and Whitt 1979 with permission).

investigations (Chapter 4). Codominant expression is an important attribute of the genetics of isozymes because it is the basis for interpreting the patterns of individual loci. Some theoretical concepts can also be proved if codominant expression holds as a rule (see Chapter 12). However, there are exceptions to codominant electrophoretic expressions. Let us consider them.

The occurrence of electrophoretically identical subunits synthesized by two different loci is observed in some salmonid fishes (Fig. 5.3.2). The genetic and evolutionary basis for such identical loci (**isoloci**) is duplication, with recent gene duplication giving rise to the electrophoretically identical gene products. Usually duplication is a result of polyploidy, and in salmonids that is tetraploidy (Wright et al. 1983; Allendorf and Thorgaard 1984). Polymorphism may occur in one locus or both loci. In either case the electrophoretic expression of isoloci complicates both the genetic interpretation of phenotypes and the statistical analysis of genotypic variability. One part of the problem is that it is often impossible to discriminate between alleles of specific loci when two (or more) isoloci are polymorphic. The problem is apparent from the phenotypes as suggested (Utter et al. 1987) and modified in Fig. 5.3.2, which shows the fractions of two isoloci where one of the loci is monomorphic and another is polymorphic.

Another complication may be due to alternative splicing, a recently discovered mechanism (Fig. 5.3.3; for the image in part A URL is www2.estrellamountain.edu/faculty/farabee/biobk/BioBookGENCTRL.html). The processing, which involves splicing, represents a potential regulatory step during gene expression. For example, several cases are known where introns present in pre-mRNAs derived from the same

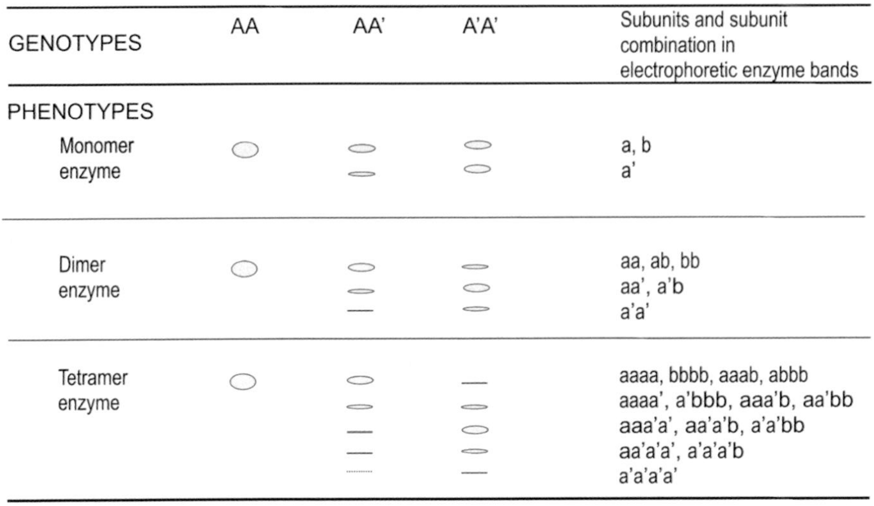

GENOTYPES	AA	AA'	A'A'	Subunits and subunit combination in electrophoretic enzyme bands
PHENOTYPES				
Monomer enzyme				a, b a'
Dimer enzyme				aa, ab, bb aa', a'b a'a'
Tetramer enzyme				aaaa, bbbb, aaab, abbb aaaa', a'bbb, aaa'b, aa'bb aaa'a', aa'a'b, a'a'bb aa'a'a', a'a'a'b a'a'a'a'

Fig. 5.3.2. Electrophoretic phenotypes when isoloci are expressed. Individuals are homozygous and heterozygous at loci coding for monomeric, dimeric and tetrameric proteins (enzymes): one locus is polymorphic (with alleles A and A' resulting from subunits a and a', respectively); and a second locus is monomorphic, coding for a subunit (b) with an electrophoretic mobility identical to that of subunit (a).

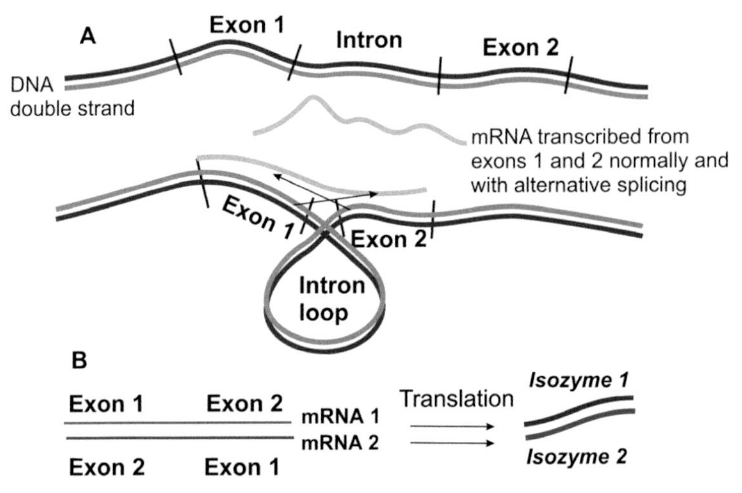

Fig. 5.3.3. The splicing (A) and alternative splicing (A, B), the mechanism for the creation of new alleles by recombination of mRNA (that gives mRNA 1 vs. mRNA 2) and as a result having different products, the isoforms or isozymes: Isozyme 1 (typical) and Isozyme 2 (new).

gene are spliced in more than one way, thereby yielding different collections of exons in the mature mRNA. This process, referred to as **alternative splicing**, yields a group of mRNAs that, upon translation, result in a series of related proteins called in proteomics the **isoforms**. In the definition given above they may be called the isozymes (see above given in italic letters). Several examples have been found in organisms ranging from viruses to fruit flies and humans. The question arises are the isoloci products of different duplicated genes, or they just the products of the combination of different exons? To discriminate between these kinds of isoforms we need a sequence analysis and inheritance pattern of the genes.

Examples given above on deviation from codominant expression are obviously not so common in higher organisms because polyploidy is restricted among them due to the limitations of chromosome sex determination. However, how frequently gene duplications occur in invertebrate animals remains unresolved. Judging by own long experience in biochemical genetic research of invertebrates, however, duplicated enzyme-coding loci are rare in these animals. Two reasons may form a basis for this: (i) duplications are rare in reality or (ii) duplications are common but diploidization events diminish their frequency due to the large mean age of the species in these groups. Even less is known about the frequency of alternative splicing as a source for isozymes.

Advances and Limitations of Electrophoretic Approach for Genetic Research

The principles outlined above for detecting genotypic variability via electrophoretic investigation of protein variability are widely applied, and have been generally recognized as a big achievement for the genetic research of wild populations (Lewontin 1974; Hartl 1980). The unmatched power of electrophoresis to detect genetic variation is enhanced by the relative ease in obtaining large samples of individuals scored according to their genotypes. Protein specimens via tissue extracts can be prepared easily and quickly. Many specimens can be run on a single gel, and multiple slices can be made from a gel after electrophoresis. Each slice is stained for different enzymes or other proteins to reveal different loci. For a trained researcher a total of upto 200–300 individuals is possible to score a day in a Lab collaboration conditions. In starch gel electrophoresis at least 6 loci can be obtained in single experimental run from each individual, because each gel can be sliced into six or more slabs, and each slab can be stained for a different type of enzyme. Usually more loci per individual can be obtained from each gel because of the expression of more than one locus under staining for an enzyme.

Nevertheless, there are obvious limitations to the technique of protein electrophoresis as genetic markers (Hubby and Lewontin 1966; Lewontin 1974; Korochkin et al. 1977). The amino acid substitutions of proteins are detected by electrophoresis indirectly via the total net charge of protein. However, (i) each base substitution does not necessarily result in a change of amino acids and (ii) amino acid substitution if they occur do not necessarily result in protein changes that are electrophoretically detectable. It has been estimated that only about 25–30% of the amino acid substitutions are detected generally via electrophoresis in most laboratories (Shaw 1965; Nei and Chakraborty 1973; Lewontin 1974). A similar estimate was also obtained under various biochemical conditions (Marshall and Brown 1975). However, sometimes the experimental estimates detect upto 50% of mutation changes in proteins (Ramshaw et al. 1979; Neel et al. 1986), and as we will see later the mutation rate itself may achieve high frequency (Altukhov 1989; see also Chapter 6). The important point is that electrophoretic identity does not necessarily mean identity of base sequences in

DNA. That is why measures of variability, for example heterozygosity, depend upon the technique used (electrophoretic data mostly comparable among themselves); this is also true for other genetic data, where alleles are inferred by phenotypes (Allendorf 1977). In recognition of the potential genetic heterogeneity within allelic classes the term **electromorph** (King and Ohta 1975) was introduced. Direct methods of DNA analysis have shown that electromorphs often include many nucleotide substitutions (see Chapter 13).

Expression of Genes in the Embryogenesis and Ontogenesis

As a result of differential gene activity different specialized cell lines are formed, which later develop into different tissues and organs. Investigations in developmental genetics for a long time were based on mutations of morphological traits in laboratory conditions, primarily on *Drosophila* lines (Korochkin 1999; Jimulev 2002). Morphological traits are formed late in ontogenesis. That is why only when methods of biochemical and molecular genetics were introduced did the analysis of early development became available.

Let us remember first a few general features of development, particularly in the ovaries. In vitelogenesis, the so called big growth of the ovum occurs. In the ovum, on the "lamp brushes", there is a very active synthesis of mRNA, rRNA, ribonucleoproteins and other proteins; mRNA is masked as an informosome. In mature ova there is no synthesis. In the zygote, after fertilization, there is also no synthesis. That is because of quick cell divisions and short interkinesises. In general, prior to the gastrula phase, the father's genes are not active. In other words, new organism on the first steps lives and develops for expense of the mother's gene products, which were accumulated in advance. This knowledge of the early developmental biology was obtained largely due to the biochemical genetic techniques and analysis of isozyme expression in embryogenesis, showing that expression of genes in males (M) and females (F) in ontogenesis is different. As stated before, it is programmed by the developmental events. Nevertheless, the fact that M genes are expressed later was established experimentally by enzyme electrophoresis. When interspecies hybrids are investigated, the different M and F-parents of the mates contribute different isozymes. In many crosses M genes were shown to be expressed later than F (Neifah and Lozovskaya 1984). For instance, in a cross between river M- and lake F-chars, the ***LDH-B**** and ***GPDH**** loci (loci with different isozymes in these two forms) first express the F isozyme (Kirpichnikov 1979; 1987).

Tissue Expression of the Enzyme Loci in Development

Enzymes and other proteins are differentially expressed in ontogenesis. There are three groups of enzymes: (i) Enzymes present in all organs and tissues throughout life, (ii) Enzymes functioning only as embryonic enzymes and (iii) Enzymes that occur at a certain stage of the life cycle and are present in only some tissues.

(i) Examples of first kind are DNA-, RNA-polymerase, adenosinetriphosphatase, ATP-synthesizes and others connected with basic cell functions. (ii) The hemoglobin

(Hb) is a good example of the proteins of the second group. You should remember the γ-Hb in human, which functions in the fetus as one kind of two kinds of subunits in the tetrameric molecule. After birth this γ-Hb is substituted with the α-Hb polypeptide chain. Similarly, for example, in salmon, out of a total of 14–18 Hb fractions, many are present only in the larval stage, and are absent after hatching. (iii) The third is the largest enzyme group. An example of this group was presented in the catostomid fish *Hypentelium nigricans* in Fig. 5.3.1. In muscle tissue only the ***GPI-B**** locus is basically expressed, while three genes are expressed in heart and gonads, and two in seven other tissues (Fig. 5.3.1). Similarly, in the sunfish *Lepomis cyanellus* ***GPI****, ***MDH**** and ***KK**** enzyme loci are duplicated, and one isozyme is present in oocytes and subsequently in the majority of tissues, but the second occurs only after hatching and only in muscle tissue. Time of expression of an enzyme locus is very similar for distant taxa, e.g., *Lepomis* and *Hypentelium* are in different taxonomic orders of fish. This suggests that regulation of ontogenesis is very strongly conserved by purifying selection, so that any deviations are removed by selection.

Functional Differences Among Isozymes

Before considering this question it is necessary to consider the connection between the duplication event and diversification of function. No diversification occurs without the duplication (Ohno 1970). Examples are available for salmon and trout in particular. LDH is duplicated: the loci ***LDH-A****, ***LDH-B****, and ***LDH-C****, have originated subsequently in time. In this tetrameric enzyme, the homomeric isozymes A4, B4, C4 differ significantly in some properties. The A4 is most thermo-resistant, most active in anaerobic conditions and most resistant to inhibitors. The C4 has all these qualities reversed, and the B4 isozyme is intermediate. LAP isozymes in mussels represent another good example. Koehn (1978) showed that its isozymes differ in activity under different salinities. The GPI isozymes of *Lepomis* differ in thermo-resistance. Ell hemoglobins differ in oxygen affinity. More details on this theme may be found in reviews (Salmenkova 1979; Hochachka and Somero 1988).

The main conclusions from Section 5.3 are as follows.

1. Isozymes or more precisely allozymes are very good markers of intra-locus or allelic variability.
2. Isozymes are the best markers to detect the tissue activity of genes.
3. Isozymes permit investigation of the time of gene activation in the ontogenesis.

5.4 EVOLUTION OF A GENOME REGULATION

Gene interaction

There are several forms of gene interaction. In the previous chapters, we encountered the terms dominance, codominance (intra-locus), epistasis and pleiotropy (inter-locus), which explain some gene interactions. In former sections of this chapter, we underlined the role of genetic regulation and interaction in the development of an individual.

This is a important topic Developmental Genetics, which is beyond our main goals. Nevertheless, it is important to recognize the great complexity of different interactions in the formation of a trait, especially a complex trait like a morphological trait or behavior pattern. The term **gene interaction** is often used to describe the idea that several genes influence a particular characteristic. This does not mean, however, that two or more genes, or their products, necessarily interact directly with one another to influence a particular phenotype. Rather, the cellular function of numerous gene products contributes to the development of a common phenotype. For example, the development of an organ such as the compound eye of an insect is exceedingly complex, and leads to a structure with multiple phenotypic manifestations, most simply described as specific size, shape, texture and color. The development of the eye, for example in *Drosophila*, can best be envisioned as a complex cascade of developmental events leading to its formation. This process exemplifies the developmental concept of **epigenesis**. This concept recognizes that each new step of development increases the complexity of a trait (organ) and is under the control and influence of one or more genes acting stepwise.

On the other hand, as mentioned in Chapter 3, in many higher organisms there are different forms or biotypes (ecotypes), which occur as differential expressions of the same basic genome in different environments. Genotype vs. environment interactions will be discussed in brief in Chapters 11–12. Here we only glance at some changes in epigenesis among simple and more complex traits.

Before considering more complex traits let us look at simple biochemical traits, such as different enzyme activity. These are quantitative traits, and it is interesting to know whether there is any regulation on this level.

Genetic Correlations Among Enzyme Activities

The enzymes are coded by well-defined structural genes, and by using balanced chromosomes and other genetic means, it is possible to examine the contributions of structural variation and background or "modifier" genes on levels of enzyme activity. In a series of papers on fruit flies, the genetic variability in the activities of 23 enzymes among a series of second chromosome and third chromosome replacement lines was examined (Laurie-Ahlberg et al. 1982). Chromosome replacement lines were kept genetically identical for all but one chromosome.

The design of the experiments was fairly elaborate, with each line tested on a number of days and rearing in repetitions. The complex design was necessary in order to determine whether the differences among lines were due to genetic differences, or to rearing effects, or to measurement errors in a particular assay. Significant variability was seen among lines in the activity of all 23 enzymes in both the second and the third chromosome replacement series, with the exception of TPI (thriose-phosphate isomerase) in the second chromosome series. Essentially the same results were obtained whether activities were expressed as reaction rates per fly, per unit weight or per unit protein basis. Wilton et al. (1982) examined the patterns of correlation among the activities of pairs of the 23 enzymes, and found that there was a tendency for positive genetic correlations in enzyme activity. A particularly striking positive correlation was seen repeatedly in the activities of the two leading enzymes in the pentose phosphate shunt, glucose 6-phosphate dehydrogenase (G6PDH) and 6-phosphogluconate dehydrogenase (6PGD). The pentose shunt directs the flux of carbon chains from glucose 6-phosphate through steps that generate NADPH, which is important in many biosynthetic steps, including lipid synthesis. Flies that lack activity of either G6PDH or both of these enzymes can survive. However, flies lacking activity for 6PGD alone cannot survive. This suggests that the intermediate metabolites generated by G6PDH may accumulate to lethal levels, analogous to many human inborn errors of metabolism. In this particular case for fruit flies the correlation of activity of enzymes G6PDH and 6PGD in the pathway suggests an adaptive (selective) mechanism for their coordinated control. In other cases, the positive correlations in activity may simply be related to effects on common substrates. For example, the enzymes hexokinase, G6PDH, GPI and PGM all share glucose 6-phosphate as a substrate, and all show positive inter-correlations in activity. The prevalence of the correlations is seen in the common patterns found among geographically isolated populations and among the second and third chromosome lines.

In our own experiments we found enzyme activity differences associated with both the biotypes and taxa (Kartavtsev and Mamontov 1983; Kartavtsev et al. 1983; 2002), as well as with joint regulation in duplicated loci in salmon (Kartavtsev et al. 1985). Differences among biotypes and differences among young, recently originated species are doubtless, while among "good species" they are either absent or not so sharp (Kartavtsev and Mamontov 1983; Kartavtsev et al. 1983; 2002). These results may support speculations on the possible role of biotypic variability in speciation; occasionally one ecotype may be fixed in a certain biotope and give rise to a new species in the future (Although, in my opinion this is hardly possible in the case of sympatry without barriers to gene flow; see more on the subject in Chapter 7).

As important as the evolutionary implications in these data are the facts that the changes in gene activities appear abruptly by a turnover mechanism and simultaneously in all individuals of a population, constituting certain ecotypes. This turnover mechanism in fish is irreversible. Landlocked salmon, for example, never become anadromous (Konovalov 1980) and vice versa. Several such examples may be enumerated. In plants the ecotypes can reverse their shape, but not completely (Stebbins 1950). Bacteria also have a kind of life form, but more simple. Changes under optimal and suboptimal environmental conditions in bacterial culture were discussed earlier (Chapter 4), and we have seen that different genes operate, even in different phases. We may conclude that the turnover mechanism that regulates epigenetic variability is evolving, and is more complex in more complex organisms. For most kinds of organisms it is still unknown, what are the specific mechanisms that turn on or turn off genes in different biotypes so dramatically changing their phenotypes. Is it operated by the same principles as in individual development and like the proliferation cascade pattern of cell lines or it is different? These are exciting questions that are necessary to solve.

5.5 TRAINING COURSE, #5

During the class these three themes and new terminology will be considered.

5.1 Heterozygosity: H_{obs}, and H_{exp}.

5.2 Estimates of Genetic Distance.

5.3 Software BIOSYS (Swofford 1981). Solve Package Examples for H_{obs}, H_{exp} and Distances.

MEAN HETEROZYGOSITY (LOCUS/INDIVIDUAL)

$H = \Sigma_{i=1}^{L} h_k/L$; $h_k = 1 - \Sigma_{i=1}^{m} p_i^2$, h_k is expected heterozygosity ($H_{exp} = H_s$), p_i is an i allele frequency;

L is number of loci. $H_{obs} = hj/n$, hj is number of heterozygotes, n is number of individuals.

GENETIC DISTANCE. **STANDARD NEI'S DISTANCE (D_N)**

$D_N = -\ln I$. I is normalized genetic identity for a random set of loci. For the locus j it is equal to: $Ij = (\Sigma_{i=1}^{k} x_i y_i)/\sqrt{\Sigma_{i=1}^{k} x_i^2 \Sigma_{i=1}^{k} y_i^2}$

For the whole set of loci I equal to: $I = Jxy/\sqrt{JxJy}$;

Where Jxy, Jx, Jy – arithmetic mean for the sums: $\Sigma_{i=1}^{k} x_i y_i$, $\Sigma_{i=1}^{k} x_i^2$, $\Sigma_{i=1}^{k} y_i^2$,

Where, x_i and y_i are allele frequencies in populations (taxa) x and y, k is number of alleles in all pairwise comparisons.

GENETIC DISTANCE. **MINIMAL NEI'S DISTANCE (Dm)**

$Dm = 1/k\, \Sigma_{i=1}^{k} d_m$; $d_m = [(\Sigma_{i=1}^{k} x_i^2 + \Sigma_{i=1}^{k} y_i^2)/2] - \Sigma_{i=1}^{k} x_i y_i$. d_m is distance for single locus, **k** is number of loci.

Multiple Molecular Forms of Enzymes (MMFE) represented by isozymes and allozymes, the genetically determined molecular forms as well as posttranslational modifications.

Allozymes are isozymes that are caused by allelic segregation at the same genetic locus.

Electrophoresis in biochemistry and biochemical genetics is a process of separation of biological macromolecules in a gel by their electric charges.

Isoloci are loci that do not differ in the electrophoretic mobility of their consequent enzymes.

6

GENES, MUTATIONS AND EVOLUTION

MAIN GOALS

SUMMARY

1. Mutations lead to the alteration of a current phenotype or to creation of variability, and by this provide a subject for genetic research. Due to this, their importance for genetics and evolution of life in general is obvious.
2. In terms of their function, genes can be classified into two groups: protein-coding or **structural genes** and RNA-coding or **regulatory genes**. Structural genes may participate in regulation too, and in this have no principal difference from the second group of genes.
3. Nucleotide substitutions can be divided into two different classes: **transition** and **transvertion**. **Transition** is the substitution of a purine for another purine (A⇔G) or the substitution of a pyrimidine for another pyrimidine (T⇔C). **Transvertions** are inverse types of nucleotide substitutions, when a purine is substituted by a pyrimidine and vice versa (G⇔T, A⇔C, C⇔G or A⇔T).
4. Classical geneticists have determined that the **rate of mutations,** that alter phenotypic characters or induce lethal effects, ***is of the order of*** 10^{-5} ***per locus per generation*** in eukaryotes such as man, *Drosophila* and corn. ***The mean rate of amino acid substitution for an average polypeptide is estimated to be***

4×10^{-7} per locus per year. Due to the evidences that electrophoresis detects only about one-quarter of all amino acid changes, ***the rate of amino acid substitutions detectable by electrophoresis is estimated to be about 10^{-7} per locus per year.***

5. Mutations play the principal role in **molecular clocks**. Mutation rate and time period since the gene pool separation of the taxonomic pair directly determines the accumulated genetic distances between these taxa during the time of separation. Thus, ***when the values of the distance and mutation rate are known, the time of taxa separation can be calculated.***

6.1 INTRODUCTION TO THE STUDY OF MUTATION

In this chapter we will deal mainly with gene mutations, their effects and connected matters. We will focus mostly on species in nature. Mutations lead to the alteration of a common phenotype or to variation and hence provide a subject for genetic research. Their importance for genetics and evolution of organisms in general is determined by their innovative role. Some organisms lend themselves to induction of mutations that can be detected easily and studied throughout reasonably short life cycles. Viruses, bacteria, fungi, fruit flies, some other invertebrates, certain plants and mice fit these criteria. Thus, these organisms have been widely used to study mutation and mutagenesis, and through other studies they have also contributed to more general aspects of genetic knowledge.

Classification of Mutations

Spontaneous Versus Induced Mutations. Mutations are classified by various schemes which are not mutually exclusive. All mutations are described as either *spontaneous* or *induced.* **Spontaneous mutations** are those that occasionally happen in nature. No specific agents are associated with their occurrence. In contrast to such spontaneous events, those that result from the influence of any artificial factor are considered to be **induced mutations**. It is generally agreed that any natural phenomenon that increases chemical reactivity in cells will induce mutations. The earliest demonstration of the artificial induction of mutation occurred in 1927, when Hermann J. Muller reported that X-rays could cause mutations in *Drosophila.*

Gametic Versus Somatic Mutations. When considering the effects of mutation in eukaryotes, it is important to distinguish whether the change occurs in somatic cells or in sex cells, gametes. Mutations in gametes are part of the germ line and are of greater concern because they are transmitted to offspring. **Dominant autosomal mutations** will be expressed phenotypically in the first generation. **X-linked recessive mutations** arising in the gametes of a heterogametic female may be expressed in hemizygous male offspring. **Autosomal recessive mutation** in the gametes of either males or females may go unnoticed for many generations.

Other Categories of Mutation. Various types of mutations are classified on the basis of their effect on the organism. The same mutation may be classified into different categories. In principle, as many types of traits exist as many kinds of mutations may be defined. (1) Those affecting morphological traits are **morphological mutations**, like coding for such garden pea characters, as smooth and wrinkle seeds, obviously changing the morphology. (2) There are **nutritional** or **biochemical mutations** that alter variation of corresponding traits. (3) The **behavioral mutations** are those that affect behavior patterns of an organism, like mating behavior, circadian rhythms in insects, etc. (4) The **regulatory mutations** can disrupt normal regulatory processes and permanently activate or inactivate a gene. (5) Viability mutations may be classified as **lethal mutations, deleterious, advanced** or **neutral**. (6) Finally, most of these categories can exist as **conditional mutations**. Even though a mutation is present in the genome of an organism, it may not be expressed under certain conditions. Among the best examples are **temperature-sensitive mutations**, found in a variety of organisms. At certain "permissive" temperatures, a mutant gene product functions normally, only to lose its functional capability at a different, "restrictive" temperature. Still another kind of mutations that are called **point mutations** can be opposed to chromosomal mutations, which as we agreed before are better classified as rearrangements.

Detection of Mutations and their Causes

Mutation expression depends upon a number of factors, which differ among kinds of organisms. That is why the techniques for the detection of mutation depend on the specificity of an organism. A method that suits bacteria is not usually good for humans or fruit flies. Detection in Bacteria and Fungi may be the easiest. Microorganisms that represent organisms nutritionally of wild types (requiring only minimal medium and the rest they can synthesized themselves) are called **prototrophs**. In this case, mutants are those individuals that require a specific supplement to the minimal medium. They are called **auxotrophs**. Nutritional mutants can be detected and isolated by their ability to grow on a complete medium and their failure to grow on a minimal medium. They can be detected on the surface of the agar-agar by simple visual inspection. All specific methods of molecular and biochemical genetics are further used for detection of small gene alterations as well.

What are the causes of mutation? Mutation occurs because of complex events. To consider all of them is beyond the scope of this course. Nevertheless, it is necessary to give few commonly known points (Jimulev 2002, pp. 213–216; Klug and Cummings 2002, pp. 283–285). The chemical basis for many mutations is the change of common nucleotides in DNA by their methylated analogs, for which repair is less perfect. The tautomeric (enole and imidasole) forms of nucleotide molecule that are less stable also occur occasionally. This leads to the formation of atypical or incorrect nucleotide pairs, like A-C and G-T, and consequently to nucleotide substitution, and hence, results in spontaneous mutation in the DNA. Repair mechanisms of DNA and RNA greatly diminish such errors and mutation. Nevertheless, errors may occur during DNA polymerase function and chemical modifications of cytosine give rise to mutation. Mistakes during DNA enzyme polymerization are common and occur at the rate 10^{-5}. Correcting 3'–5'exonuclease activity of DNA polymerases decreases this frequency down to 10^{-10} per nucleotide. Other repair mechanisms apply later and may lower this frequency even more. However, some fraction of changes still remains in genes and populations, which forms the pool of spontaneous mutations.

6.2 STRUCTURE AND FUNCTIONS OF GENES

We have considered the concept of the gene and its action in previous chapters. Our knowledge of gene functioning and its relation to morphogenesis and regulation is still very limited. Before considering the appearance of mutations and their rate it is necessary to review some basics on the genetic code and what types of genes we generally deal with.

Types of Genes

In terms of their function, genes can be classified into two groups: protein-coding or **structural genes** and RNA-coding genes. Protein-coding genes control protein synthesis via translation mechanisms and their derivatives. RNA-coding genes are those that produce tRNA, rRNA and some others, like small nuclear RNAs (snRNA). These RNAs are the final products of these genes. Ribosomal RNAs are certain components of the ribosomes; specifically, they are a part of transcription and protein synthesis. In contrast tRNAs, although part of translation, transfer the genetic information from mRNAs into amino acid sequences of proteins (see Chapter 3). The function of snRNAs is not completely clear, but they are essential for mRNA transcription and processing. Some structural genes and most RNA-coding genes participate in regulatory functions: regulation of replication, transcription, translation and the life cycle, including cell division and ontogenesis. They are called **regulatory genes**, although these are not distinctly different from the rest of the genes, as we see.

The genetic code for genes is in the general "universal" for viruses, prokaryotes and eukaryotes, although a few differences exist. Thus, in the mitochondrial genes, codes are slightly different. The nuclear genetic code is given in Table 6.2.1. For each three letters of the codon code, the three-letter codes for amino acids are represented (Table 6.2.1).

There are $4^3 = 64$ possible triplet codons for the four different nucleotides uracil (U), cytosine (C), adenine (A) and guanine (G). Three of them (UAA, UAG, UGA) are, however, amino acid terminating codons (or **nonsense codons**), and do not determine any amino acid. Each of the remaining 61 codons (**sense codons**), are codes for a particular amino acid. However, there are only twenty amino acids used for making proteins. Thus, it is obvious that code is redundant, and there are several codons for the same amino acid. Codons determining the same amino acid are called **synonymous codons**, and nucleotide substitutions here are synonymous.

In the genetic code, codon AUG codes for methionine, but this codon is also used as the initiation codon. Therefore, every mRNA has this codon at the beginning when transcription occurs. Later, upon translation, the methionine encoded by the initiation codon is removed from the polypeptide. As noted above, there are slight differences in code for mitochondrial genomes, which is illustrated for mammals mitochondrial genes in Table 6.2.2. In the mitochondrial genetic code, AUA, which codes for isoleucine in the nuclear code, is used for methionine. The codon UGA is not a termination codon but codes for tryptophan. On the other hand, codons AGA and AGG are terminating codons in these organisms instead of being coding for arginine. The genetic code of yeast

Table 6.2.1. The coding dictionary.

First codon letter	Second codon letter				Third codon letter
	U	C	A	G	
U	Phe	Ser	Tyr	Cys	U
	Phe	Ser	Tyr	Cys	C
	Leu	Ser	Stop	Stop	A
	Leu	Ser	Stop	Trp	G
C	Leu	Pro	His	Arg	U
	Leu	Pro	His	Arg	C
	Leu	Pro	Gln	Arg	A
	Leu	Pro	Gln	Arg	G
A	Ile	Thr	Asn	Ser	U
	Ile	Thr	Asn	Ser	C
	Ile	Thr	Lys	Arg	A
	Met	Thr	Lys	Arg	G
G	Val	Ala	Asp	Gln	U
	Val	Ala	Asp	Gln	C
	Val	Ala	Glu	Gln	A
	Val	Ala	Glu	Gln	G

Note. U—uracil, C—cytosine, A—adenine and G—guanine. AUG encodes methionine, which initiates most polypeptide chains. All other amino acids except tryptophan, which is encoded only by UGG, are represented by two to six triplets. The triplets UAA, UAG and UGA are termination signals and do not encode any amino acids.

Table 6.2.2. The genetic code for mammalian mitochondrial genes (from Nei 1987).

Codon	Amino Acid	Codon	Amino Acid	Codon	Amino Acid	Codon	Amino Acid
UUU	Phe	UCU	Ser	UAU	Tyr	UGU	Cys
UUC	Phe	UCC	Ser	UAC	Tyr	UGC	Cys
UUA	Leu	UCA	Ser	UAA	Ter	UGA	Trp
UUG	Leu	UCG	Ser	UAG	Ter	UGG	Trp
CUU	Leu(Thr)	CCU	Pro	CAU	His	CGU	Arg
CUE	Leu(Thr)	CCC	Pro	CAC	His	CGC	Arg
CUA	Leu(Thr)	CCA	Pro	CAA	Gin	CGA	Arg
CUG	Leu(Thr)	CCG	Pro	CAG	Gin	CGG	Arg
AUU	He	ACU	Thr	AAU	Asn	AGU	Ser
AUC	He	ACC	Thr	AAC	Asn	AGC	Ser
AUA	Met(Ile)	ACA	Thr	AAA	Lys	AGA	Ter(Arg)
AUG	Met	ACG	Thr	AAG	Lys	AGG	Ter(Arg)
GUU	Val	GCU	Ala	GAU	Asp	GGU	Gly
GUC	Val	GCC	Ala	GAC	Asp	GGC	Gly
GUA	Val	GCA	Ala	GAA	Glu	GGA	Gly
GUG	Val	GCG	Ala	GAG	Glu	GGG	Gly

Note. The codons for yeast are identical with those for mammals except the ones in parentheses.

mitochondrial genes is slightly different from that of the mammalian mitochondrial code (Table 6.2.2). In yeast, CUU, CUC, CUA and CUG code for threonine instead of leucine, and AUA codes for isoleucine instead of methionine.

Universality of the genetic code supports much of modern molecular genetics. Thus, many studies involving recombinant DNA technology reveal that bacterial genes can be inserted into eukaryotic cells and transcribed and translated and vice versa. However, as new data increase some more exceptions to the universality of the genetic code become evident. In some protozoans, such as *Paramecium* and *Tetrahymena,* UAA and UAG do not appear to be termination codons, even in nuclear genes, but code for glutamine. For one of the most simple prokaryotic organisms, *Mycoplasma capricolum*, the usual termination codon UGA is used for encoding of tryptophan.

The Triplet Nature of the Code

In the 1950s, G.A. Gamov (Cit. after Ratner 1998) and later Brenner et al. (1961) first argued on theoretical grounds that the code must be a triplet because three-letter words represent the minimal use of four letters to specify 20 amino acids. A code of four nucleotides, taken two at a time, for example, provides only 16 unique code words (4^2). A triplet code provides 64 words (4^3)—clearly more than the 20 needed—and it is much simpler than a four-letter code (4^4), which specifies 256 words. Experimental evidence supporting the triplet nature of the code was subsequently obtained by F. Crick and his colleagues. Using phage T4, they studied **frameshift mutations**. *The **frame of reading** is the start and finish of a translation*, and the space from the initiating codon to the terminating codon is the Open Reading Frame (ORF). Frameshift mutation results from the addition or deletion of one or more nucleotides within a gene and consequently the mRNA transcribed by it. The gain or the loss of one or more letters shifts the frame of reading during translation. Crick and his colleagues found that the gain or loss of one or two nucleotides caused such a mutation. When three nucleotides were involved, the frame of reading was reestablished (Fig. 6.2.1, I and II). This would occur only if the code is a triplet. The triplets are also called codons. This work also suggested that most triplet codes are not blank, but rather encode amino acids, supporting the concept of a degenerate code.

Degeneracy of the code, its nonoverlapping and the Wobble Hypothesis. When inspecting the coding dictionary, most evident is that the code is redundant or degenerated, as the early researchers predicted. That is, almost all amino acids are specified by two, three or four different codons. Three amino acids (serine, arginine and leucine) are each encoded by six different codons. Only tryptophan and methionine are encoded by single codons (Table 6.2.1).

The code is nonoverlapping, i.e., each codon specifies one amino acid at a time; then the next codon defines the next amino acid without blanks along ORF, starting from a fixed initiating point. Most often in a set of codons specifying the same amino acid, the first two letters are the same, with only the third differing. Crick (1966) observed a pattern in the degeneracy at the third position and postulated the **wobble hypothesis**, which explains the ability of substitutions in third codon position without altering the amino acid coded.

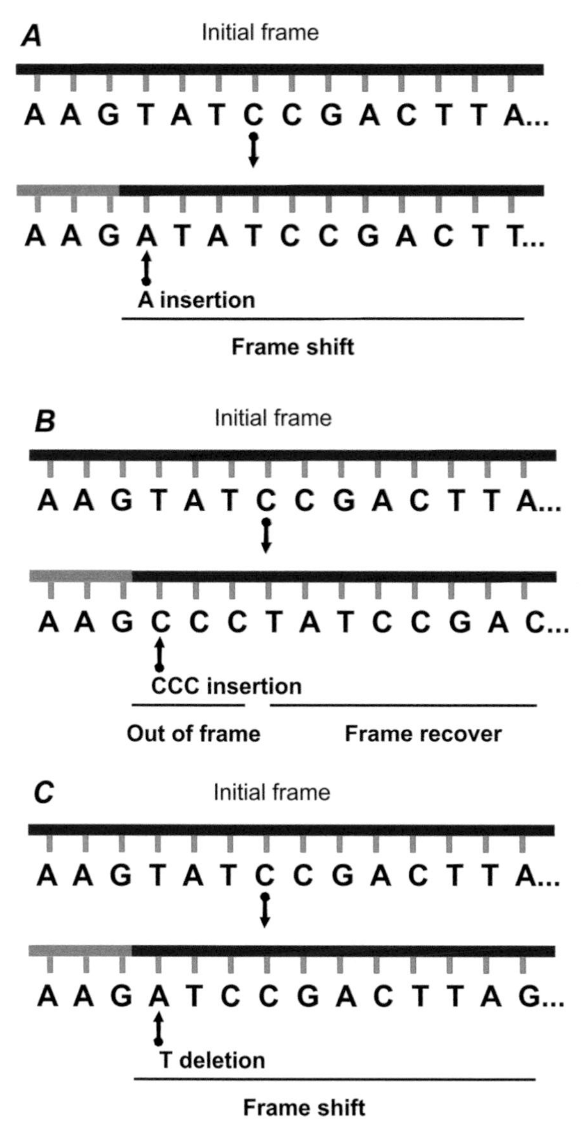

Fig. 6.2.1. The effect of frameshift mutations on a DNA sequence repeating after the triplet sequence AAG (shown in color for changed frame). (A) The insertion of a single nucleotide shifts all subsequent triplet reading frames, (B) The insertion of three nucleotides changes only two triplets, but the frame of reading is then reestablished to the original sequence. (C) Example in frame shift due to deletion.

6.3 TYPES OF MUTATIONS ON DNA LEVEL

Substitution, Deletion, Insertion and Inversion

All morphological, biochemical, physiological and other characters of organisms are controlled by genes, located in DNA, so any mutational changes in these characters are due to some change in DNA molecules. Four main modes in DNA changes may be defined: (1) **substitution** of a nucleotide for another nucleotide in a sequence, (2) **insertion** of nucleotides into an existing sequence (see Fig. 6.2.1, II), (3) **deletion** of nucleotides from a sequence (see Fig. 6.2.1, II) and (4) **inversion** of nucleotide orders in a sequence. Insertion and deletion may occur with one or more nucleotides as a unit, and inversion with two or more (see examples in Fig. 6.2.1). Insertion and deletion may shift the reading frames of the nucleotide sequence, leading to partial or complete loss of sense in a coding gene sequence (see Fig. 6.2.1).

Transition and Transvertion

Nucleotide substitutions can be divided into two different classes, i.e., **transition** and **transversion**. **Transition** is the substitution of a purine for another purine (A↔G) or the substitution of a pyrimidine for another pyrimidine (T ↔ C). Inverse types of nucleotide substitutions, i.e., G↔T, A↔C, C↔G or A↔T are called **transversions** (Fig. 6.3.1).

Molecular genetic data indicate that insertion and deletion occur quite often, especially in non-coding regions of DNA. The number of nucleotides in an insertion or in a deletion varies from a few nucleotides to several thousand. Short insertions or deletions are apparently caused by errors in DNA replication. Long insertions or deletions seem to be due mainly to an unequal crossover or DNA transposition (see Chapter 16).

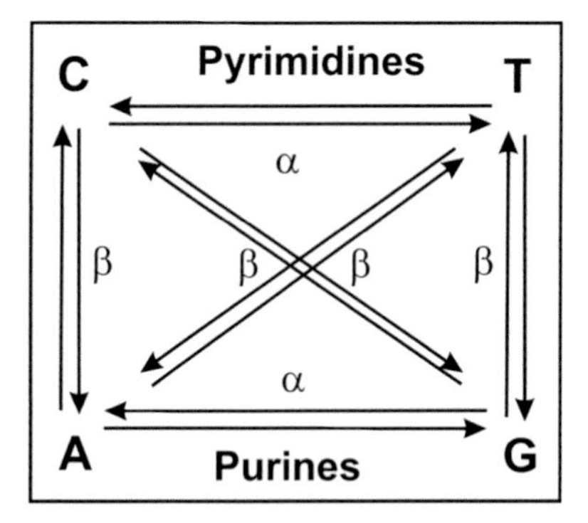

Fig. 6.3.1. Transitional, α (A↔G and T↔C) and transversional, β (four others) nucleotide substitutions (from Nei 1987 with permission).

Pattern of Nucleotide Substitution

There are many different types of genetic changes, and we will deal with some later. However, if we consider only a sequence order of DNA, the majority of the changes are nucleotide substitutions. If mutation occurs at random among the four nucleotides A, T, C and G, we would expect twice as many transversional changes as transitional changes (see Fig. 6.2.1; 4β/2α). In practice, however, transitional changes occur more often than expected (Fitch 1967; Vogel 1972). Thus, if we compare pseudogenes, which are nonfunctional or silenced genes and believed to have originated by a duplication of functionally close genes, it is possible to evaluate the correct substitution rate, depending here only upon mutation rate (Nei 1987, Chapter 13). In the research by Gojobori et al. (1982) and Li et al. (1984), the relative substitution changes among the four nucleotides by using pseudogenes comprising transitional changes (59.30%) was much higher than the expected proportion (33.3%) (Table 6.3.1).

However, the four types of transitional changes do not occur with equal frequency, the frequency of C→T being higher than that of the other changes. The high frequency of C→T change seems to be partially related to methylation of cytosine in the CG dinucleotides in the genome, but other factors are also involved. It is interesting to note that the eight types of transversional changes occur with more or less the same frequency, though the G→T and C→A changes tend to be higher. The reader who is interested in the biochemical reasons for the unequal mutation rates among the four nucleotides should consult Topal and Fresco (1976) and Gojobori et al. (1982). For our purpose, it is sufficient to keep in mind that the mutation rates among the four nucleotides are not equal.

Table 6.3.1. Relative mutation rates among the four nucleotides A, T, C and G in pseudogenes in mammals (from Li et al. 1984 with modification).

Mutant nucleotide	Original nucleotide			
	A	T	C	G
A		4.4 ± 1.1	6.5 ± 1.1	**20.7 ± 2.2**
T	4.7 ± 1.3		**21.0 ± 2.1**	7.2 ± 1.1
C	5.0 ± 0.7	**8.2 ± 1.3**		5.3 ± 1.0
G	**9.4 ± 1.3**	3.3 ± 1.2	4.2 ± 0.5	

Note. Substitution rates for transitions are shown in bold values.

6.4 MUTATIONS AND AMINO ACID SUBSTITUTIONS

Because the amino acid sequence of a polypeptide is determined by the nucleotide sequence of a structural gene, any change in amino acid sequences is determined by a mutation occurring in DNA. However, a mutational change in DNA does not necessarily result in a change in the amino acid sequence because of the degeneracy of the genetic code and the possibility of synonymous codon usage. Experimental detection of mutations, such as in enzyme electrophoresis (see Chapter 5), can lead to the underestimation of the number of mutations.

If nucleotide changes occur at random, the proportion of nonsynonymous mutations can be computed by using the genetic code. If we assume that all codons are equally frequent in the genome and the probability of substitution is the same for all pairs of nucleotides, the proportion of nonsynonymous mutations in nuclear genes becomes 71 percent, excluding nonsense mutations (Nei 1975). That is, about 29 percent of the mutations occurring at the nucleotide level cannot be detected by examining amino acid sequences. This percentage does not change seriously, even if we make more realistic assumptions about the probabilities of nucleotide substitutions and the codon frequencies.

Based on the genetic vocabulary in Tables 6.2.1 and 6.2.3, it may be seen that synonymous codons occur mainly because of the redundancy of the third nucleotide position. Calculation shows that nearly 72% of nucleotide substitutions at the third position lead to synonymous codons. In contrast, at the first nucleotide position, synonymous changes occur but the proportion of them is smaller and close to 5 percent. Lastly, at the second nucleotide position there are no synonymous substitutions at all.

As outlined in previous chapters, in population genetics and molecular taxonomy, protein differences are often studied by electrophoresis. However, as noted, this technique detects only about 25–30% of mutations (Chapter 5). It should be noted, however, that electrophoretic detectability of protein differences is greatly enhanced by heat treatment (Ayala 1975) and by the application of various electrophoretic conditions (Singh et al. 1976; Ramshaw et al. 1979; Coyne 1982; McLellan 1984; Ayala 1984). In practice, these techniques are not used for large-scale population surveys because they decrease the speed of sampling from populations. A method that can be used for a population survey to detect more protein variation than standard electrophoresis is isoelectric focusing electrophoresis. However, it is not suited for enzymes, the main markers. Another method applied in genetics is two-dimensional electrophoresis. In this method, as we explained also in Chapter 3, many different proteins can be examined on the same electrophoretic gel, so it is suitable for a large-scale survey of genetic variability. By changing the content of buffers during each phase of electrophoresis, it is even possible to detect separate peptides, thus, increasing resolution up to 200–10,000 fractions-dots in fingerprinting. Goldman et al. (1987) were able to study 383 polypeptides in humans and apes. However, the proportion of polymorphic loci detectable by this method is lower than that detectable by ordinary electrophoresis (Aquadro and Avise 1981; Ohnishi et al. 1983; McLellan et al. 1983; Ayala 1984). The reason for this seems to be that the proteins studied by this technique are mainly structural proteins such as histones, actin, myosin, etc., and these proteins are less variable because of stronger functional constraints (Kimura 1983a; Nei 1987).

6.5 MUTATION RATE

Direct Methods of Estimation

The rate of spontaneous mutation is one of the most important population genetic parameters, which requires reliable estimates. Nevertheless, this important parameter has not been measured by molecular markers with appropriate accuracy in a sufficient variety of organisms. One of the main problem in getting representative estimates is a very low mutation rate in general, and so a huge number of individuals must be sampled and several loci studied. Classical genetic methods determined that the rate of morphological or viability mutations inducing lethal effects is close to 10^{-5} per locus per generation for some higher eukaryotic organisms such as corn, *Drosophila*, mice and humans. It is not possible to extend these findings to mutations at the DNA level. Another problem is that most morphological mutations are deleterious, if not lethal, and usually contribute little to the evolutionary history of a species. For comparison, the rate of non-deleterious mutations in morphological traits of barley was estimated not to exceed 8.6×10^{-9} per locus per generation, with the probability level of 95% (Kahler et al. 1984).

Many estimates were obtained on the basis of biochemical and molecular genetics in recent years (Lewontin 1974; Limborskaya 1981; Vogel and Motulsky 1986; Nei 1987; Ajityxob 1989). Extensive studies ware carried out on *Drosophila* by Mukai with collaborators (Mukai and Cockerham 1977; Voelker et al. 1980). They investigated spontaneous mutations for seven allozyme loci in *D. melanogaster*, maintaining 1,000 lines during 211–224 generations. These authors examined 3,111,598 gene-generations (the number of generations multiplied by the number of genes examined). Voelker et al. (1980) estimated that the electrophoretic mobility and null mutations rates are 1.28×10^{-6} and 3.86×10^{-6} per locus per generation, respectively. It was found later that lines used by these authors carry mutator genes that increased the rate of lethal mutations nearly ten times; however, it is not clear whether these genes influence the mutation rate for enzyme loci. In any case, the estimates obtained make clear that the mutation rate at the molecular level may be lower than the rate for some morphological characters or lethal genes. As noted above this conclusion was supported by data obtained for barley (Kahler et al. 1984). These authors also examined five enzyme loci for 841,260 gene generations, but found no mutant allele. On the basis of this observation, the authors calculated that the mutation rate per locus per generation was below 3.6×10^{-6} with the 95% confidence level. Vast electrophoretic studies on mutation rate were also conducted in human populations (Harris et al. 1974; Neel et al. 1980; Altukhov 1989). Although a total of 522,119 gene generations were examined in the first of two papers, no mutation affecting electrophoretic mobility was found. Based on these data, Neel et al. (1980) concluded that the mutation rate must be lower than 6×10^{-6} per locus per generation with 95% confidence level. The estimates in this paragraph, it seems, are closer to the situation with the selectively neutral loci. In some other cases of molecular markers the situation may be different, and mutation rate, or precisely the intragenic recombination rate, for allozyme genes may reach values 6.8×10^{-4} (Altukhov et al. 1983).

Indirect Methods of Estimation

Several authors (Neel 1973; Nei 1977a; Neel and Rothman 1978; Bhatia et al. 1979; Altukhov 1989) have attempted to estimate the mutation rate for electrophoretic alleles indirectly, considering the balance among mutation, selection and genetic drift. For example, Nei (1977a) obtained an estimate of 2.3×10^{-6} per locus per generation, using data on 29 protein loci from Japanese macaques. Extending Nei's work, Kimura (1983c) estimated the proportion of deleterious mutation among the total mutations. His results suggest that 80 to 96% of new mutations in electrophoretic loci are deleterious. In the case of Japanese macaques, he estimated that the rate of neutral mutation is 1.65×10^{-7} per locus per generation. An original approach was suggested by Altukhov and coworkers (Altukhov et al. 1983; Altukhov 1989). By this technique the rare allelic variants at enzyme markers were studied at two stages of the life cycle in common pine, and detection depended on specified monomorphic and polymorphic loci groups. The "pure" mutation rate was estimated in the pine example as high as

1.9×10^{-4} (Altukhov 1989, p. 182); however for group of polymorphic loci it was 8.7×10^{-4}. As above, it also was emphasized that most of new mutations at protein-coding loci are deleterious. Coefficients of natural selection were estimated to be near 70% ($s = 0.71$; Altukhov 1989, Table 5.8). Although these estimates are quite reasonable, they are dependent on the assumption of population equilibrium. There is no guarantee for the validity of this assumption, so these estimates should be regarded as tentative.

Estimates of Mutation Based on the Rate of Gene Substitution

If we assume that most gene substitutions occur by the random genetic drift of neutral mutations, the rate of mutations can be estimated from data on gene substitution. This is because the rate of gene substitution (α) for neutral mutations is equal to the mutation rate (μ) (Nei 1987, Chapter 13). In reality, most new mutations are deleterious, as follows from the previous discussion, but they are quickly eliminated from the population by normalizing or purifying selection. Thus, if we estimate the mutation rate under the assumption of $\mu = \alpha$, it would give an underestimate of the total mutation rate. Exclusion in this respect refers to pseudogenes, which are subject to no selection. However, as noted by Nei (Nei 1987, p. 32), "this estimate can be used for testing the "null hypothesis" of neutral mutations, because in the neutral theory of molecular evolution only those mutations that contribute to gene substitution or polymorphism are considered (Kimura 1983a,b)".

In the genomes of many eukaryotic taxa, pseudogenes are very frequent. Because pseudogenes do not have any detected biological function, the total mutation rate can be estimated from the rate of nucleotide substitution, if the time of function loss is known (Nei 1987, p. 147). Some estimates of the rate of nucleotide substitution for pseudogenes (b) are given in Table 6.5.1, together with the rates for functional genes.

From data presented in Table 6.5.1, it is clear (i) that the rate of nucleotide substitutions is highest at the third position in a codon, and (ii) that nucleotide substitution rates are at least twice as great in pseudogenes than in even the third nucleotide position. The average rate of nucleotide substitutions in pseudogenes is 4.7×10^{-9} per nucleotide site per year (Table 6.5.1), which consequently may be considered as the rate of neutral mutations. The mammalian α globin genes consist of 423 nucleotides (141 codons), so that the total mutation rate per gene (coding regions only) is estimated to be 2.0×10^{-6} per year (Nei 1987). During the last one million

Table 6.5.1. Times since gene duplication (Td), times since non-functionalization (Tn) and rates of nucleotide substitution per site per year (b) for various globin pseudogenes; a_1, a_2 and a_3, denote the rates of nucleotide substitution for the first, second and third nucleotide positions of codons in functional genes, respectively (from Li et al. 1981 adopted from Nei 1987).

Gene	Td (MY)	Tn (MY)	b ($\times 10^{-9}$)	a_1 ($\times 10^{-9}$)	a_2 ($\times 10^{-9}$)	a_3 ($\times 10^{-9}$)
Mouse $\Psi\alpha 3$	27	23	5.0	0.75	0.68	2.65
Human $\Psi\alpha 1$	49	45	5.1	0.75	0.68	2.65
Rabbit $\Psi\beta 2$	44	43	4.1	0.94	0.71	2.02
Gout $\Psi\beta^x$ and Ψ^z	46	36	4.4	0.94	0.71	2.02
Average			4.7	0.85	0.70	2.34

years, the average generation time for the human lineage was probably about 20 years. Therefore, the total mutation rate for the coding region of the human α globin gene is estimated to be $\mu = 4.0 \times 10^{-5}$ per generation, and sometimes, as we stressed above, it may even reach $\mu = 1.9 \times 10^{-4}$. These rates are quite high. We must remember however, that in functional genes, most of these mutations are eliminated by purifying selection at early stages of the life cycle and they are not involved in further evolution in populations. Hence, the rate of the rest of the mutations should be lower.

The above result suggests that synonymous nucleotide substitutions occur more frequently than non-synonymous substitutions, possibly at the same rate as the pseudogene rate. The rate of synonymous substitution can be estimated by the methods that we will consider later. With such methods it was found (Li et al. 1984) that the rate of synonymous substitution is quite similar for many genes, like α- and β-globins, growth hormones, etc., and that the average rate for mammalian taxa is 4.6×10^{-9} per nucleotide site per year (Nei 1987, Table 5.6). This rate is identical to the rate for pseudogenes, suggesting that synonymous substitutions are obviously subjected to no purifying selection; although there are still conflicting data, which may result from biased estimates. Still, if the former is the case, the rate of neutral mutations can also be estimated from synonymous substitutions. In the above paragraphs, only nuclear genes were considered. However, the mutation rate in mitochondrial genes in mammals and other groups is known to be higher than that in nuclear genes (Brown et al. 1979; 1982; Brown 1983; Avise 1994). The rate of silent nucleotide substitution has been estimated to be 4.7×10^{-8} per site per year (Brown et al. 1982). This is nearly one order higher than that of nuclear pseudogenes. Very high rates of point (nucleotide) mutation were also observed in bacteria and RNA viruses (Air 1981; Holland et al. 1982; Nei 1983; Gojobori and Yokoyama 1985; Alikhanyan et al. 1985; Lewin 1987; Russel 1998). In bacteria, the spontaneous mutation rate is $10^{-6} – 10^{-7}$ per locus per generation (Lewin 1987). In the research on the nucleotide substitutions in the influenza A virus it was suggested that the mutation rate must be on the order of 0.01 per nucleotide site per year to fit the divergence pattern between virus lineages. This is nearly 6–7 times as high as the rate for nuclear genes in eukaryotes. In hot spots of *E. coli,* the spontaneous rate of transitions and stop or amber mutations is 2, 3, 10 and even 100 per *lac* gene (Russel 1998). It is believed that these high rates of mutation in non-nuclear genes are caused by the lack of or weakness of the repair mechanism of mutational damage of DNA or RNA (Holland et al. 1982; Lewin 1987; Jimulev 2002). For evolutionary genetics, it is important to know, as we stated earlier, the rate of mutations that contribute to amino acid substitutions in proteins and so determine their polymorphisms in populations. When neutrality of mutations is assumed, this mutation rate can be estimated from that rate for amino acids (Table 6.5.2). The mutation rate for a given polypeptide is given by the substitution rate per amino acid site multiplied by the number of amino acids in the polypeptide. This rate varies sufficiently among different polypeptides; the example of the distribution of the rate among the 41 polypeptides obtained by Chakraborty et al. (1978) shown here (Fig. 6.5.1). The distribution is roughly the gamma distribution, with mean equal to 2.47×10^{-7} and standard deviation 2.51×10^{-7}. Variation in the substitution rate per polypeptide depends upon two causes, the difference in the number of amino acids or/and in the substitution rate per site.

Table 6.5.2. Rates of amino acid substitutions per amino acid site per 10^9 years ($\times 10^9$) in various proteins (from Dayhoff 1978, adapted from Nei 1987).

Protein	Rate	Protein	Rate
Fibrinopeptides	9.0	Thyrotropin beta chain	0.74
Growth hormone	3.7	Parathyrin	0.73
Ig kappa chain C region	3.7	Parvalbumin	0.70
Kappa casein	3.3	Protease inhibitors, BP1 type	0.62
Ig gamma chain C region	3.1	Trypsin	0.59
Lutropin beta chain	3.0	Melanotropin beta	0.56
Ig lambda chain C region	2.7	Alpha crystallin A chain	0.50
Complement C3a anaphylacoxin	2.7	Endorphin	0.48
Laccalbumin	2.7	Cytochrome b;	0.45
Epidermal growth factor	2.6	Insulin (exc. guinea pig and coypu)	0.44
Somatotropin	2.5	Calcitonin	0.43
Pancreatic ribonuclease	2.1	Neurophysin 2	0.36
Lipotropin beta	2.1	Plastocyanin	0.35
Haptoglobin alpha chain	2.0	Lactate dehydrogenase	0.34
Serum albumin	1.9	Adenylate kinase	0.32
Phospholipase Az	1.9	Triosephosphate isomerase	0.28
Protease inhibitor, PST1 type	1.8	Vasoactive intestinal peptide	0.26
Prolactin	1.7	Corticotropin	0.25
Pancreatic hormone	1.7	Glyceraldehyde 3-P04 dehydrogenase	0.22
Carbonic anhydrase C	1.6	Cytochrome C	0.22
Lutropin alpha chain	1.6	Plant ferredoxin	0.19
Hemoglobin alpha chain	1.2	Collagen (exc. nonrepetitive ends)	0.17
Hemoglobin beta chain	1.2	Troponin C, skeletal muscle	0.15
Lipid-binding protein A-II	1.0	Alpha crystallin B chain	0.15
Gastrin	0.98	Glucagon	0.12
Animal lysozyme	0.98	Glutamate dehydrogenase	0.09
Myoglobin	0.89	Histone H2B	0.09
Amyloid AA	0.87	Histone H2A	0.05
Nerve growth factor	0.85	Histone H3	0.014
Acid proteases	0.84	Ubiquitin	0.01
Myelin basic protein	0.74	Histone H4	0.010

Note that the mean substitution (mutation) rate is approximately equal to the product of the average substitution rate (1×10^{-9}) in Table 6.5.2 and the average number of amino acids per polypeptide (240). The difference in the substitution rate per site is considered to reflect the variation in the intensity of purifying selection.

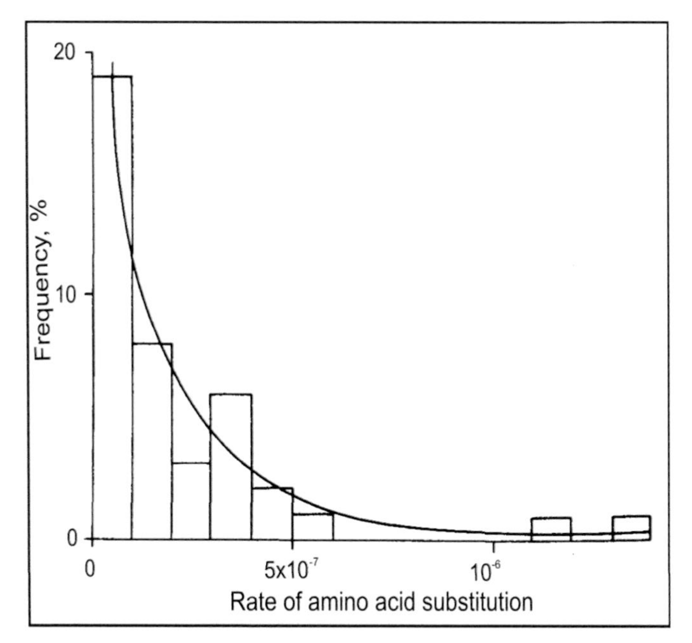

Fig. 6.5.1. Distribution of the rate of amino acid substitution per polypeptide per year. The number of polypeptides used is 41 (from Nei 1987 with permission).

It should be noted that the proteins of which the substitution rate was studied are generally of small size, the average number of amino acids per polypeptide being 240. Most proteins used for electrophoretic studies are larger. Molecular weights for 119 protein subunits in mammalian species (Darnell and Klotz 1975) fit a gamma distribution (Fig. 6.5.2). The mean and standard deviation of this distribution are 45,102 and 24,531, respectively. The average molecular weight of an amino acid is about 110, so the average polypeptide seems to have about 410 amino acids. If we assume that the rate of amino acid substitution per site is the same for both groups of polypeptides in Figs. 6.5.1 and 6.5.2, ***the mean rate of amino acid substitution for an average polypeptide is estimated to be 4×10^{-7} per locus per year***. Previously, we noted that electrophoresis detects only about one-quarter of all amino acid changes. ***Therefore, the rate of amino acid substitutions detectable by electrophoresis is estimated to be about 10^{-7} per locus per year***. Under the "null hypothesis" of neutral mutations, this is equal to the mutation rate.

Mutation Rate and Time of Divergence

As mentioned earlier, the rate of amino acid or nucleotide substitution seems to be constant per year rather than per generation. Why the mutation rate at the DNA level is constant per year, understandable from the viewpoint of the neutral theory (Kimura and Ohta 1971; Nei 1975). However, classical genetics has established that the mutation rate is generally constant per generation rather than per year. What is the reason for this difference? One possible explanation is that the mutations studied by

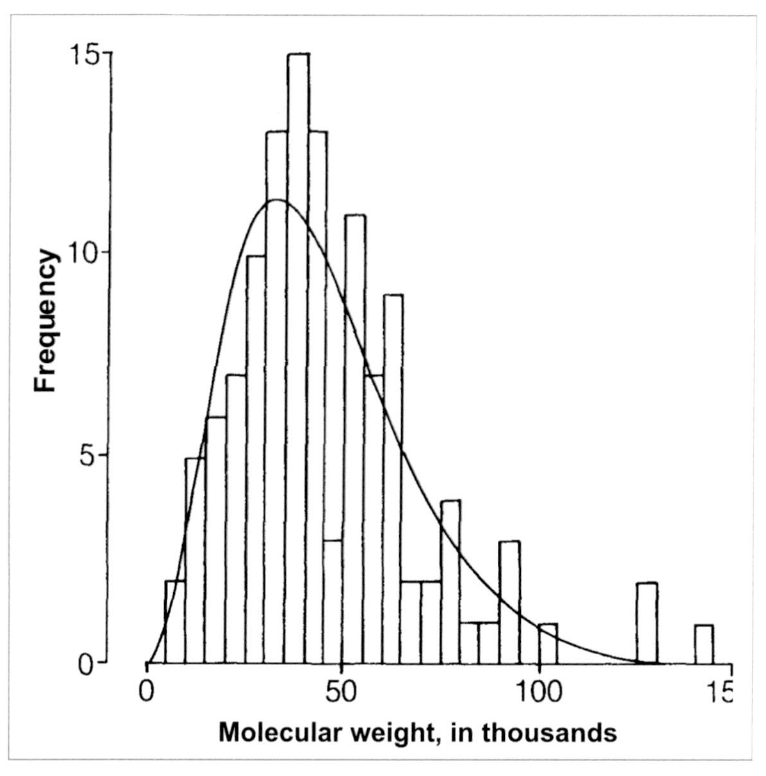

Fig. 6.5.2. Distribution of molecular weights of protein subunits in mammalian species. The total number of subunits used is 119. The gamma distribution fits the data very well ($\chi^2(9) = 6.43$; $P > 0.65$; from Nei 1987 with permission).

classical geneticists are different from those contributing to amino acid or nucleotide substitution. In classical genetics, the mutation rate was estimated mostly by using traits of external morphology or viability mutations that are deleterious or lethal. It is possible that a majority of these mutations are due to large changes, like big deletions, insertions or at the molecular level, the frameshifts that lead to loss of gene function. Thus and Magni (1969) showed in yeast that most deleterious mutations are frameshifts that occur at the time of meiosis. Müller (1959) also showed that the majority of lethal mutations in *Drosophila* occur at the meiotic stage. It may be expected that the rate of deleterious mutations is constant per generation rather than per year. In contrast, small alterations or mutations that occur at other, less selectively significant genes, are a small part of the total mutations and may appear constantly and independent of generation length. Data supporting this view have been obtained for T4 phage (Drake 1966) and for *E. coli* (Novick and Szilard 1950). However, this question, still needs further clarification. The mutation frequency must depend both on DNA replication cycles and on absolute time. For example, Kubitschek (1970) found that replication-dependent mutations include those caused by caffeine and some other chemicals, whereas time-dependent mutations include those induced by ultraviolet rays.

Nevertheless, the approximate constancy of nucleotide or amino acid substitution for both prokaryotes and eukaryotes, as observed with 5S RNA (Hori and Osawa 1979), fibrinopeptide, hemoglobin and cytochrome-b (Dickerson 1971), suggests that there are mutations that occur in proportion to chronological time. They could be due to ultraviolet radiation or to background radiation. Unfortunately, we know very little about the real mechanisms of mutation in nature. It is important no note for ourselves that the total mutation rate (μ_T) comprises of several components: $\mu_T = \mu_N + \mu_D + \mu_R$. Here: μ_N is the rate of neutral mutations, μ_D the rate for deleterious mutations and μ_R the rate for favorable mutations. Favorable mutations occur quite rarely, but once they appear they quickly become common and then fixed in the species, i.e., variation at the locus is lost. As noted above, deleterious mutations are removed from populations by purifying selection. In other words, molecular variation and polymorphism in populations at this level have to be maintained due largely to neutral mutations (Kimura 1983a; Nei 1987).

Time of Divergence. Suggestions of the neutrality of molecular variation lay in the basis of time reconstruction based on the mutation rate μ_N. Time in this case is measured via accumulated mutation number since isolation of a pair of taxa and corresponding separation of single gene pool into two independent ones, where molecular clocks through time are turn on. In the case of protein markers, absolute time of divergence (T) is estimated as proportional to the standard genetic distance (D_n, Nei 1972) that inverse to the mutation rate for an average protein gene μ_N ($\mu_N = 10^{-7}$, for electrophoretically detected loci, see above). Thus (Nei 1975; 1987),

$$T = D_n \, 1/\mu_N. \qquad \textbf{(6.1)}$$

Estimates of divergence time on the basis of sequence of amino acids or nucleotides are based on the same principle, i.e., the accumulated number of mutations for a pair of sequences. Such an approach was considered in brief in Chapter 4 for amino acid sequences, and will be further developed for nucleotide sequences in Chapter 13. In estimating T there are many difficulties. Frequently it is necessary to correct mutation rate for the particular set of loci because those coding for proteins with more functional importance exhibit lower mutation rates; big difference may exist for them in this property (see text above and Table 6.5.2). That is why in some papers the μ_Ns accepted may be different from value given above (Carson 1975, $\mu_N = 2*10^{-6}$; Avise et al. 1980, $\mu_N = 5*10^{-6}$, $\mu_N = 2*10^{-6}$; Kartavtsev et al. 1983; Kartavtsev and Mamontov 1983, $\mu_N = 5*10^{-6}$; Chichvarkhin et al. 2000, $\mu_N = 10^{-7}$; Kartavtsev et al. 2000, $\mu_N = 10^{-6}$). In the cases of sequence divergence data for a certain gene it is particularly important that the substitution (mutation) rate is constant among phyletic lines. However, the mechanism for the constant rate of molecular evolution remains an unresolved problem at the present time. Furthermore, the constant rate is an approximate property, and there are many exceptions, as will be discussed later. Finally, it should be noted that even if the mutation rate is generation-dependent, it is possible to explain the linear relationship between molecular evolution and chronological time by using the theory of slightly deleterious mutations (Kimura 1985), though the explanation is somewhat contrived. The mechanisms of mutations are variable and many of them new, discovered recently (Jimulev 2002; Klug and Cummings 2002). It should also be noted that the production of mutations is an extremely complicated process, involving DNA lesion and repair, and it is difficult to make a priori predictions about the mutation frequency from the molecular mechanism alone (Kondo 1977).

We conclude this chapter by remembering the phenomenon of **transposable genetic elements**, sometimes referred to as **transposons**, genetic units that can move or be "transposed" within the genome. Because the movement of genetic units from one place in the genome to another often disrupts genetic function and results in phenotypic variation, transposons fit a broad definition of mutation. Recent data show their ability as enhancers of mutation rate. They also may create regulatory change, which can stimulate genesis cancer in a normal cell. All these activities are beyond the scope of this chapter. The role of transposons in evolution will be briefly considered in Chapter 15. In one of the above sections, it was stated that favorable mutations are rare and quickly became shared by all individuals of a population through direct natural selection. However, it must be understand that persistence of some mutations is supported by balancing natural selection or by selection with changing direction of action (Lewontin 1974). These mutations may stay for many generations in populations. What proportion of mutations is neutral and what is supported by balancing selection is a major question of long-term discussion, which was mentioned earlier, and the precise answer to which is still unknown. Information on the impact of natural selection on molecular polymorphism will be considered again in Chapters 8–10 and 12.

All the material of the current chapter may be summarized in three points.

1. Mutations have a complex nature. Occurrence of even a simple point mutation determines by a complex of chemical events, e.g., for a nucleotide mutation, two events could act, one altering its substitution type and another involving its reparation.
2. From the total mutations $\mu_T = \mu_N + \mu_D + \mu_R$, only μ_N and some of μ_R are conserved as variations. In other words, molecular polymorphisms in populations are maintained basically by neutral mutations and other portion of it by balancing natural selection.
3. Mutations play a principal role in molecular clocks. Mutation rate and time since gene pool separation of a taxonomic pair directly determine the value of accumulated genetic distance between the pair. Thus, when the values of the distance and mutation rate are known, the time of separation of the taxa can be calculated.

6.6 TRAINING COURSE, #6

6.1 Solve these three questions in class to deepen knowledge on mutation.

1) What is the difference between a chromosomal mutation and a gene mutation? Between a somatic and a gametic mutation?
2) Why do you suppose that a random mutation is more likely to be deleterious than neutral or beneficial?
3) Most mutations in a diploid organism are recessive. Why?

6.2 Visit Internet sites on genetics and learn one point you like on mutations that was not pointed out in Chapter 6.

6.3 Consider a few new terms.

- **Spontaneous mutations** are those that occasionally happen in nature. No specific agents are associated with their occurrence. In contrast to such spontaneous events, those that result from the influence of any artificial factor are considered to be **induced mutations**.
- **Nutritional** or **biochemical mutations** are mutations that alter variation of corresponding traits.
- Mutations that result in synonymous codons are called **synonymous** or **silent mutations**, whereas others are called **nonsynonymous** or **amino acid altering mutations**.
- Viability mutations may be classified as **lethal mutations, neutral**, **advanced** or **deleterious**.

7

GENETIC ASPECTS OF SPECIATION

MAIN GOALS

7.1 General Genetic Approach: Advances and Limitations
7.2 What Data are Necessary? What is the Data Base?
7.3 Species Concept. Review of Literature Data
on Heterozygosities and Genetic Distances
7.4 Speciation Modes (SM): Population Genetic View
7.5 TRAINING COURSE, #7

SUMMARY

1. A population genetic approach increases the validity of our conclusions on the current species status.
2. Application of the approach proposed in this chapter permits us to define modes of speciation in a taxon.
3. On the basis of molecular genetic data, phylogenetic reconstructions by kinship, as well as dating the time of divergence are possible.

7.1 GENERAL GENETIC APPROACH: ADVANCES AND LIMITATIONS

The problem of biological species and speciation are the main focus of this chapter. These problems have attracted researchers' attention since the establishment of biology as a science. Most popular now among biologists is the Synthetic Theory of Evolution (STE), which includes the Biological Species Concept (BSC). The basis of the STE and it systematic description were presented in the fundamental books by Haldane (1932), Dobzhansky (1937; 1943; 1955), Huxley (1954), and Mayr (Mayr 1947; Mayr 1968). A popular summary of the STE is the book by Timofeev-Resovsky with co-authors (1977). The concept requires further development and constructive ideas on some STE themes were developed by Vorontsov (1980), Harrison (1998) and others.

One weak point in the STE is the frequent inability to test experimentally a key criterion of the BSC—i.e., reproductive isolation of the species in nature. There are a lot of other criticisms addressed to BSC that were summarized, for example, by King (1993). Nowadays, an important controversy is taking place between the BSC and the Phylogenetic Species Concept (Avise and Wollenberg 1997). The theory of speciation is also not well developed in the STE. Strictly speaking, there is no quantitative theory of speciation at all. Reflecting this limitation is the series of new theoretical and experimental developments (Avise and Wollenberg 1997; Arnold and Emms 1998; Harrison 1998; Howard 1998; Templeton 1998; 2004; Wu and Hollocher 1998; Kartavtsev et al. 2002; Kartavtsev 2009a,b; Kartavtsev 2011a,b).

It should be emphasized, nevertheless, that many directions of the STE and the genetics of speciation have been developed. There have been diverse analyses performed to understand the genetic and conceptual bases of speciation (Fox and Morrow 1981; Grant 1985; King 1993; Harrison 1998; Templeton 1998; De Qeiros 1998). The genetic basis for creation of reproductive isolation has also been subjected to analysis (Leslie 1982; Templeton 1981; Nei et al. 1983; Coyne 1992; Templeton 1998; Butlin 1998). The possibility of sympatric speciation (Bush 1975; Kondrashov et al. 1998; McCune and Lovejoy 1998), the role of saltations or genetic revolutions in evolution (Altukhov and Rychkov 1972; Carson 1975; Altukhov 1983; 1997), and the genetic differentiation during formation of living forms and taxa (Ayala et al. 1974; Avise 1976; Avise and Aquadro 1982; Nevo and Cleve 1978; Thorpe 1982; Nei 1975; 1987) have also been considered. What, in general, are the advances and limitations in the contemporary genetic approach?

Advances.

- Data reduction down to genotypic codes (values) give us a possibility to use genetic theory in the analysis.
- A comparative investigation of variability among structural and regulatory elements of the genome and genetic divergence of taxa is possible.
- Investigations on protein divergence in species from nature discovered a "molecular clock".
- A new possibility of a phylogenetic reconstruction developed: (1) not by similarity, but by kinship and (2) by dating the divergence.

Limitations.

- Deduction is limited by genotypic descriptions and genetic theory.
- Analysis is connected with preliminary laborious experimental investigation (with its own limitations).
- Investigation of a species from nature is frequently limited by unique or rarely repeated events (phenomena).
- Genotypic effects of the marker loci on phenotype are weak.
- The theory is not sufficiently developed in some directions.

7.2 WHAT DATA ARE NECESSARY? WHAT IS THE DATA BASE?

In general we need:

- Data that support (or reject) the central dogma of Neo-Darwinism, i.e., evolution can occur only on the basis of genetic change.
- Data on variability at different levels of biological organization in genetic terms (per locus quantitative genotypic values, AA, ...), i.e., single-dimensioned data tables (DT).
- Data on genotypic values of an individual at the set of loci (genotype AA Bb...), i.e., multi-dimensional DT.
- Complementary data: Morphological traits, data on abiotic variability (at least as an expert estimate, e.g., grouping variables), etc.

The keystone of the STE may be represented by Dobzhansky's scheme (Fig. 7.2.1), in which genetic divergence is a key to speciation. If one provides a fact that evolution is possible without genetic change in lineages, then the evolutionary genetic paradigm and STE in particular can be rejected.

What database is available to operate with? Among the most representative now are data on the variation of the isozyme genes. We will consider below such data that are based on allozyme variability. An example of such an intra- and interspecies variability is phosphoglucomutase in dace (Fig. 7.2.2).

With the biochemical genetic techniques and some population genetic formulae it is easy to calculate both the genetic variability and similarity (or distance) of gene pools between individuals from populations of the same and different species. A few equations that are useful for this are given below, and calculation procedures were considered in detail in Training Course #5.

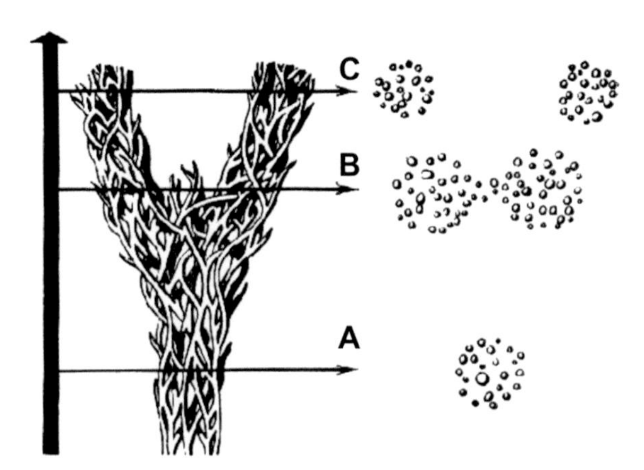

Fig. 7.2.1. A schematic representation of species divergence in time (From: Dobzhansky 1955). (A) Single species population, (B) Initial phase of divergence (subspecies), (C) Different species. Circles on the right side represented the gatherings of separate demes or populations having exchange by gene flow with each other up to phase B.

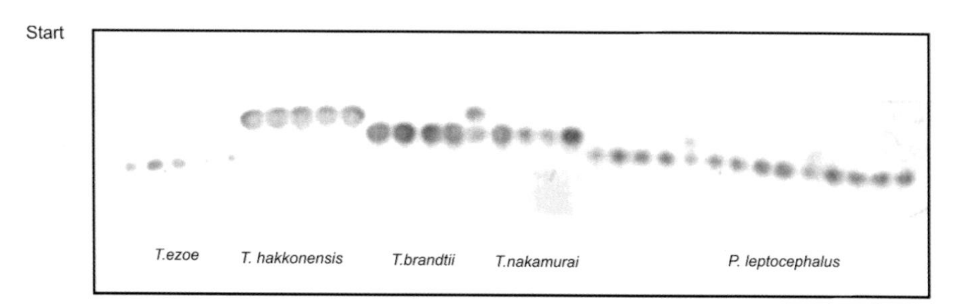

Fig. 7.2.2. Enzyme electrophoresis of phosphoglucomutase (**PGM***) locus in the dace, cyprinid fishes (*Pisces, Cyprinidae, Tribolodon*) (Kartavtsev et al. 2002). Both the genetic variability and similarity can be estimated through the gel inspection.

Mean Heterozygosity

$H = \Sigma_{i=1}^{L} h_k / L$, H is mean expected heterozygosity per locus or per individual.

$h_k = 1 - \Sigma_{i=1}^{m} p_i^2$, h_k is expected heterozygosity ($H_{exp} = Hs$), p_i is an *i* allele frequency; L, number of loci. Observed heterozygosity is $H_{obs} = hj/n$, hj is the number of heterozygotes, *n* is *the* total number of individuals in a sample.

Genetic Distance. Standard Nei's Distance (D_n)

$D_n = -\ln I.$

I is the normalized genetic identity for a random set of loci. For the locus ***j*** it is equal to:

$I_j = (\Sigma_{i=1}^{k} x_i y_i) / \sqrt{\Sigma_{i=1}^{k} x_i^2 \Sigma_{i=1}^{k} y_i^2}.$

For the whole set of loci ***I*** is equal to:

$I = Jxy / \sqrt{JxJy};$

Where ***Jxy, Jx, Jy***—arithmetic mean for the sums: $\Sigma_{i=1}^{k} x_i y_i$, $\Sigma_{i=1}^{k} x_i^2$, $\Sigma_{i=1}^{k} y_i^2$,

Where, x_i and y_i are allele frequencies in populations (taxa) *x* and *y*, ***k*** is number of alleles in all pair-wise comparisons.

Mean heterozygosity per individual (per locus) has been recognized as the best measure of variability (Lewontin 1978; Nei 1987; Zhivotovsky 1991). The most popular measure of taxa divergence during evolution at this level is standard Nei's distance, D_n and the inverse measure, normalized similarity, I (Nei 1972; Nei 1975; 1987).

An Example of Evolutionary Genetic Research

Let us first see an example of research that summarized the allozyme data for six cyprinid fish (Kartavtsev et al. 2002). We will include for comparison our mtDNA sequence data as well. Genetic variability is typical for these cyprinid species (Table 7.2.1) in comparison with other fish (see data below).

Based on the allele frequencies, several distance measures were calculated between species; two of them, Rogers' D_r and Nei's D_n, are shown below (Table 7.2.2).

Table 7.2.1. Genetic variability at 30 loci in 6 cyprinid fish species (Pisces, Cyprinidae).

Вид	**N (SD)**	**N_A (SD)**	**P (%)***	**H (SE)**	
				H_{obs}	**H_{exp}(H–W)****
(1) *T. hakonensis*	24.1 (1.4)	1.3 (0.1)	23.3	.036 (.016)	.043 (.017)
(2) *T. brandti*	29.2 (1.5)	1.2 (0.1)	16.7	.018 (.010)	.021 (.013)
(3) *T. sachalinensis*	13.4 (0.4)	1.2 (0.1)	16.7	.045 (.020)	.067 (.029)
(4) *T. nakamurai*	5.0 (0.0)	1.0 (0.0)	3.3	.007 (.007)	.007 (.007)
(5) *P. leptocephalus*	14.0 (0.0)	1.1 (0.1)	10.0	.017 (.010)	.016 (.009)
(6) *L. waleckii*	8.0 (0.0)	1.1 (0.1)	13.3	.029 (.014)	.042 (.022)
Mean (± *SE*), species *1–4*	17.9 ± 5.4	1.2 ± 0.06	15.0 ± 4.2	.026 ± .008	.034 ± .013
Mean (± *SE*), species *1–6*	15.6 ± 3.8	1.2 ± 0.04	13.9 ± 2.8	.025 ± .006	.033 ± .009

N. Mean sample size per locus, N_A. Mean number of alleles per locus, P. Proportion of polymorphic loci, H. Mean heterozygosity per individual, SD. Standard deviation, SE. Standard error.
* Locus is considered as polymorphic, if predominate allele frequency is less than 0.99.
** H–W. Notation explains heterozygosity, which expected under Hardy-Weinberg equilibrium, unbiased estimate (Nei 1978).

Table 7.2.2. Genetic distances among 6 cyprinid fish species (Pisces, Cyprinidae) at 30 protein loci.

Species	**1**	**2**	**3**	**4**	**5**	**6**
(1) *T. hakonensis*	–	.145	.243	.217	.300	.810
(2) *T. brandti*	.146	–	.177	.145	.266	.743
(3) *T. sachalinensis*	.242	.189	–	.284	.290	.762
(4) *T. nakamurai (U-ugui)*	.212	.146	.271	–	.223	.769
(5) *P. leptocephalus*	.272	.242	.276	.207	–	.760
(6) *L. waleckii*	.569	.538	.557	.547	.542	–

Below, the diagonal Rogers' distances, D_R (Rogers 1972) and above, diagonal standard Nei's genetic distances, D_n (Nei 1972) are given.

The main outcome from a distance-similarity matrix is a dendrogram or tree, which can illustrate how the taxa join or aggregate with each other. The main purpose of the tree diagram is to convert numerical information from the matrix into a form that is easy to see. Representation of the data in a dendrogram is not the only way, however. Indeed, when there are many units of comparison, especially in intraspecific investigations, it much better to use other methods of cluster analysis or ordination (Rao 1980). For example, the matrix of genetic distances can be summarized as multi-dimensional scores and then plotted in two or three dimensions. In our case, the tree constructed indicates that the dace species are closest to each other and the other two taxa are external to that group (Fig. 7.2.3 and Fig. 7.2.4). Differences in the

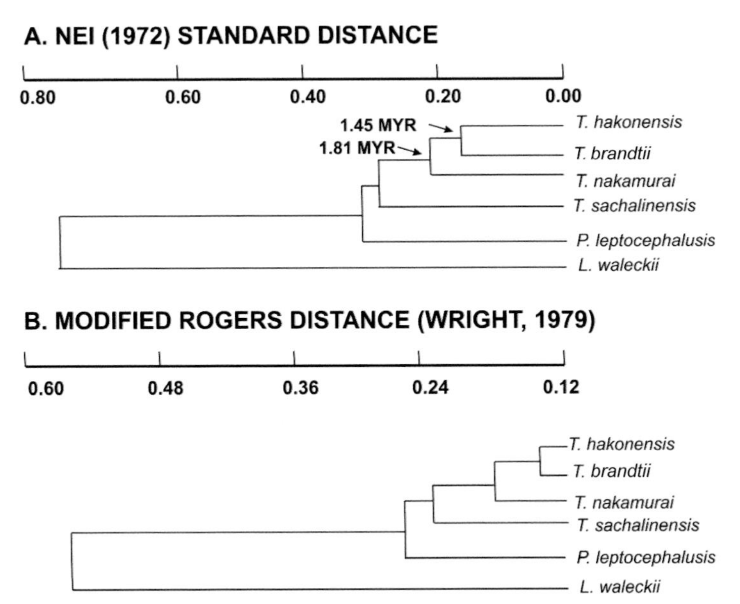

Fig. 7.2.3. Dendrograms that illustrate clustering via UPGM method among cyprinid fishes based on allele frequencies at 30 allozyme loci, using Nei's D_n (Nei 1972) and modified Rogers' distance (D_r) (Wright 1978). The same clusters were obtained using Nei's D_m. Thus, clustering and tree topology only weakly depend on the distance measure. Coefficient of cophenetic correlation: r = 0.99.

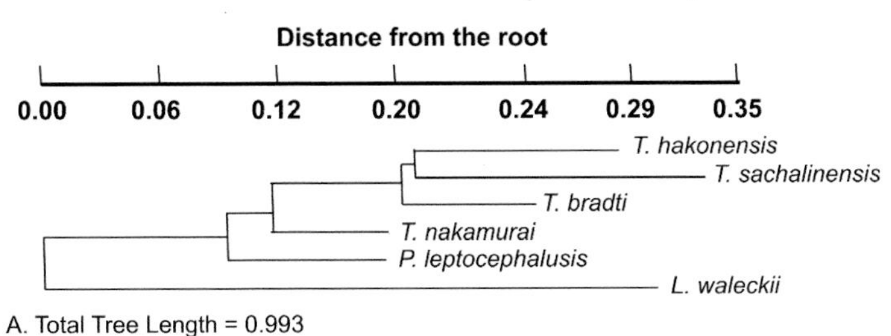

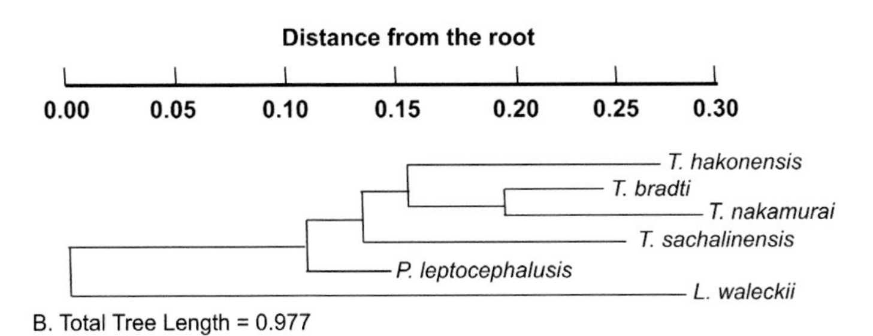

Fig. 7.2.4. Variants of reconstruction of phylogenetic relations in Cyprinidae, which were obtained using Rogers' distance (D) (Rogers 1972). A, B are two possible phylogram, constructed by Wagner's method. The coefficients of cophenetic correlation are: A, r = 0.979, B, r = 0.982.

tree shape depend on several things (Bailey 1970; Rao 1980; Nei 1987): the number of traits (loci), diversity of taxa, technique of cluster formation and distance measure (When distances among taxa are large, the type of measure has less influence on the tree topology). The DNA sequence data on mitochondrial cytochrome-b gene gave relationships similar to those for the allozymes, i.e., dace species are very similar genetically as provided in Sasaki and Kartavtsev et al. (2007) (see Fig. 7.2.5).

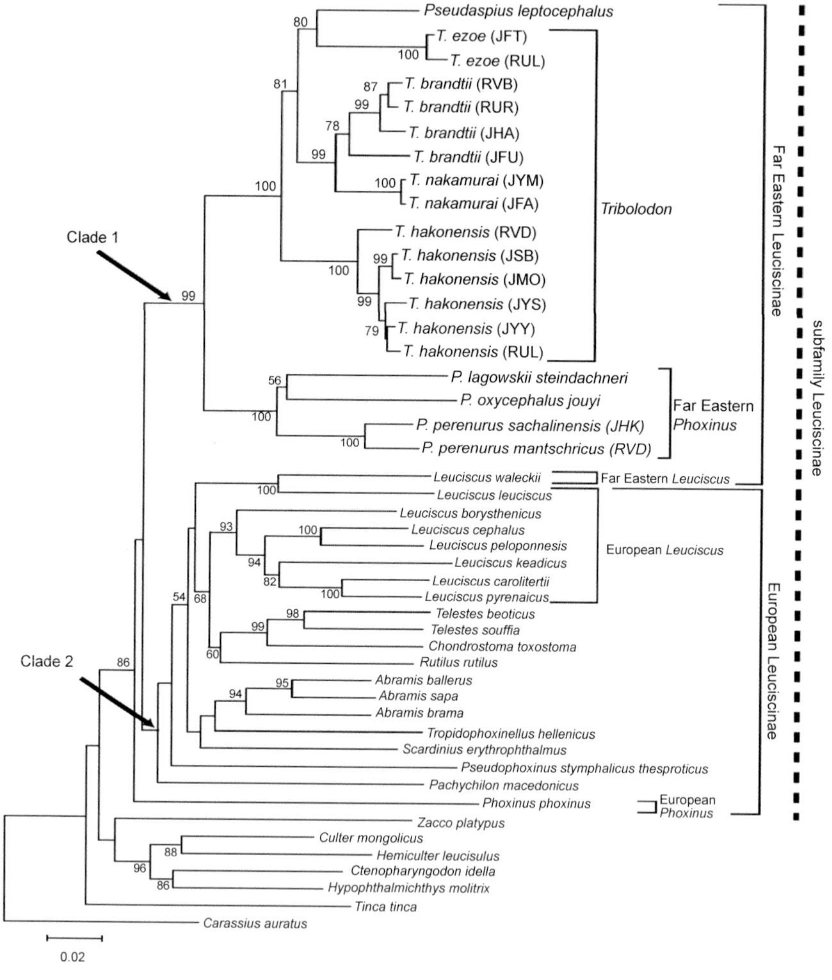

Fig. 7.2.5. Neighbor-joining tree based on mitochondrial cytochrome-b sequences and K2P distances (Kimura 1980). Number at nodes represents percent bootstrap support in 1000 replicates. The scale bar shows 0.02 base substitutions per nucleotide site. Abbreviations: *T. ezoe*, JFT: Japan, Fukushima, Tadami River, RUL: Russia, Sakhalin, Uliker River, *T. brandti*, RVB: Russia, Vostok Bay, RUR: Russia, Sakhalin, Ulike River, JHA: Japan, Hokkaido, Abira River, JFU: Japan, Fukushima, Ukedo River, *T. nakamurai*, JYM: Japan, Yamagata, Mogami River, JFA: Japan, Fukushima, Aga River, *T. hakonensis*, RVD: Russia, Vostok Bay, JSB: Japan, Shiga, Biva Lake, JMO: Japan, Miyazaki, Oydo River, JYS: Japan, Yamagata, Sakata Port, JYY: Japan, Yamagata, Mamigasaki River, RUL: Russia, Sakhalin, Ulike River, *P. perenurus*, JHK: Japan, Hokkaido, RVD: Russia, Vladivostok Area.

We noted earlier that the tree branching depends on many things, and the true tree topology is definitely known only in a theory and in some special cases (Nei 1987; Hillis et al. 1996). To construct the most convenient tree, it is quite frequently necessary to use several approaches, and then combine the phylogenetic signal into a consensus tree. For our cyprinids, the consensus tree combined both the genetic and ecological information (Fig. 7.2.6).

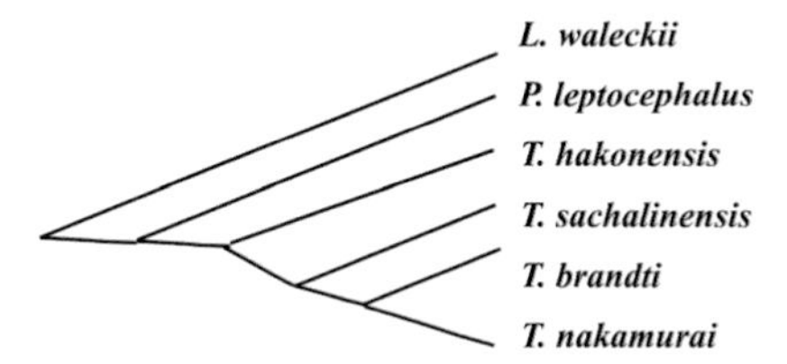

Fig. 7.2.6. Consensus dendrogram for the studied species of the Leuciscinae subfamily of cyprinid species.

7.3 SPECIES CONCEPT. REVIEW OF LITERATURE DATA ON HETEROZYGOSITIES AND GENETIC DISTANCES

We will start the review with the inspection of the main concepts on species origin that are relevant to some genetic forces causing the process. Developing main ideas of the STE and Dobzhansky's scheme in particular (see Fig. 7.1.1), Bush (1975) presented a scheme, which shows the roles for a break in gene flow in different modes of speciation (Fig. 7.3.1).

The barriers to gene flow are able to create Reproductive Isolation Mechanisms (RIM) or Reproductive Isolating Barriers (RIB), which in their turn further lead to origin of species; under different situations in nature, different modes of speciation act (Fig. 7.3.1). Neither the scheme above nor the paper itself (Bush 1975) answers many fundamental questions of speciation. For instance, it is unclear, what mode is most frequent and whether gene flow is the sole primary factor that alter gene pools or if there are others. To continue discussion it is necessary to define what a species is.

Species Definition

A species is a biological unit, which is reproductively isolated from other units, consisting of one to several more or less stable populations of sexually reproducing individuals that occupy a certain area in nature (my definition). In principal points, this is the definition of the BSC. In one of the original BSC definitions, "***A species is a reproductive community of populations (reproductively isolated from others) that occupies a specific niche in nature***" (Mayr 1982, p. 273). We will accept the BSC for further discussion, although we will keep in mind that it is restricted mainly to bisexual organisms (Mayr 1968; Timofeev-Resovsky et al. 1977; Templeton 1998). The BSC is closest to the theory of population genetics, and that is why I prefer to use this definition in this and subsequent chapters; it may be restricted, but it is closer to

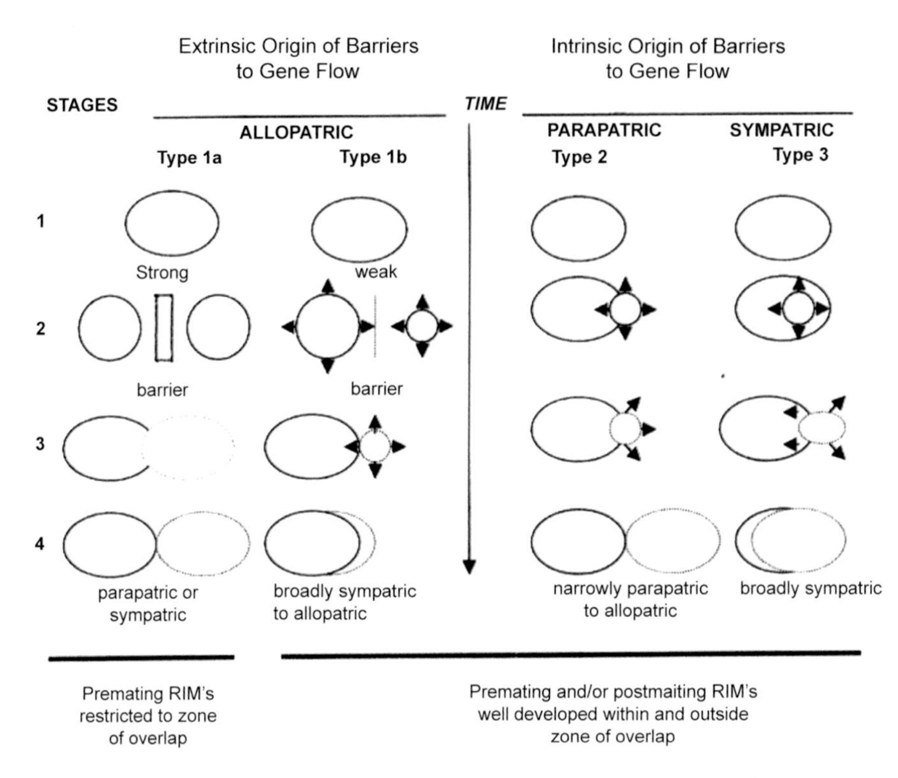

Fig. 7.3.1. Diagram of main modes of species origin (from Bush 1975 with permission from publisher).

the theory of population genetics than other definitions. Several other species concepts, with their own advances and limitations, have been critically reviewed (King 1993; Krasilov 1977). The list of these concepts is given below.

- The Linnaean Species;
- Biological Species Concept;
- The Biological Species Concept (BSC) (Mayr 1947; 1968);
- BSC Modification II (Mayr 1982);
- The Recognition Species Concept (Paterson 1978; 1985);
- The Cohesion Species Concept (Templeton 1998);
- Evolutionary Species Concept;
- Simpson's (1961) Evolutionary Species Concept;
- Wiley's (1981) Evolutionary Species Concept;
- The Ecological Species Concept (Van Valen 1976);
- The Phylogenetic Species Concept (Cracraft 1983).

From the genetic point of view on the speciation we need to know the level of genetic variation both among the intraspecific gatherings and among the taxonomic groups (demanded by the BSC paradigm; Mayr 1968; Timofeev-Resovsky et al. 1977), as well as their relative gene pool divergence, that represent supposedly different stages of species origin (in accordance to the BSC and other paradigms; see Fig. 7.2.1, Fig. 7.3.1 and Mayr 1968; Timofeev-Resovsky et al. 1977; King 1993). Firstly, let us look at genetic variability, which is best represented by means of heterozygosity, ***H*** (see above). A simple review of the literature for enzyme loci give: ***H*** = 10.2% for plants, ***H*** = 11.8% for invertebrates and ***H*** = 7.3% for vertebrates (Fig. 7.3.2).

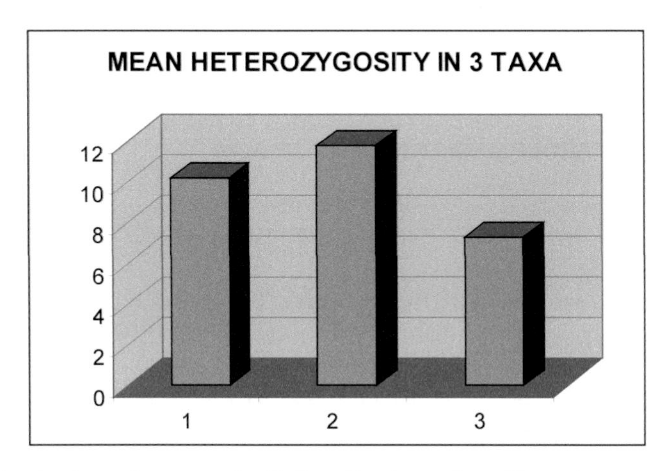

Fig. 7.3.2. Levels of mean heterozygosity (H, %, ordinate) as estimated by nuclear enzyme markers in three different taxa: (1) plants, (2) invertebrates, (3) vertebrates (Literature data are reviewed in the Ph.D. thesis by Kartavtsev 1978).

In a different review, it was reported that vertebrate animals have ***H*** = 5%, invertebrates have ***H*** = 9%, and plants are intermediate to these two values (Nevo et al. 1984). Other review sources gave similar figures; for example, for mammals ***H*** = 5% (Makarieva 2001), and for fish ***H*** = 5.1 ± 0.3 (Ward et al. 1992). Taking into account that only 25–50% of variation of protein gene markers can be detected by electrophoresis (Lewontin 1974; Nei 1987 and see discussion in Chapter 5), we can conclude that there is a large amount of genetic variability in wild species. This fact supports the ability, which is accepted by the BSC, to accumulate genetic changes through the gene pool's stages of isolation. In stages 1–4 in Type 1a of Bush's scheme (Fig. 7.3.1), for example, this may lead to the consequent origin of RIMs by pleiotropic effects of new mutations accumulated and thus lead to species origin.

Comparison of some other variability parameters shows that genetic variability may depend on many factors, so it is not simply related to a certain general evolutionary vector like time. For example, the level of genetic diversity in two taxonomically close salmon genera, which are ecologically divergent and differ in mean population size (Altukhov et al. 1997, Table 2.5), is significantly different (Fig. 7.3.3). These data support the idea that larger population size causes higher ***H*** (~G_{st}, see development in

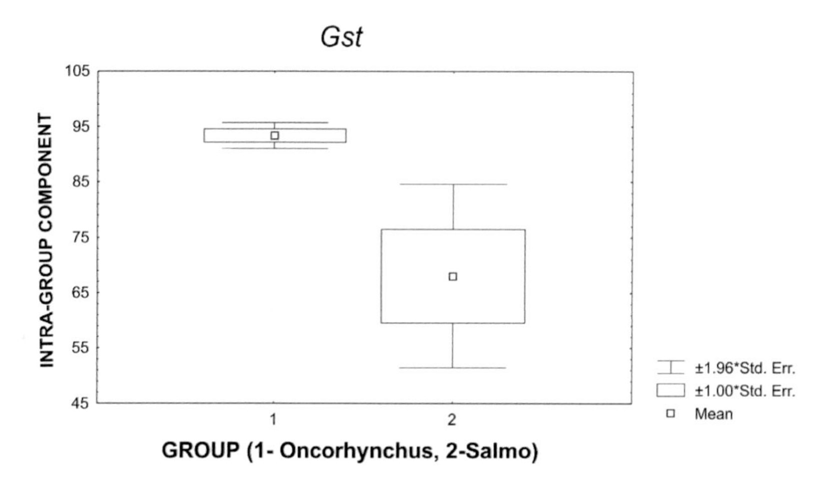

Fig. 7.3.3. Comparison of the genetic diversity in two salmon genera in the scale of standardized allele frequencies ($G_{st} = F_{st}$). Judging from the SE, the G_{st} differences are statistically significant.

Chapter 9), but do not support the time concept of two gene pools separation, because *Salmo* is a more ancient group than *Oncorhynchus* (Neave 1958).

In other words, size or more precisely, genetically effective size of a population (N_e, see Chapter 8), can be the key factor, which determines ***H***. The effective sizes of populations differ among animal groups; for example, it is generally less in vertebrates compared to invertebrates, and especially marine invertebrates (Fig. 7.3.4; ANOVA shows that differences are significant statistically).

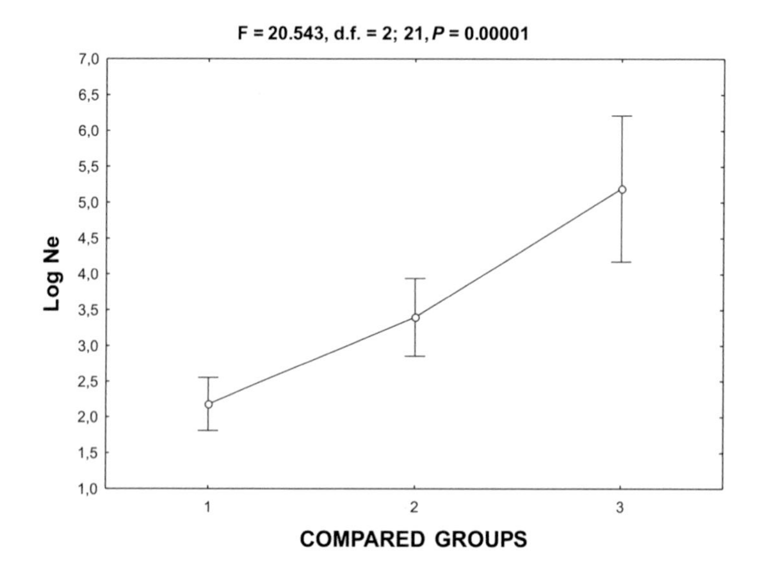

Fig. 7.3.4. Plot of effective size (N_e) values across three different animal groups: (1) Vertebrates, (2) Land Invertebrates, (3) Marine Invertebrates. Vertical lines show confidence intervals (95%).

In accordance with the latest findings, we may interpret the difference in ***H*** between the vertebrate and invertebrate animals (see Fig. 7.3.2) largely as the difference between the values of their N_e. ***H*** and N_e are functionally related according to the infinite mutation model (Nei 1987). This model, which is principally the neutral mutation's model, is not incompatible with the non-neutral statement that marker loci can have pleiotropic effects on RIMs during the speciation process. However, many scientists oppose the solely neutral interpretation for enzyme loci variability (Lewontin 1974; Altukhov 1989), whereas others accept neutrality as most appropriate for molecular evolution (Kimura 1969; Nei 1987). We will not debate this complex matter in this book, although, we will present some facts that disagree with both the extreme views. For now, let us look at data on genetic similarity or its converse, dissimilarity (distance).

In the hierarchy of the animal taxa, subspecies have values in the range ***I*** = 0.6 to 1.0 with the mode near 0.8, semispecies and sibling species have values of ***I*** = 0.5 to 1.0 with the mode at 0.7, species have ***I*** = 0.0 to 1.0 with the mode also 0.7 and between genera ***I*** = 0.0 to 1.0 with the mode 0.2 (Fig. 7.3.5).

Three of four distributions are uni-modal, although they are clearly far from normality, while the one for the semispecies and sibling species group is bi-modal (Fig. 7.3.5). On average there is an obvious trend that shows the decrease of genetic similarity with increasing rank of taxa (Fig. 7.3.6).

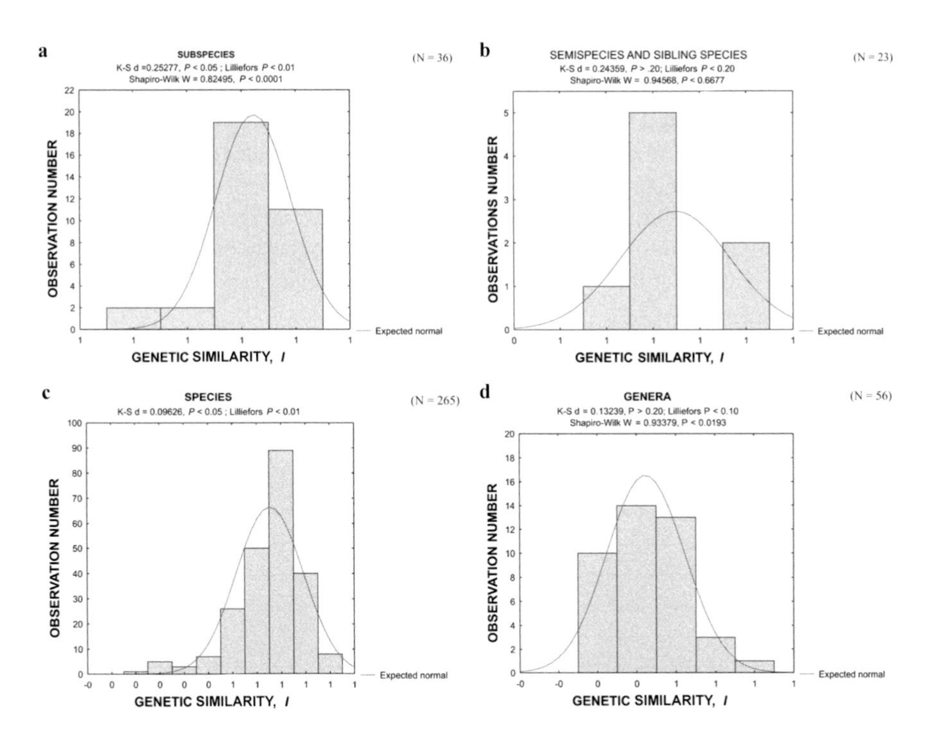

Fig. 7.3.5. Frequency distribution of the similarity coefficient among the representatives of different taxa in animals. Data are from Kartavtsev (2005b). Bulk of Is are the normalized similarity indices (Nei 1972). N is the number of taxa in comparisons.

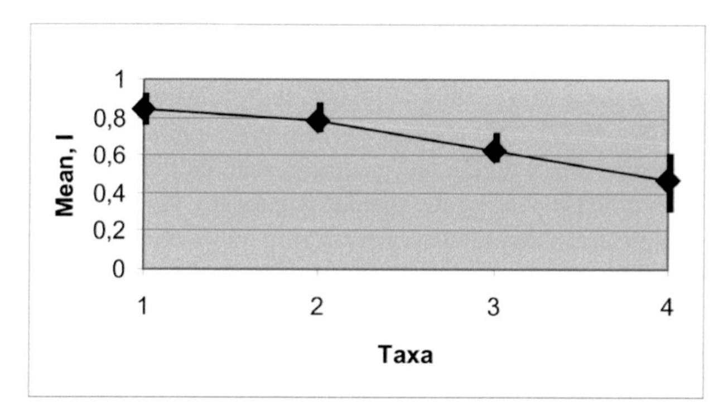

Fig. 7.3.6. Genetic similarity in animal taxa of different rank: mean for the groups. (1) Subspecies, (2) Semispecies and sibling species, (3) Species, (4) Genera (Kartavtsev 2005b).

Intraspecies genetic distances have been measured for many groups of organisms (Lewontin 1974; Nei 1987; Altukhov 1989). Mean genetic similarity on this level is near $\boldsymbol{I} = 0.95$ (see details in Chapter 8). Thus, the available data make it possible to guess (i) that one of the basic requirements of the STE and BSC, the ability to maintain separate gene pools and accumulate genetic differences in time, holds true, and (ii) that the geographic mode of speciation obviously prevails in the animal world, as a Type 1a model, because the long-time accumulation of genetic changes prevails in our data from natural populations.

Do the data presented assume that speciation always follows the Type 1a mode? A few examples below indicate that the answer is no. Two sympatric forms of trout were known in Sweden, and it was unknown whether their gene pools were isolated. Genetic research (Ryman et al. 1979) detected two fixed alleles, which helped to classify all individuals into the two groups. When unambiguously detected, these two taxa appear to have other loci that show the gene pools differ (Table 7.3.1).

Table 7.3.1. Genetic differences among sympatric taxa of the trout *Salmo trutta* in Sweden (from Ryman et al. 1979 with modification).

LOCUS (ALLELE)	FORM I *LDH-1 (100/100)*	FORM II *LDH-1 (240/240)*	Chi^2
AGP-2 (100)	0.976	1.000	4.82*
CPK-1 (100)	0.562	0.293	26.04***
EST-2 (100)	0.967	0.886	3.37
LDH-5 (100)	0.872	1.000	12.63***
MDH-4 (100)	0.571	1.000	51.08***
SDH-1 (100)	0.891	1.000	10.58***
SOD (100)	0.989	1.000	0.01

Note. Asterisks denote: * $P < 0.05$, *** $P < 0.001$.

This table illustrates the idea that sometimes very few structural gene changes can create separate biological unities. In this particular case, the genetic distance (D_n) is only 0.02 (Ryman et al. 1979), which is normal for intraspecies levels of genetic differentiation. Many other such examples are available for salmon (Fig. 7.3.7).

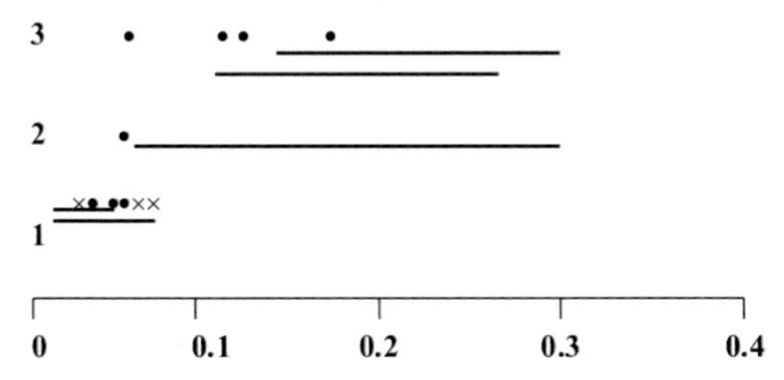

Fig. 7.3.7. Genetic distances in taxa of salmon fishes (from Kartavtsev 1995). (1) Populations within species, (2) Subspecies, (3) Species.

This plot supports the notion that in salmon very little structural gene change can create separate biological species in a short time, which suggests an alternative mode of speciation (we will return to this point in Section 7.4).

Other facts also preclude simple conclusions about speciation. It is obvious, that "weight" of a species, say in the D_n scale, may be different in different animal taxa. For instance, in Amphibia within genus average $D_n = 1.1$, which is an order of magnitude larger than that in Aves, where $D_n = 0.1$ (Avise and Aquadro 1982). The rate of molecular genetic changes, or the speed of morphological evolution (or both), may be different in separate lineages, creating different mean values of traits, like for example D_n at the rank of genus in birds and frogs as noted above. A similar signal is known for nucleotide divergence at *Co-1* and *Cyt-b* loci for fish (Kartavtsev 2009; 2011a). Other sources may add some uncertainty as differential levels of attention are paid to some groups. For example, studies in birds and mammals are much more common, which may lead to overestimation in the rank of taxa compared to other, less studied groups. We should be careful in defining any simple, universal criteria for the species' rank, like, for example, the value of D_n.

We may accept now that speciation is not necessarily linked with structural gene changes. This raises the possibility that changes in a regulatory gene may be also the factors causing the diversity of taxa (Wilson 1976). Such data are not common in the literature, because of the complexity of research and uncertainty of the border between the structural and regulatory gene (Wilson 1976). If we simply look at differences in the activity of enzyme markers as another aspect of genetic change, then some useful information may be gathered for understanding speciation. In particular, we were able to find reasonably large regulatory differences between two sibling species of chars, for up to 32% divergent loci, even though their $D_n = 0.08$ is near the typical value for intraspecies comparisons (Kartavtsev et al. 1983). Similar results have been obtained for the group in *statu nascendi*, the Whitefishes and Grayling forms in Baikal Lake, with a range of genetic distances of $D_n = 0.01$ to 0.03 (Table 7.3.2; Kartavtsev and Mamontov 1983).

Table 7.3.2. Comparison of isozyme activity in three Whitefish forms (Coregonidae) and Grayling (Thymallidae).

LOCUS and FORM	DIFFERENCES IN ENZYME ACTIVITY (EXPRESSION)			
	C. autumnalis* and *C. lavaretus pidschian	***C. autumnalis* and *C. lavaretus baikalensis***	***C. lavaretus pidschian* and *C. lavaretus baikalensis***	***T. arcticus* (Black form) and *T. arcticus* (White form)**
*GPDH-1**	–	–	–	++
*GPDH-2**	–	+	++	–
*MDH-2**	–	–	–	+
*ME**	–	–	+	–
*6PGD**	++	–	+	+
*IDH-1**	–	–	+	–
*IDH-2**	+	–	–	–
*PGM-2**	++	+++	++	–
*FUM**	+	–	+	–
*ACPH-1**	–	–	–	+++
RATIO (E)	18.2 ± 8.2	9.1 ± 6.1	27.3 ± 9.5	17.4 + 7.9

Note. E = n/N, %. N is the total number of loci, n is number of loci, for which different activity was detected by site. N is for Whitefish – 22, for Grayling – 23. Signs denote: "–", Activity does not differ significantly, "+", Repetitive activity difference, "++", two-fold difference, "+++", three-fold or greater difference.

7.4 SPECIATION MODES (SM): POPULATION GENETIC VIEW

We mentioned in the preceding section that speciation theory in evolutionary genetics lacks exact scientific meaning in its predictive inability, e.g., to detect future change using the theory. In this case, this is to predict the origin of species, or at least discriminate among several modes of speciation on the basis of some quantitative parameters or their empirical estimates. Attempts made in this direction (Avise and Wollenberg 1997; Templeton 1998) do not fit the above criteria. That is why we attempted to test the discrimination of modes of speciation on the basis of some key population genetic measurements available in the literature (Kartavtsev 2000; Kartavtsev et al. 2002; Kartavtsev and Lee 2006; Kartavtsev 2009a,b; Kartavtsev 2011a,b). This approach may lay the framework for a genetic speciation concept and the possible future genetic theory of speciation. Before considering the concept let us review data on divergence of DNA markers in a hierarchy of taxonomic categories that are relevant to speciation genetics as explained in previous sections of this chapter.

7.4.1 Genetic Divergence within Species and in a Hierarchy of Taxonomic Categories

The Biological Species Concept, BSC, implies that a species is an isolated reproductive unity. The molecular data, especially pertaining to mtDNA, show that, on the one hand, natural hybridization between species may lead to introgression of genes from one gene pool to the other one. On the other hand, sequences of individual genes exemplify that the variability of DNA markers increases with the rank of the taxon (Johns and Avise 1998; Hebert et al. 2002a; Ward et al. 2005). Hence, I believe it is expedient to compare the data on nucleotide divergence for several genes, from several data sources, and, in addition, to substantiate both the variability and distance parameters. The latter is important for understanding the essence of estimating divergence of DNA or other markers and its connection to species identification and to the speciation process. Such complications as saturation with mutations and unequal rates of substitutions among sites can obscure real DNA variability. There may be other hidden factors too. In particular, various genes may encode different functional properties of phenotypes (macromolecules firstly), and this obviously affects estimates of distance (Graur and Li 1999; Kartavtsev 2009a; 2011a,b).

If the nucleotide sequences for a particular set of loci or alleles in a population sample are known, then DNA polymorphism can be assessed in several ways. The best measures of DNA sequence divergence are π and p or their derivatives (see details in Nei 1987; Nei and Kumar 2000; Kartavtsev 2009a; 2011a). Numerical simulations showed that when p-distances are small, $< 20\%$, then different substitution models give similar scores of diversity in time (Nei and Kumar 2000, p. 41). It is also useful to remember that because of the heterogeneity of substitution rates among sequences and in different parts of genes, an important correction of p-distance is the gamma-correction (e.g., Nei and Kumar 2000; Felsenstein 2004).

Divergence of DNA Nucleotide Sequences at Intra- and Interspecies Levels

As measures for comparison, one may employ uncorrected p-distances, distances of the two-parameter Kimura model (K2P) or other indices used in the literature for the genes *Cyt-b, Co-1* and others (Kartavtsev 2009a; 2011a). The possibility of their use follows from theory and from numerical simulation, as noted in Nei and Kumar (2000, p. 41) and outlined earlier (Kartavtsev 2009a; 2011a).

Pair-wise K2P comparisons for sequences of the *Cyt-b* gene, presented, for instance, in a review of data on vertebrate animals, show a far from normal distribution (Johns and Avise 1998). This creates additional problems of analyzing this and other genes, in which the distance distributions also seem to deviate from normality. The analysis of their distribution is based on the data table including 20,731 species (Kartavtsev 2011a) for the five groups of comparison (1–5) and for genes *Cyt-b* and *Co-1*. The five groups of comparison are as follows: (1) populations within species; (2) subspecies, semispecies and sibling species; (3) morphologically distinct species within genera; (4) genera within a family and (5) families within an order. Indeed the original data showed great variability and different patterns of distributions, both for the two genes and for groups of comparison (Kartavtsev 2009a; 2011a). In such cases, means of the estimates generally provide more satisfactory distributions as was indeed obtained (Kartavtsev 2009a; 2011a,b).

A one-way ANOVA (model with random effects for groups of the same size) showed that mean distances in the five groups analyzed were significantly different for each of the two genes: *Cyt-b*, $F = 2048.60$, d.f. = 4, 3138; $P < 0.0001$; *Co-1*, $F = 9876.80$, d.f. = 4, 19089; $P < 0.0001$. Pooling data in a two-way MANOVA (see scheme below) for the two genes produced a statistically significant increase in differences among *p*-distances in the hierarchy of the comparison groups: $F = 124.15$, d.f. = 4, 279, $P < 0.000001$ (Fig. 7.4.1, top). Interaction of factors in this data set was not statistically significant: $F = 1.82$, d.f. = 4, 279, $P = 0.1258$. However, this pooling is not quite correct for all the mtDNA sequences compared, because it includes heterogeneous groups of different size. Consequently, categorized representation of mean values with weighting an individual score on a sample size (n) for each gene is more correct (Fig. 7.4.1, bottom). However, both approaches showed that the distance for two genes increases with the rank. Mean unweighted distances (%) for the five

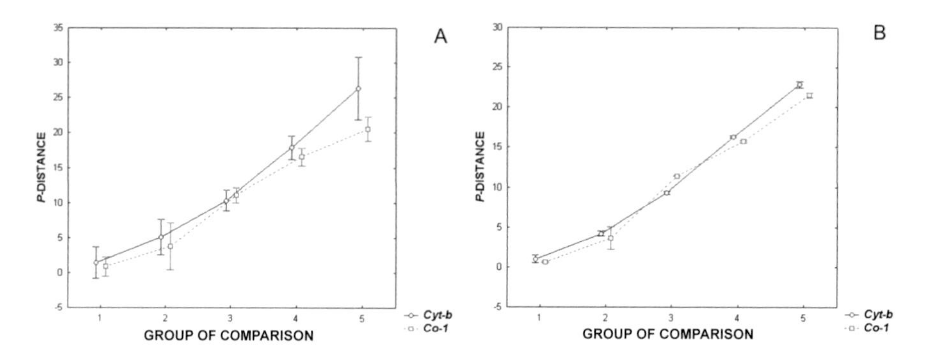

Fig. 7.4.1. Graphs from the two-way ANOVA and mean p-distance values at five levels of differentiation in animal species (From Kartavtsev 2011a; reusage by publisher's policy). Part **A** is variation among comparison groups without weighting the distance scores on the species number (n). Main effect, i.e., the differences among five comparison groups is evident: $F = 124.15$, d.f. = 4, 279, $P < 0.000001$. Difference in mean distance score for the two genes is also statistically significant, but interaction of two factors is non-significant (see text). Part **B** is variation among five comparison groups with weighting the distance scores on the n. Factorial effects for the group of comparison and the interaction between gene & taxa are significant. For the interaction effect, i.e., the differences among the distance scores at two genes in five comparison groups, $F = 101.05$, d.f. = 4, 22227, $P < 0.0001$. Bars are confidence intervals for mean (95%). All p-distances and means for five comparison groups at each of two genes Cyt-b and Co-1 are presented in original paper (Kartavtsev 2011a). Comparison groups: (1) Within species, among individuals of the same species; (2) Within sibling species (plus semispecies, and subspecies), (3) Within genus, among morphologically distinct species of the same genus; (4) Within family, among genera of the same family; (5) Within order, families of a certain order.

groups were as follows: *Cyt-b* (1) 1.38 ± 0.30, (2) 5.10 ± 0.91, (3) 10.31 ± 0.93, (4) 17.86 ± 1.36, (5) 26.36 ± 3.88 and *Co-1* (1) 0.89 ± 0.16, (2) 3.78 ± 1.18, (3) 11.06 ± 0.53, (4) 16.60 ± 0.69, (5) 20.57 ± 0.40 (Kartavtsev 2011a,b).

Taking into account variation in sample size (n) for each i-th distance measure in comparison groups (Kartavtsev 2011a,b), we performed a two-way MANOVA with *p*-distances weighted by n (factor 1, comparison groups: 1 to 5, see above; factor 2, genes: *Cyt-b* and *Co-1*; also, a model with random effect of factors was applied) (Fig. 7.4.1, bottom). In this MANOVA, the effect of factor 1 (i.e., group of comparison) was significant $F = 4964.01$, d.f. = 4, 22227; $P < 0.000001$. The effect of factor 2 (mean *p*-distance differences for two genes) proved to be non-significant: $F = 1.15$, d.f. = 1,

22227; $P = 0.2842$. The interaction between factors 1 and 2 was also significant too: $F = 101.05$, d.f. = 4, 22227; $P < 0.000001$. The categorized graph of the distribution of mean weighted *p*-distance values supported the earlier conclusion on the increase of distances with the rank of the groups compared. Fit of the bivariate distribution (the distance score, "*p*" against the taxa rank, "taxa") showed that there is accordance with the linear regression model, although factorial impact is moderate, at 67–70%: for *Cyt-b*, taxa = 1.9002 + 0.1201**p* ($r_p = 0.84$; $R^2 = 0.7091$; $t = 97.93$, d.f. = 3141, $P < 0.0001$); for *Co-1*, taxa = 1.4332 + 0.1471**p* ($r_p = 0.82$; $R^2 = 0.6717$; $t = 198.96$, d.f. = 19092, $P < 0.0001$). Thus, it is possible to conclude that there is little, if any, impact of saturation on either gene up to the order level in our data set. The lower graph in Fig. 7.4.1 clearly shows the meaning of the factor interaction: the *p*-distance values or its derivatives of these two genes differ among some of the five comparison groups; i.e., the substitution rates are different for *Cyt-b* and *Co-1* at least in some of the groups of animal taxa compared. This conclusion, on an extended data set, validates the same conclusion made earlier for these two genes (Kartavtsev and Lee 2006; Kartavtsev 2009a).

The data presented in Fig. 7.4.1 demonstrate that both genes show a trend of increasing mean *p*-distances with increasing rank of the groups compared, from populations to orders. Average *p*-distance values for two genes and different taxa levels are summarized for convenience in the Table 7.4.1. Because of the importance of this conclusion, the data presented in Fig. 7.4.1 were additionally tested using nonparametric Kruskall–Wallis ANOVA. In this case, unweighted scores were used to have more conservative estimation. For gene *Cyt-b*, $H = 57.01$, d.f. = 4, N = 85, $P = 0.0001$. For gene *Co-1*, $H = 74.05$, d.f. = 4, N = 103, $P = 0.0001$. Thus, the comparative analysis of the data for nucleotide sequences of genes *Cyt-b* and *Co-1*, performed for groups with increasing taxonomic rank for each of the genes separately, demonstrates (with a probability of error $P < 0.0001$) that in animals, genetic divergence increases with the taxon rank. Heterogeneity of the rate of gene evolution, also significant in our data for the two genes (Fig. 7.4.1, bottom), is widely known in literature (e.g., Li 1997; Machordom and Macpherson 2004), which was noted previously. An applied outcome comes from the difference between levels 1 vs. 3, i.e., an order of magnitude difference that exists in an average *p*-distance value between the intra- and interspecies levels suggest that most species may be easily discriminated by these mtDNA markers using only few specimens. In other words, the vast data reviewed here provide theoretical and good empirical support for per individual species identification or DNA barcoding.

The differences in estimates of *p*-distance between the two genes can have the following interpretations. Firstly, the substitution rate may in fact be different in the two genes but hidden somehow. For instance, the data on taxonomic groups from the most representative sources (Johns and Avise 1998; Hebert et al. 2002a,b), which can differ in divergence level, may be differently represented in our database. Actually, heterogeneity of K2P values at *Cyt-b* gene was found for the vertebrate groups examined: amphibians and reptiles have the highest, and birds the lowest variability (Johns and Avise 1998). Significant heterogeneity of the nucleotide diversity was obtained for *Co-1* among flatfish genera (Fig. 7.4.2; Kartavtsev et al. 2008). Interspecies heterogeneity of estimates of nucleotide diversity at *Cyt-b* can be found

Table 7.4.1. Grand means of genetic distances in animal species for two mtDNA genes (*Cyt-b* and *Co-1*) in five comparison groups of the increased categorical (taxa) ranks.

Mean distance score (%)*), ***p*-distance, etc.**	**Standard error of mean, SE**	**Number of pair-wise comparisons when calculating mean, k**	**Species or taxa number, n**
Cyt-b Within species, among individuals of the same species (1)			
1.38	0.30	16	160
Within genus, among sibling species, semispecies and subspecies (2)			
5.10	0.91	12	305
Within genus, among morphologically distinct species of the same genus (3)			
10.31	0.93	36	964
Within family, among genera of the same family (4)			
17.86	1.36	27	1557
Within order, among families of the same order (5)			
26.36	3.88	4	186
Co-1 Within species, among individuals of the same species (1)			
0.89	0.16	42	2338
Within genus, among sibling species, semispecies and subspecies (2)			
3.78	1.18	7	16
Within genus, among morphologically distinct species of the same genus (3)			
11.06	0.53	70	15229
Within family, among genera of the same family (4)			
16.60	0.69	48	1184
Within order, among families of the same order (5)			
20.57	0.40	27	327

Note. *) Grand means were calculated on the basis of data in Kartavtsev 2011a (Appendix). Here unweighted means are shown as explained in the text.

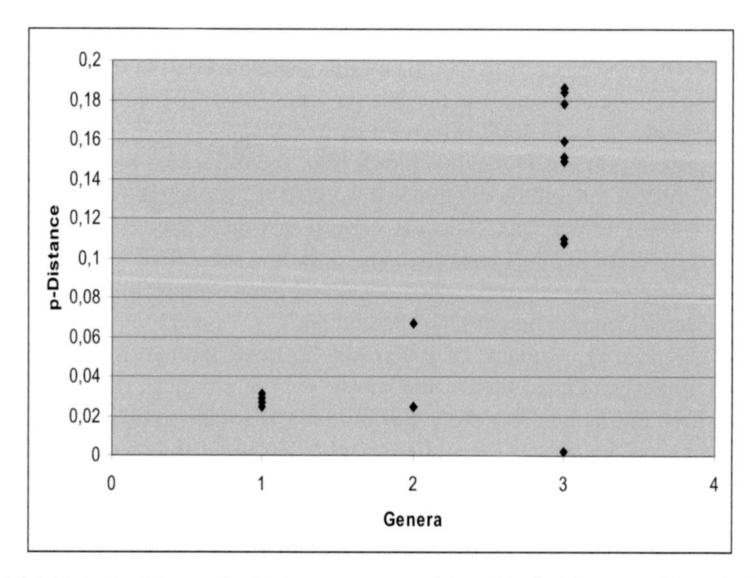

Fig. 7.4.2. Plot of p-distances for Co-1 gene sequence data within flatfish genera (Numerical data are from Kartavtsev et al. 2008). On the x-axis are 3 flatfish groups: 1. Pseudopleuronectes, 2. Verasper + Hippoglossoides, 3. Cynoglossus. On the y-axis are the p-distance scores for intragroup comparisons.

even within a single fish genus (Garcia-Machado et al. 2004). Secondly, in the two most representative works on *Co-1,* several different measures were used (Hebert et al. 2002a; 2002b). In addition, instead of K2P and other similar measures (expected distance), non-corrected *p*-distance (observed distance) was employed in many studies. In general, a shortcoming of analysis of such data arrays is high biological heterogeneity of the material and presence of some unknown or not identifiable components of the estimates (some of them were mentioned above). For instance, *p*-distances and other distance measures can differentially represent one and the same group of comparison. However, non-weighted *p*-distances in the most numerous comparison group (morphologically distinct species within a genus) did not significantly differ statistically between two groups, (1) *p*-distance and (2) other distance estimates (K2P, GTR, TrN, etc.; Kartavtsev and Lee 2006; Kartavtsev 2009a). However, unmodified *p*-distance must undergo homoplasy faster, i.e., be smaller than the expected values of K2P, GTR, TrN, etc. (Nei and Kumar 2000, p. 41). The differences between these groups are also non-significant, when n is used as a covariate in ANOVA of the distance scores. However, the differences between the groups are significant if the distance scores are weighted by n: *Cyt-b*, $F = 231.38$; d.f. = 1, 943; $P < 0.01$; *Co-1*, $F = 207.60$; d.f. = 1, 13888; $P < 0.01$ (Kartavtsev 2009a). The latter differences are apparently caused by unequal representation of taxa in compared groups and also their different numeric representation. This effect is still obscure, e.g., there was no correlation detected between the distance score and n: $r_p = -0.0122$, N = 289, $P = 0.6836$ (all comparison groups included) and $r_p = 0.0336$, N = 106, $P = 0.7320$ (only genera included). For the *Cyt-b* gene, all groups consist almost exclusively of vertebrates, which may on average have differed in *p*-distances compared to invertebrates that were mostly tested on *Co-1* (Kartavtsev 2009a; 2011a,b).

7.4.2 Genetic Variability, Divergence and Introduction of an Operational Criterion for Delimiting a Speciation Mode in Genetic Terms

Below I briefly compare molecular genetic and biochemical genetic data, and draw some conclusions from these and the evidence from previous sections that are relevant to speciation genetics.

An outcome from Biochemical Genetic Data and their Comparison to Nucleotide Divergence

Evidence on variability of structural protein-coding genes as assessed by electrophoretic allozyme analysis is also useful to measure genetic divergence, as noted in previous sections. Although they are now not very popular, they give quite a representative view of variability of nuclear genes. Examination of genetic diversity in wild natural species requires analysis of both the heterozygosity (diversity) and the distances (differences), assessing different aspects of variability, which is not always taken into account. Note, however, that *p*-distance and π can be used as measures of both variability and distance.

A number of surveys give similar data on heterozygosity per individual H, with total mean H = 0.076; for vertebrates, H = 0.054, for invertebrates, H = 0.100 (Aronshtam et al. 1977; Hedgecock and Nelson 1981; Nei and Koehn 1983; Nevo et al. 1984; Hedgecock 1986; Ward et al. 1992; Kartavtsev 2009b, etc.). The H value underestimates the actual genetic diversity by approximately one-third, owing to technical restrictions of protein electrophoresis, which is commonly used to estimate variability at that level (Lewontin 1978; Nei 1975; 1987; Altukhov 1989; 1999; Chapter 5). Coefficients of intraspecific genetic similarity or divergence (distance) have been estimated in many groups of animals. The mean genetic similarity at this level is I = 0.95 (Lewontin 1978; Nei 1987; Altukhov 1983; 1989; Kartavtsev 2009b). According to our database, which comprises of more than 300 populations of 80 animal species, average I = 0.94 ± 0.01 (Kartavtsev 2005b; 2009b; Kartavtsev and Lee 2006; see also the Section 7.3). Data on protein loci similarity in the hierarchy of animal taxa were summarized before (Fig. 7.3.6 and represented in Avise and Aquadro 1982; Nei 1987; Thorpe 1982; Kartavtsev 2005b; 2009b; Kartavtsev and Lee 2006). This means that genetic similarity significantly decreases with increasing rank of the taxon, and conversely, distance increases with increasing taxon rank (Kartavtsev 2005b; 2009a,b; Kartavtsev and Lee 2006). In other words, the allozyme data along with mtDNA markers are supportive of long-time accumulation of genetic differences through typical allopatric speciation, and in this the BSC.

Thus, in general, the current molecular genetic evidence and the results of analysis of protein marker genes support, firstly, the basic idea of the BSC, that taxon formation necessarily requires isolation of gene pools followed by their gradual genetic divergence. Within species the differentiation may be reversed, while it is normally irreversible at upper divisions or at taxa levels. Secondly, correlated evidence suggests that the geographic, allopatric or divergent (D1, Templeton 1981) speciation mode prevails in nature, implying a common rule of gradual accumulation of small genetic differences after separation of gene pools with increase of distance in a hierarchy of taxa levels. This may be considered as the main gate of species origin in the animal world, as follows from current analysis and classical views (Mayr 1968). Yet, there are facts warning against simplified conclusions on modes of speciation using distance scale alone, as noted in the Section 7.3, in the literature (Avise and Aquadro 1982) and as shown for *Co-1* marker divergence within flounder genera (Fig. 7.4.2). Other examples of such disparities can be found (Avise 1994). The range of nucleotide diversity also shows that some animal taxa display a high divergence level among the species, while others are characterized by a very low value of this measure. As already noted above, avian taxa are substantially less differentiated at *Cyt-b* than amphibians and reptiles (Johns and Avise 1998; see also Kartavtsev 2009a; 2011a,b and data in the Section 7.3). For three main geographic phyletic groups of *Orizias latipes*, the nucleotide diversity of *Cyt-b* was found to be comparable to the within-genus divergence: $p = 11.3–11.8\%$ (Takehana et al. 2003). For the other gene, *Co-1*, the species within the genus *Cnidaria* have $p = 1\%$, while in crustaceans $p = 15.4\%$ (Hebert et al. 2002b). The difference for flatfish genera in *Co-1* sequences was demonstrated too (see Fig. 7.4.2), and in a wider context the heterogeneity among animal taxa at the level of the genus for both *Cyt-b* and *Co-1* occurs. One-way parametric ANOVA and K-W ANOVA support this conclusion: for the *Cyt-b*, $F = 265.08$, d.f. = 3, 10654, $P < 0.01$;

K-W $H = 10.87$, d.f. = 3, n = 32, $P = 0.01$ and for *Co-1*, $F = 196.91$, d.f. = 3, 13886, $P < 0.01$; K-W $H = 12.11$, d.f. = 3, n = 43, $P = 0.007$ (Kartavtsev 2009a).

It is possible to conclude that in animals, the prevalence of the mode of D1 means that morphologically different, "good" species usually are quite distinct from intraspecies categories in a genetic distance scale, which allows the successful genotypic identification of a species (DNA barcoding). Nevertheless, the absence of a universal distance scale means there is a need for a complex approach to the species definition and delimitation of speciation mode (Davis and Nixon 1992; Brower 1999; Templeton 2001; Sites and Marshall 2004; Kartavtsev 2000; 2009a; 2011a,b). One such genetically based approach is exemplified below.

A scheme and an algorithmic approach to distinguish speciation modes (models) on the basis of key population genetic parameters and their estimates have been developed (Kartavtsev 2000; Kartavtsev et al. 2002; Kartavtsev 2009a,b; Kartavtsev 2011a,b). As a basis for the evolutionary genetic concept of speciation, verbal descriptions of seven speciation modes were used (Templeton 1981). Consequently, a classification scheme for seven modes of speciation was developed (Kartavtsev et al. 2002; Kartavtsev 2005b; 2009a,b; 2011a,b). Here in this section, a revised scheme, updated for sequence data, is presented (Kartavtsev 2011a,b). In brief, the main elements of this scheme from all 7 types represent: D1–D3 (divergent speciation) and T1–T4 (transformative or transilience speciation) (Fig. 7.4.3). This approach leads to a relatively simple, logical and experimental scheme, which allows us to (1) organize an investigation of speciation in various groups of organisms, based on a focused approach with defined genetic terms and (2) obtain analytic expressions (equations) for each of the speciation modes (Fig. 7.4.4). Using the proposed scheme (Fig. 7.4.3), one can determine two kinds of conditions required for speciation, (1) the necessary conditions and (2) the sufficient conditions, which denote requirement for mechanisms that control species origin in genetic terms, and that are sufficient for the formation and recognition of a species. Importantly, in addition to the general definition of the sufficient conditions, six (1–6) experimentally measured descriptors are introduced (their number including mtDNA and nDNA molecular markers can be increased up to appropriate level) to clarify how, and in which form, these conditions are manifested in a particular case of speciation or in a potential model. For instance, the divergent type of speciation D1 explains classic geographic (or allopatric) speciation (see Fig. 7.4.3).

According to the BSC, the D1 model implies that large populations are isolated (disruption of the gene flow) and evolve separately, accumulating mutations, while reproductive isolating barriers (RIBs) are caused by pleiotropic effects. The longer the time elapses from the isolation event, the greater the distances between the corresponding taxa. Accordingly in this notation, distance descriptors are introduced: 1. $D_T > D_S$ and 4. $p_T > p_S$ (where subscripts T and S indicate genetic distances at structural genes/sites in the putative parental taxon and in conspecific populations or at the higher and lower levels of the taxonomic hierarchy *in statu nascendi* situations). Likewise, since upon implementation of the D1 mode, no significant genetic diversity differences appear in either structural genes or the regulatory part of the genome (because the initial and derived taxa have large effective size, N_e and thus small rate of reduction of diversity), the following parameters are introduced: 2. $H_D = H_P$, 3. $E_D = E_P$ and 5. $\pi_D = \pi_P$ (differences in heterozygosity/diversity and gene expression between the daughter

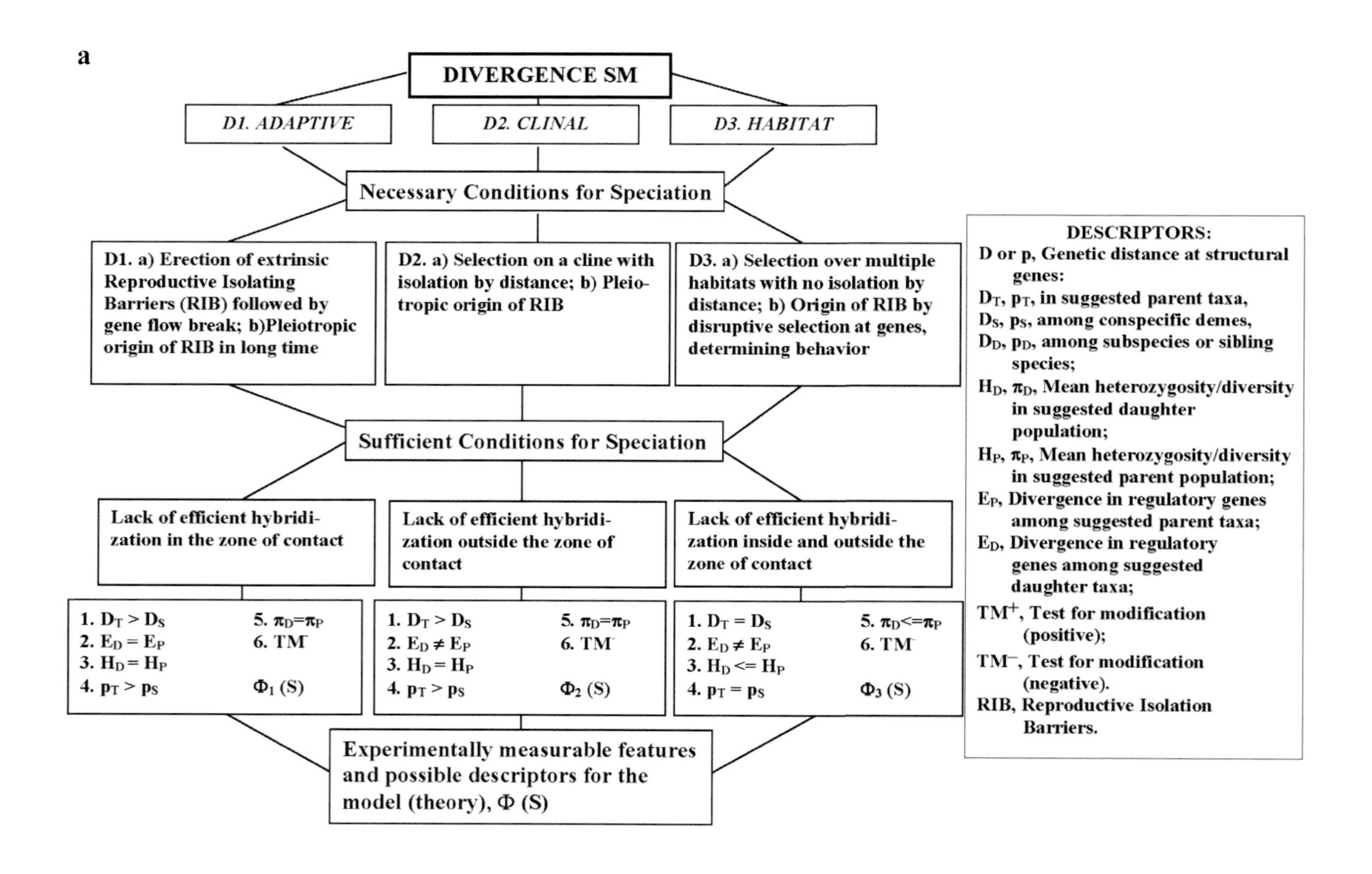
a
DIVERGENCE SM
D1. ADAPTIVE
D2. CLINAL
D3. HABITAT
Necessary Conditions for Speciation
D1. a) Erection of extrinsic Reproductive Isolating Barriers (RIB) followed by gene flow break; b)Pleiotropic origin of RIB in long time
D2. a) Selection on a cline with isolation by distance; b) Pleiotropic origin of RIB
D3. a) Selection over multiple habitats with no isolation by distance; b) Origin of RIB by disruptive selection at genes, determining behavior
Sufficient Conditions for Speciation
Lack of efficient hybridization in the zone of contact
Lack of efficient hybridization outside the zone of contact
Lack of efficient hybridization inside and outside the zone of contact
1. $D_T > D_S$ 2. $E_D = E_P$ 3. $H_D = H_P$ 4. $p_T > p_S$ 5. $\pi_D=\pi_P$ 6. TM^- Φ_1 (S)
1. $D_T > D_S$ 2. $E_D \neq E_P$ 3. $H_D = H_P$ 4. $p_T > p_S$ 5. $\pi_D=\pi_P$ 6. TM^- Φ_2 (S)
1. $D_T = D_S$ 2. $E_D \neq E_P$ 3. $H_D <= H_P$ 4. $p_T = p_S$ 5. $\pi_D<=\pi_P$ 6. TM^- Φ_3 (S)
Experimentally measurable features and possible descriptors for the model (theory), Φ (S)
DESCRIPTORS:
D or p, Genetic distance at structural genes:
D_T, p_T, in suggested parent taxa,
D_S, p_S, among conspecific demes,
D_D, p_D, among subspecies or sibling species;
H_D, π_D, Mean heterozygosity/diversity in suggested daughter population;
H_P, π_P, Mean heterozygosity/diversity in suggested parent population;
E_P, Divergence in regulatory genes among suggested parent taxa;
E_D, Divergence in regulatory genes among suggested daughter taxa;
TM^+, Test for modification (positive);
TM^-, Test for modification (negative).
RIB, Reproductive Isolation Barriers.

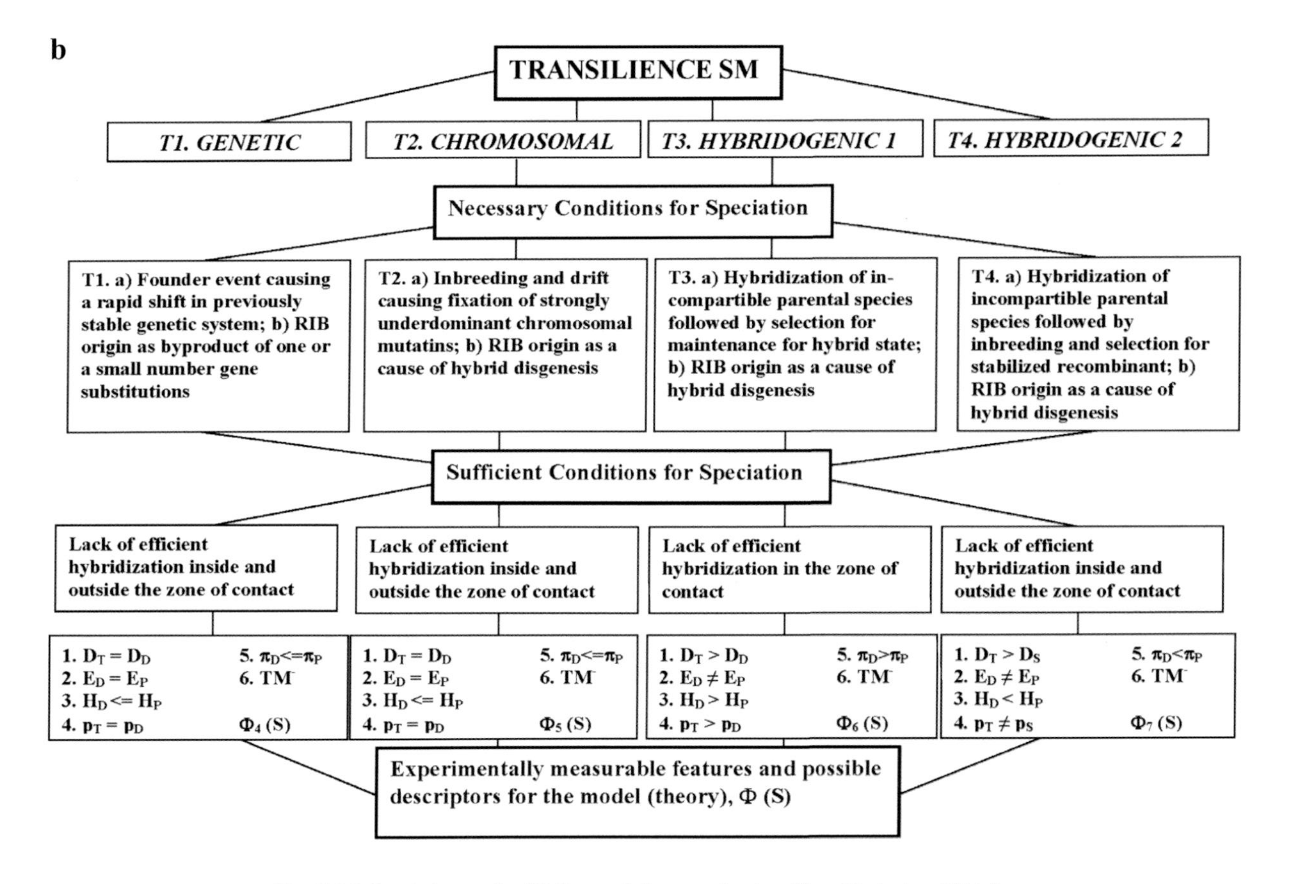

Fig. 7.4.3. Speciation modes (SM): population genetic view (from Kartavtsev 2011a).

$$\Phi_1 (S) \in \{(D_T > D_S) \subset (E_D = E_P) \subset (H_D = H_P) \subset (p_T > p_S) \subset (\pi_D = \pi_P) \subset TM^-\} \quad \text{(D1)}$$

$$\Phi_2 (S) \in \{(D_T > D_S) \subset (E_D \neq E_P) \subset (H_D = H_P) \subset (p_T > p_S) \subset (\pi_D = \pi_P) \subset TM^-\} \quad \text{(D2)}$$

$$\Phi_3 (S) \in \{(D_T = D_S) \subset (E_D \neq E_P) \subset (H_D <= H_P) \subset (p_T = p_S) \subset (\pi_D <= \pi_P) \subset TM^+\} \quad \text{(D3)}$$

$$\Phi_4 (S) \in \{(D_T = D_D) \subset (E_D \neq E_P) \subset (H_D <= H_P) \subset (p_T > p_D) \subset (\pi_D <= \pi_P) \subset TM^-\} \quad \text{(T1)}$$

$$\Phi_5 (S) \in \{(D_T = D_D) \subset (E_D = E_P) \subset (H_D > H_P) \subset (p_T > p_D) \subset (\pi_D > \pi_P) \subset TM^-\} \quad \text{(T2)}$$

$$\Phi_6 (S) \in \{(D_T > D_D) \subset (E_D \neq E_P) \subset (H_D > H_P) \subset (p_T > p_D) \subset (\pi_D > \pi_P) \subset TM^-\} \quad \text{(T3)}$$

$$\Phi_7 (S) \in \{(D_T > D_S) \subset (E_D \neq E_P) \subset (H_D < H_P) \subset (p_T > p_S) \subset (\pi_D < \pi_P) \subset TM^-\} \quad \text{(T4)}$$

Fig. 7.4.4. Analytical description of seven types of speciation modes. Descriptors for equations are given in Fig. 7.4.3. D1–D3 are divergence speciation modes, T1–T4 are transilience speciation modes.

and the parental taxa are absent). Finally, upon some types of speciation, not only variability and genetic distances, but also some quantitative loci (polygenes) are of major importance, which could not be distinguished at the molecular level, but lead to the formation of RIBs. Hence, the next descriptor is introduced: 6. TM (TM^+ vs. TM^-, an experimental test to detect scores for modifications or RIB-important differences in quantitative traits in nature. This test also allows us to distinguish between epigenetic variation and taxonomically significant difference.

Do all data presented above imply that speciation always corresponds to the D1 type? Apparently not. One nice example supporting this answer was mentioned for two trout (*Salmo trutta*) (Ryman et al. 1979). There are other examples of bursts of fish evolution, documented by molecular markers (Rutaisire et al. 2004; Duftner et al. 2005). These, as well as other data, for instance from our database of coefficients of similarity, indicate that sometimes very small differences in structural genes may result in the appearance of RIBs (and thus reproductively isolated biological entities). In the case of the trout mentioned above, the genetic differences between the two forms $Dn = 0.02$ (Ryman et al. 1979), which corresponds to the level of intraspecies genetic differentiation. There are many other examples for salmonid fishes (Kartavtsev 2009a,b; Kartavtsev 2011a,b), supporting the view that in these fishes, small changes can generate biological species during a short period of time. This evidence also suggests an alternative speciation mode, such as the transformational (T1) or other T-modes (Figs. 7.4.3–7.4.4), though in general, the D1 speciation mode prevails in this group too.

The original developments described above are similar to other directions in evolutionary genetics. For instance, the method of distance scaling along phyletic lines was suggested by Avise and Walker (1999). It was designed for the normalization of taxa weights, and as an outcome the unification of systematics is expected. The estimation of cohesion of gene trees was suggested by Templeton (2001) to decide on species boundaries. The approach includes the notion of genetic exchangeability and/or ecological interchangeability among lineages belonging to the same species (Templeton 2001). Both approaches are operational for species delimitation but these techniques will not solve the "rigidity" of species and species boundaries without formalization of a species notion. Some authors have reached similar conclusions on the basis of independent analysis of different characters and approaches for species delimitation (Ferguson 2002; Wiens and Penkrot 2002; Sites and Marshall 2004). In particular, the latter authors emphasize the idea of diffuse peculiarities of the species concept and species boundaries and, consequently, the necessity and applicability of several sets of operational criteria in a multi-faceted approach for species identification (Sites and Marshall 2004). This is also emphasized in the approach suggested here (Figs. 7.4.3–7.4.4 and relevant text). The scheme presented in the current paper was designed originally to define a mode of speciation. However, it is also contains the logical criteria of whether species have originated yet. Thus, this approach is quite suitable for delimiting species, as it has both a theoretic and empirically operational approach. It has a weakness, however, it is common to all current methods, the non-tree based and tree-based (Sites and Marshall 2004), i.e., in some cases, the approach will require researchers to make qualitative judgments because of the variety of ways for species to originate. Potentially the approach is close to the Population Aggregation Analysis (PAA) in Davis and Nixon's (1992) because it is based on population based parameters like D_T, H_D, etc. (see Figs. 7.4.3–7.4.4). However, this PAA1 approach could easily be converted to the mode PAA2 as defined by Brower (1999). Developed notations (see Figs. 7.4.3–7.4.4, and see also Sites and Marshall 2004) may even have properties of the tree-based method (see below). As in PAA2, it is suggested to use not only genotypic scores (character states) but other suitable descriptors (qualitative and quantitative traits: QT, QTL, etc.); they could be represented as per individual sets of the records or as vector-scores for implementing a multi-dimensional analysis (Canonical, PCA, PAA, etc.) with the aims of (a) testing a null hypothesis (H1) of the absence of vectors' gatherings and if rejected, the alternative hypothesis H2 will be tested for discrimination among them and taking the solution in the frame of logic suggested (Fig. 7.4.3), and (b) determine whether vectors' genetic (=phylogenetic) unity is present, both as a distance value and a coalescent signature obtained from a tree; again solving H1 and H2. To obtain the phyletic signal it will be necessary to develop new descriptors in the Fig. 7.4.3 scheme and introduce them in the set of equations D1–T4 (and others when developed) in Fig. 7.4.4. These special descriptors, like the branch length or the parsimony outcomes to the current OTU (Operational Taxonomic Unit) at a consensus tree built at several gene sequences, could be operational criteria among others. The approach is basically empirical but different from others reviewed (Sites and Marshall 2004), as having (1) a general genetics and population genetics theoretical basis and (2) having formalization as equations of set theory. Such an approach has its own limitations as well as benefits. One limitation is that it is restricted to sexually reproducing species, for which basic population genetic principles are more or less fit. The other limitation is that generalizations (deductions) are possible only in a framework of the genetic terms defined. But individuals comprising species are phenotypes. Thus, genotype/phenotype correspondence should be defined in an appropriate form, and genotype-and-environment interaction or ecological interchangeability in Templeton's (2001) sense should also be introduced somehow, partly, it is supposed, when QTL or other complimentary descriptors are introduced. An advantage is that this approach is wider than many others suggested for delimiting species (see Sites and Marshall 2004), in its ability to define different speciation modes (or take into account the differences in species types). Also, by weighting the members of equations in a specific way it is possible to develop the approach further as framework for developing a genetic theory of speciation.

Obviously, one can enter other descriptors, derived for example from DNA and immunogenetic markers. Most important, however, is that the scheme summarized in formal equations (Fig. 7.4.3) is operational, i.e., it allows classification of speciation modes. It was tested for cyprinids (Kartavtsev et al. 2002), and it explains well our own earlier data on salmonids (Kartavtsev and Mamontov 1983; Kartavtsev et al. 1983). Both the testing of the scheme presented and its theoretic background must be further developed.

7.5 TRAINING COURSE, #7

1. Software (POPULUS 1.4): "POPULUS"—Random Drift. Selection and Gene Drift. Test Two Different Modes with Two Sets of Parameters (N = 10 and N = 999; p = 0.5).
2. Internet support. Visit and learn about species and speciation in the website: //:http/www.biology.ucsc.edu/people/barrylab/public_html/classes/evolution/genetic.htm

8

POPULATION GENETIC STRUCTURE OF A SPECIES

MAIN GOALS

8.1 Population Structure in Bisexual Species
8.2 Modes of Population Structure
8.3 Principles of Intraspecific Genetic Structure Study
8.4 A Field Population Genetic Study
8.5 TRAINING COURSE, #8

SUMMARY

1. The main parameters that determine genetic dynamics of populations are (1) ***p*, allele frequency**, (2) **N_e, effective size of population**, (3) ***m*, coefficient of migration,** defining a gene flow, and (4) *s*, **coefficient of natural selection**.
2. Research on **intraspecific genetic diversity** requires a **complex approach**, taking into consideration a complexity of factors that cause variability and differentiation within a species.

8.1 POPULATION STRUCTURE IN BISEXUAL SPECIES

Let us consider first what a species is and what a population is.

Species

Here and in the following we will consider species as it was defined in Chapter 7 or according to the Biological Species Concept (BSC). The species in the BSC is most naturally supported by population genetic theory, and that is why we prefer this concept despite the limitations, which were outlined earlier (Chapter 7). Populations that form such a species may be defined in genetic terms, and we will learn some fundamental qualities of these populations in this chapter and the next.

Species, as you may know, is a taxonomical category as well. Species and subspecies, if present, are the lowest taxonomic categories (Fig. 8.1.1; other examples are available On-Line, e.g., http://www2.estrellamountain.edu/faculty/farabee/biobk/BioBookDivers_class.html).

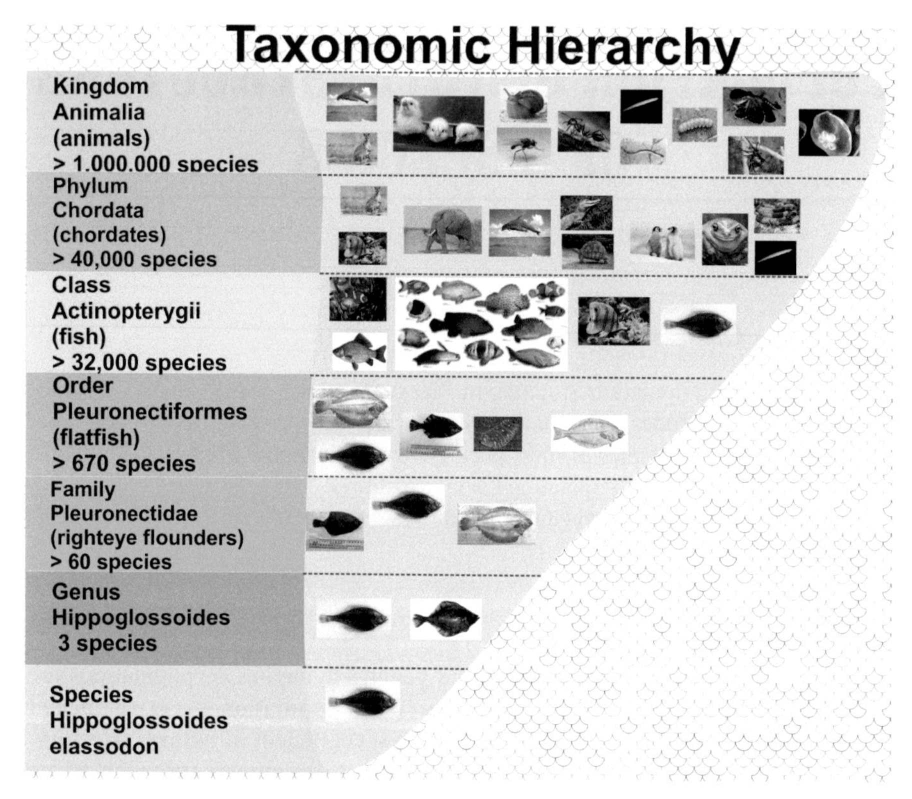

Fig. 8.1.1. Examples of the main taxonomic categories: Kingdom, Phylum, Class, Order, Family, Genus and Species.

Population

Population may be defined in several ways. Lebedev (1967) used the name the **elementary population** to define the smallest self-reproducing unit of a species. Yanulov (1962) used the term **stock** as a self-reproducing unit, and Marr (1957) believed that a self-reproducing unit is a **sub-population**. Altukhov (1974), defined a **local stock (local population) as a system of sub-populations, which is theoretically** *equivalent to a subdivided population,* ***and is the smallest stable reproductive unit*** of a species. More recent definitions have also occurred: Mina (1986), Yablokov (1987), etc. I think that in its applicability to genetics the following definition is very satisfactory: "*A* ***population*** *is a reproductive unit of organisms which share a common* *gene pool*" (Dobzhansky 1970). Such a population is also frequently called a **deme** or **Mendelian population**. Local stock and sub-population fit this definition. Altukhov and his colleagues (Altukhov and Rychkov 1970; Altukhov 1974; 1989) provide strong evidence that a subdivided population, like local stock, is more stable both in space and in time. Before we go too deep, I'd like to stress that there are two usages of the term population: one is in Dobzhansky's sense, and another in a more common sense, like any group of individuals without assuming a genetic perspective. In the latter sense the temporary age gatherings, eco-groups and races of a species may be named populations. It is better to avoid this usage.

An overview of some population genetic notions and species problems are given in Altukhov (1974; 1983; 1989), Yablokov (1987) and Ivankov (1998).

To summarize, within a species we may observe three genetically meaningful gatherings or divisions (Fig. 8.1.2).

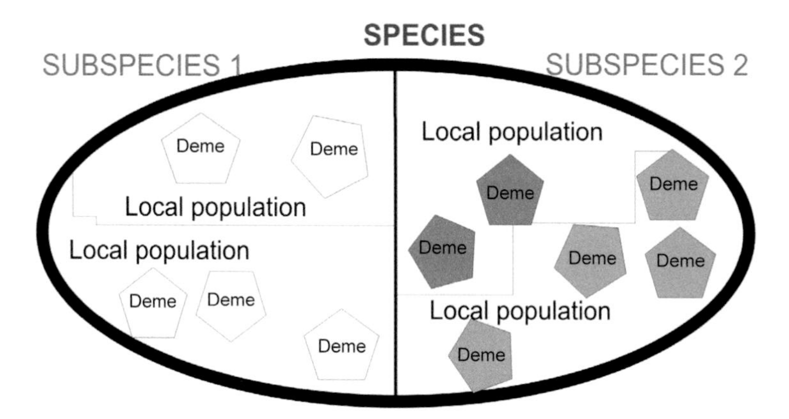

Fig. 8.1.2. Main intraspecies divisions: (1) Subspecies (geographic race), (2) Local population (stock), (3) Deme (subpopulation, Mendelian population).

8.2 MODES OF POPULATION STRUCTURE

There are two different approaches to a description of population structure.

Phenomenological Approach

There is an example of a phenomenological approach in the book by Kirpichnikov (1979). Under this approach there are four modes of population structure: (1) Eel-pout mode, (2) Wild-carp mode, (3) Pacific salmon mode and (4) Tuna mode.

What are the reasons for mode one? Eel-pout is a demersal fish with low mobility and low fecundity, and it is viviparous. Thus, under this concept the first mode comprises small, well isolated, populations of the species. The second mode is most typically attributed to wild carp, with large isolated populations inhabiting separate rivers or river systems. The third mode's typical representatives are pacific salmons (genus *Oncorhynchus*). Each river is inhabited, according to Kirpichnikov's theory, by a separate, isolated local stock, the integrity of which is supported by homing (an instinct of returning to the river of birth). Finally, the fourth mode is typical of tuna, with their huge mobility, large population size and high fecundity; all these factors give very large and weakly differentiated populations. What is wrong with such an approach? Let us consider mode 3 as an example. Even among the six most abundant species of pacific salmons, there is no unity in a sense of population structuring. Homing in pink salmon is small, while such species as sockeye and masou have very perfect homing. Pink salmon stocks may reach sometimes millions of fish, however for the latter two species a normal number of fish in the stock is around one to ten thousand. These differences lead to weak genetic differentiation in pink and strong differentiation in masou and sockeye. The contrast in within-species differentiation is even greater between salmon species of *Salmo* and *Oncorhynchus* genera. We will see that later.

In other words, the phenomenological approach is not precise and we need another approach to describe population structure. It may be called a scientific approach.

Scientific Approach

The main ideas of this approach are based on the theory of population genetics; it thus has precise scientific solutions. However, it is necessary to stress that these solutions are sometimes difficult to achieve because of limited data or, less frequently, because the theory itself is not well developed.

8.3 PRINCIPLES OF INTRASPECIFIC GENETIC STRUCTURE STUDY

Major Factors of Population Genetic Dynamics

The factors that influence the dynamics of populations may be random (or stochastic) or systematic. For example, random genetic drift is one of the best known factors of genetic differentiation, especially in small populations. It is a stochastic factor by definition. Natural selection and migration are systematic factors. Major population genetic parameters that determine the degree of differentiation are: ***(i) p, allele frequency (we defined it before), (ii) N_e, genetically effective size of population, (iii) m, coefficient of migration, (iv) s, coefficient of natural selection.*** We will consider more of the theoretical background of these and other factors in the next chapter.

For now, let us only illustrate the principal action of these factors, starting with the first pair of parameters. Consider in time genetic drift for one population of a small size. Genetic drift occurs due to stochastic errors in the transfer of genes from one generation to the next. Let $N_e = N = 10$ and $p_A = 0.5$, which are the population size (number of individuals) and an allele (say *A*) frequency in an initial generation (t_0). We may calculate the standard error for p_A in this case (as we did in Training Course #4 on the basis of equation 4.3): SE = $\sqrt{[(0.5*0.5)/20]}$ = 0.16. So, in the next generation p_A must be 0.5 ± SE. In other words, some of the next generation gene pool samples may have p_A = 0.7 or p_A = 0.3 (33% probability by a normal approximation, see Chapter 11 for more explanation), and even 0.5 ± 2SE, p_A = 0.9 or p_A = 0.1

(5% probability by a normal approximation), with some simplifications for the sake of clearness. Obviously, very soon in such a small population of 10 individuals (N = 10), an allele will become either fixed **($p_A = 1$)** or lost **($p_A = 0$)**. In POPULUS software we have seen such examples in the dynamics for several generations and for different ***p*** and N_e. Above are the theoretical considerations. The experimental results support these ideas (Fig. 8.3.1).

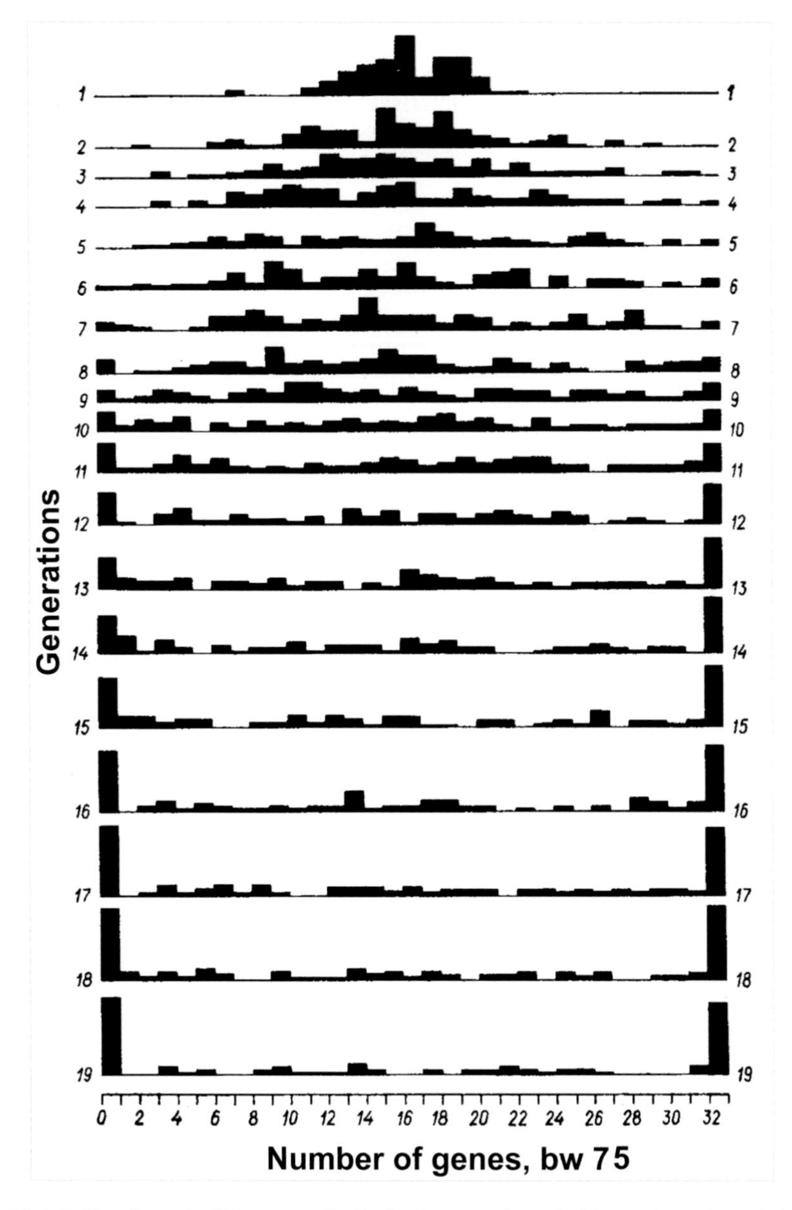

Fig. 8.3.1. Drift and gene bw[75] frequency distribution in generations of 100 experimental populations of *Drosophila melanogaster* (from Buri 1956; after Altukhov 1989).

All our conclusions hold true for the case of spatial differentiation through genetic drift (Wright 1943; Li 1955; 1978).

How does migration or gene flow influence the allele frequency, or the parameter $\boldsymbol{p}$? When dynamics are dependent on N_e (drift), initial value of the frequency has direct impact on the number of generations until fixation or loss of an allele. This is very vividly seen when numeric simulations are done in POPULUS or other software. In the numerical example above, if allele frequency at t_0 is $\boldsymbol{p}_A = 0.9$, then with N = 10, and SE = 0.07 the allele frequency in the next generation may be, correspondingly, $\boldsymbol{p}_A = 0.97$ or $\boldsymbol{p}_A = 0.17$ (33% probability) or $\boldsymbol{p}_A = 1.0$ or $\boldsymbol{p}_A = 0.24$ (5% probability). Thus, when the initial frequency is 0.9, fixation of the allele ***A***, i.e., its change to a monomorphic condition, may occur even in the first generation.

Migration is the next of the four most important factors that influence population structure. We will consider the impact of the parameter ***m***, as well as the parameter ***s***, in more detail in the next chapter. Migration influences not only the integrity of a population, but also increases the effective size, by uniting the populations (Fig. 8.3.2).

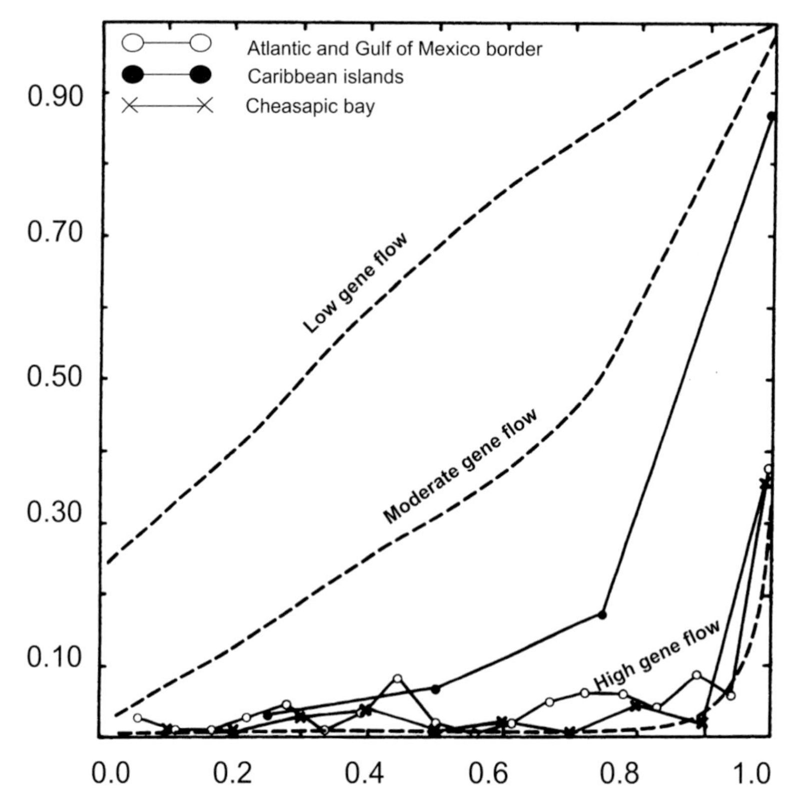

Fig. 8.3.2. Graph of distribution of a conditional mean allele frequencies (p_i) within some species of oysters and pearl clam that illustrate the role of gene flow in creating high N_e.
On ordinate is p_i, on abscissa is p_i that is normalized on sample size. Broken lines show levels of gene flow that correspond to certain p_i. Circles, dots, crosses correspond to 2 oyster and 1 Pearl Clam data sets (Modified from Buroker 1984).

Data on Fig. 8.3.2 for marine invertebrate populations correspond to high values of migration or gene flow (Buroker 1984; Durand and Blank 1986). Mathematically $\boldsymbol{m}$ and N_e are proportional to each other (see Chapter 9, Section 9.3, Equation 9.23).

The influence of natural selection is very diverse. Selection can increase or hinder practically all processes in populations. The impact of genetic drift can be greatly decreased by normalizing selection, while with directional selection fixation of alleles during drift can occur sooner. We will deal with these situations, making numerical simulations with allele frequencies by changing the $\boldsymbol{s}$ parameter during Training Course #8. There are many developed models of different modes of selection. Some of them will be considered in Chapters 9 and 12. Those who want to obtain more information are advised to see other sources (Hartl and Clark 1989; 1997).

Measure of Genetic Differentiation

As a measure of a degree of genetic differentiation between populations, the statistics F_{st} (standardized variance in allele frequency) and D_m (minimal genetic distance) are useful. Using F_{st} (or G_{st}) is preferable for population analysis, as we will see in Chapter 9 together with more details on the properties of this statistic. The simplest way to understand what F_{st} means is to define it in terms of heterozygosity (diversity): $\boldsymbol{F_{st} = (H_t - H_s)/H_t}$. Here H_s is the expected heterozygosity within a subpopulation (sample), and H_t is expected heterozygosity in the whole population (in the set of samples; the formula for calculating **H** was given in Chapter 7, Section 7.2). In this definition $F_{st} = G_{st}$ (Nei 1977b). Obviously under this definition the intrapopulation portion of variability is given as $1 - F_{st}$.

Simplest Model of Population Genetic Structure

We have already dealt with this model, that is the H-W ratio: $\mathbf{(p + q)^2 = p^2 + 2pq + q^2}$. That is the equilibrium model for large (infinite in theory), randomly mating populations, which explains the stability of the allele and, to a lesser extent, genotype frequencies without the pressure of selection and migration.

Complex Nature of Population Genetic Dynamics

In nature neither of the factors act separately. They commonly interact in different fashions, and vary in space and in time, from generation to generation, from young to adult (in ontogenesis), and even in seasons within the same cohort. A classic example for the wild population (which nobody has repeated since the mid-seventies) is the two-locus polymorphism in sockeye salmon, which undergoes the influence of two factors (Fig. 8.3.3). Polymorphism at the ***PGM**** locus is better explained by the curve obtained under some selection pressure along with variability of the impact of N_e and $\boldsymbol{m}$ (Fig. 8.3.3). Polymorphism at ***LDH**** may be considered as selectively neutral because it is well explained by only N_e and $\boldsymbol{m}$.

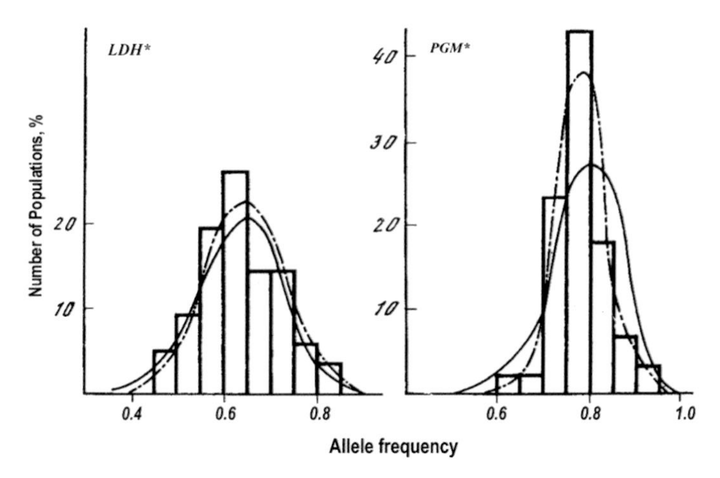

Fig. 8.3.3. Allele frequency distribution at two loci in spawning subpopulations of sockeye (*O. nerka*) in Azabachee Lake (Kamchatka) (from Altukhov et al. 1975). Histogram represents observed frequencies. With the dotted line the expectations of the island model of population structure are shown for $\mathbf{N_e}$, **m**, and **s** (drift, migration and selection; the solid line illustrates the action of only $\mathbf{N_e}$ and **m** (drift and migration).

Classification of the Modes of Intraspecies Structures by Using Population Genetic Tools

Three different principal modes are possible to define, based on population genetic tools: (i) Panmictic population, (ii) Subdivided population and (iii) Continuous population. We will consider here in brief (and in Chapter 9 in more detail) the first and the second modes. The third mode is mathematically more complex to present in the context of this course, and more importantly it is rare in nature.

i) The panmictic population mode. This mode is theoretically described by the H-W model. Within such a population no genetic differentiation exists. Thus, $F_{st} = 0$ when no action of selection is occurring. Species may be represented in the area by one or several such populations, which are completely or nearly completely isolated from each other.

ii) Subdivided population mode. This mode of population structure is theoretically described by the **island model** (Wright 1943; see approximation by this model in Fig. 8.3.3), or by the **stepping stone model** (Kimura and Ohta 1971) of population differentiation. A differentiation in a subdivided population at equilibrium state, without selection, is described by approximate equation: $Fst = 1/[4Nm + 1]$.

Peculiarity in Population Organization of Fish and Shellfish Species

There are some peculiar features of intraspecific structure in the marine realm, which is connected with generally large (numerous) populations that have vast migration and wide distribution (active or passive). However, all genetic principles hold true for these species as well as for others. For example, there are large and small mean

F_{st} values for separate species, which are connected principally with inversely related genetic drift and migration (see Fig. 8.3.2, Fig. 8.4.5). Let us consider some general data on variability and genetic differentiation, and then turn to specific research on a representative shellfish, the pink shrimp.

Genetic Variability and Differentiation

Levels of mean heterozygosity in different taxa vary in wide limits in plant and animal taxa (see Chapter 7, Fig. 7.3.2). The overall mean for both is **H** = 0.076, while for vertebrate animals it is **H** = 0.054, and for invertebrates, **H** = 0.100 (Nevo et al. 1984). The frequency distribution of the genic loci with respect to similarity coefficients among species is U-shaped; that is the reverse of the expectation under neutrality (Ayala et al. 1974; Ayala 1975; Fig. 8.3.4).

Nevertheless, for some protein loci the similarity indices have shown the distribution that is close to normal (Fig. 8.3.5). This is another type of distribution that may agree with differentiation through drift.

Some results suggest that natural selection is the most plausible explanation for the distribution of the individual heterozygosity at allozyme loci (H_0) and its association with the variation of morphological traits in a population-and-environment gradation (Fig. 8.3.6).

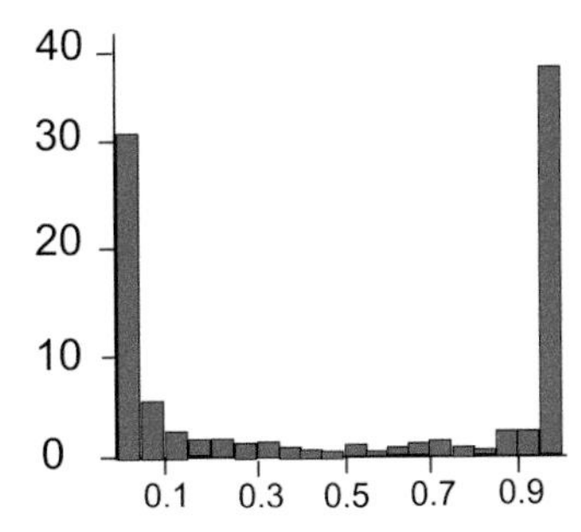

Fig. 8.3.4. Similarity coefficients plot for the sibling species in *Drosophila willistoni* species group. Y-axis, Frequency of loci (%), X-axis, Similarity index, I. Most loci show either fix or zero similarity, while the rest exhibit moderate to high levels of similarity. The U-shaped distribution does not agree with drift or the neutral theory of differentiation in time. Rather it should be close to normal distribution (Modified from Ayala et al. 1974).

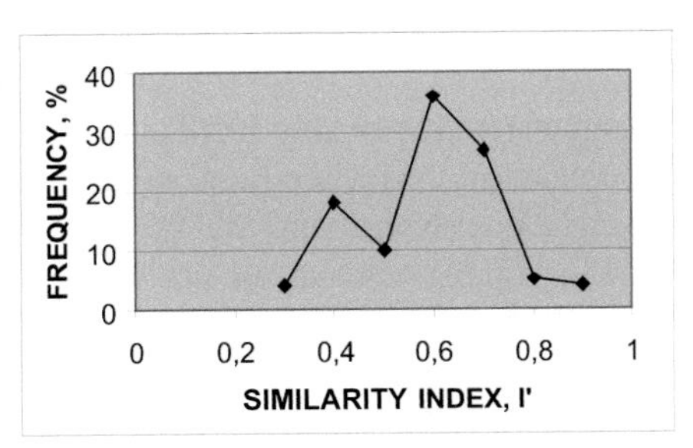

Fig. 8.3.5. Frequency distribution of a similarity index at duplicated hemoglobin loci in different salmon species (from Kartavtsev 1995).

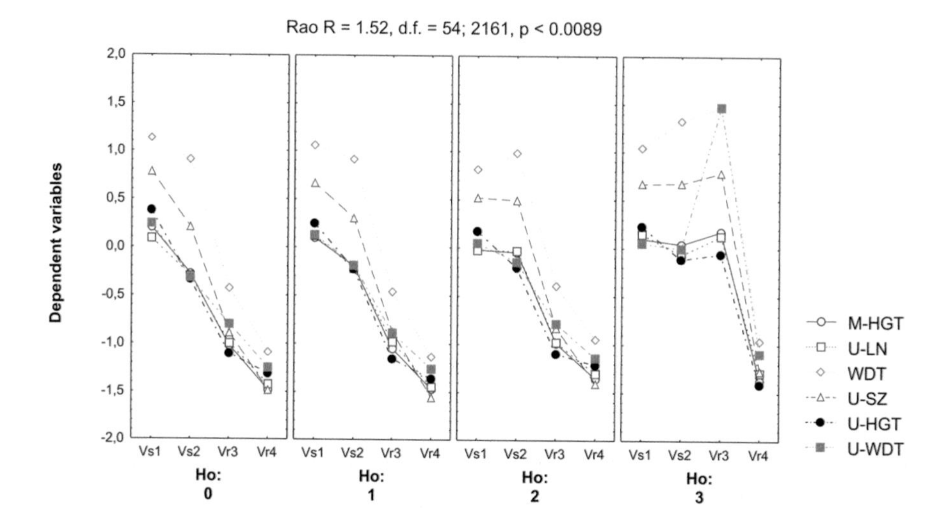

Fig. 8.3.6. Plot of means illustrating the results of two-factor variance analysis: the H_o and morphology traits variability in dogwhelk (*Nucella heyseana*), in Vostok and Nakhodka bays (Peter the Great Bay, Sea of Japan). Increase of size in the group of tri-heterozygotes ($H_o = 3$, on the right) in two samples requires a selection as an explanation of these results (Built from data in Kartavtsev and Svinyna 2003).

As was demonstrated earlier, the sources of genetic variation in the populations are complex (Figs. 8.3.3–8.3.6). Sometimes it is evident that neither selection nor neutrality could be considered as the explanation of variation, but rather heterogeneity of the data is the cause (Fig. 8.3.7).

Obvious non-normality is evident for both distributions (Fig. 8.3.7). The heterogeneity in the bottom graph is due to a low similarity among several island populations of the lizards investigated (I = 0.6–0.7 and less; Kato et al. 1978) (Fig. 8.3.7). These data suggest that there is heterogeneity among population groups, which represented differently among compared intraspecies groups.

If homogenized population groups are compared, the normality may be found (Fig. 8.3.8).

Data that illustrate levels of intraspecies genetic differentiation in salmon show that there is no unity for members of the genera *Salmo* and *Oncorhynchus*. However, in general, for both groups, the intrapopulation component prevails, while F_{st} (G_{st}) is much lower (Fig. 8.3.9).

Data in Fig. 8.3.9 illustrate: (i) a reasonably small interpopulation differentiation, G_{st} comparative to intrapopulation divergence (upper figure) and (ii) the difference between *Salmo* and *Oncorhynchus*, on one hand, and species' differences on the other (middle and bottom figures). Populations and species of trouts (*Salmo*) in total are more differentiated genetically than pacific salmons (see also Fig. 7.3.3, Chapter 7).

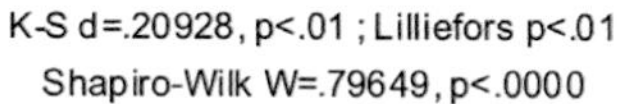

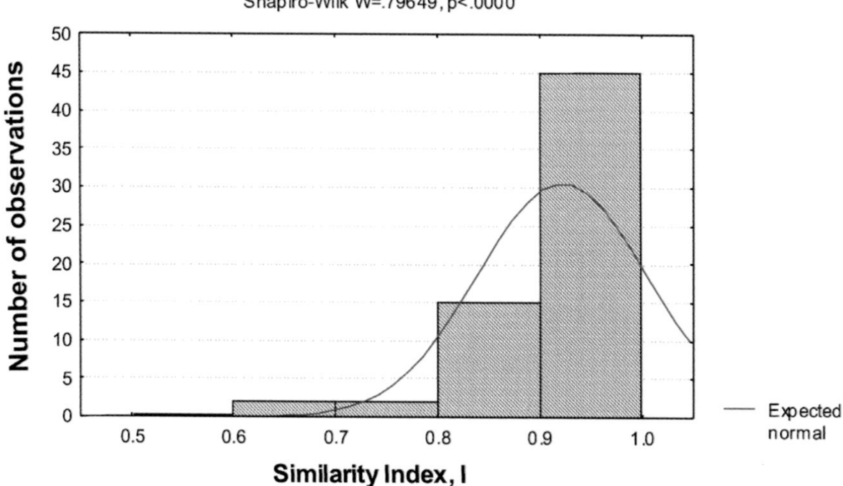

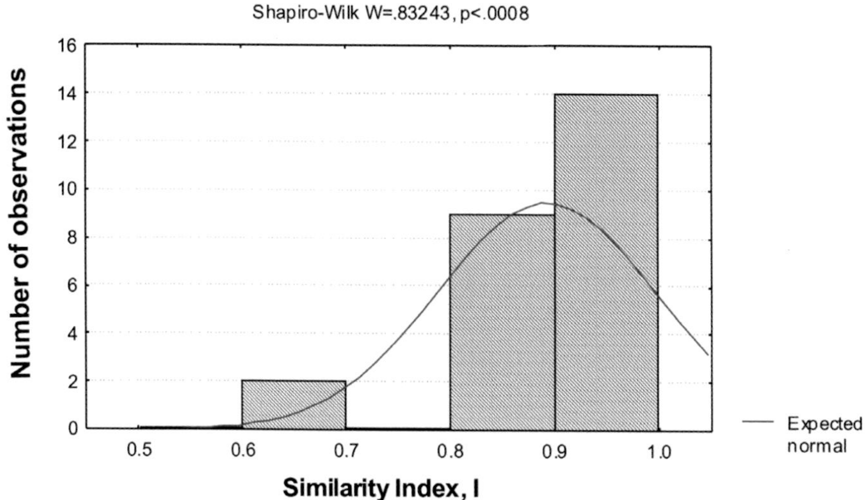

Fig. 8.3.7. Frequency distribution of a similarity index (I) within species (Data of the author, unpublished). Upper figure shows I distribution for the whole database, while the bottom figure for only one reptile genus *Eumeces* (from Kartavtsev 1995).

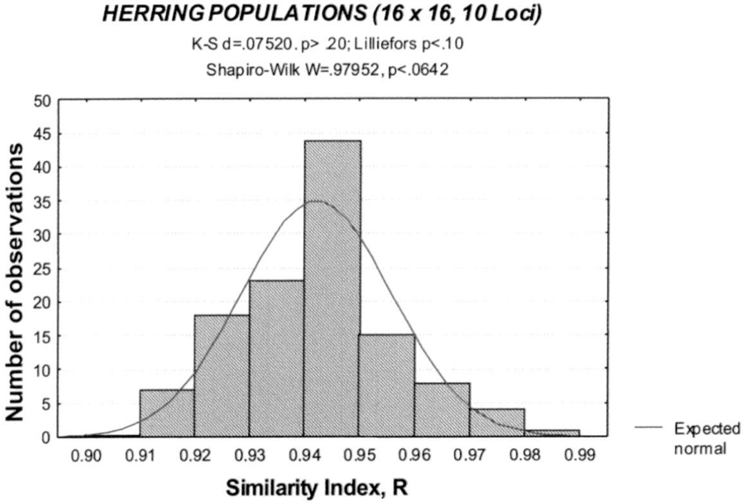

HERRING POPULATIONS (16 x 16, 10 Loci)

K-S d=.07520, p> .20; Lilliefors p<.10

Shapiro-Wilk W=.97952, p<.0642

Number of observations

50
45
40
35
30
25
20
15
10
5
0

0.90 0.91 0.92 0.93 0.94 0.95 0.96 0.97 0.98 0.99

Similarity Index, R

— Expected normal

Fig. 8.3.8. Frequency distribution of a similarity index (R_c) within the species of Pacific herring, *Clupea pallasii* (Calculated and plotted from data in Kartavtsev and Rybnikova 1999).

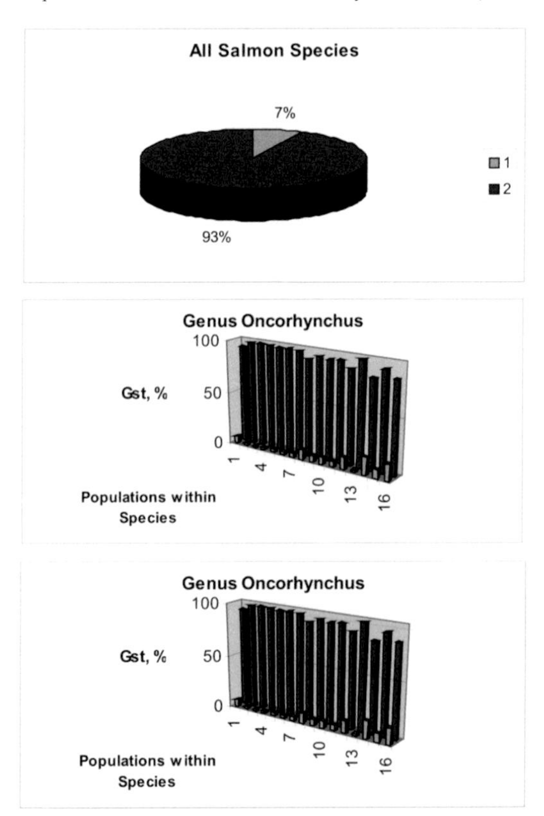

Fig. 8.3.9. Values of G_{st} as measurements of intraspecies differentiation in salmon (from Kartavtsev 2005b). 1. Inter-, 2. Intra-population components.

There are several possible reasons for different levels of intraspecies genetic differentiation. One of the important reasons may be an average difference in genetically effective size, N_e, which is higher in pacific salmons (see Chapter 7). The larger N_e is, the smaller is random genetic drift, and consequently the smaller is genetic differentiation of populations and the species in total (see details in Chapter 9). At least, for vertebrates and marine invertebrates, effective size is found to be different (see Fig. 7.3.4, Chapter 7).

8.4 A FIELD POPULATION GENETIC STUDY

The Investigation of Intraspecies Population Structure in Pink Shrimp, *Pandalus borealis*

The investigation was based on sample collections that covered nearly the whole species distribution (Fig. 8.4.1; Kartavtsev 1994).

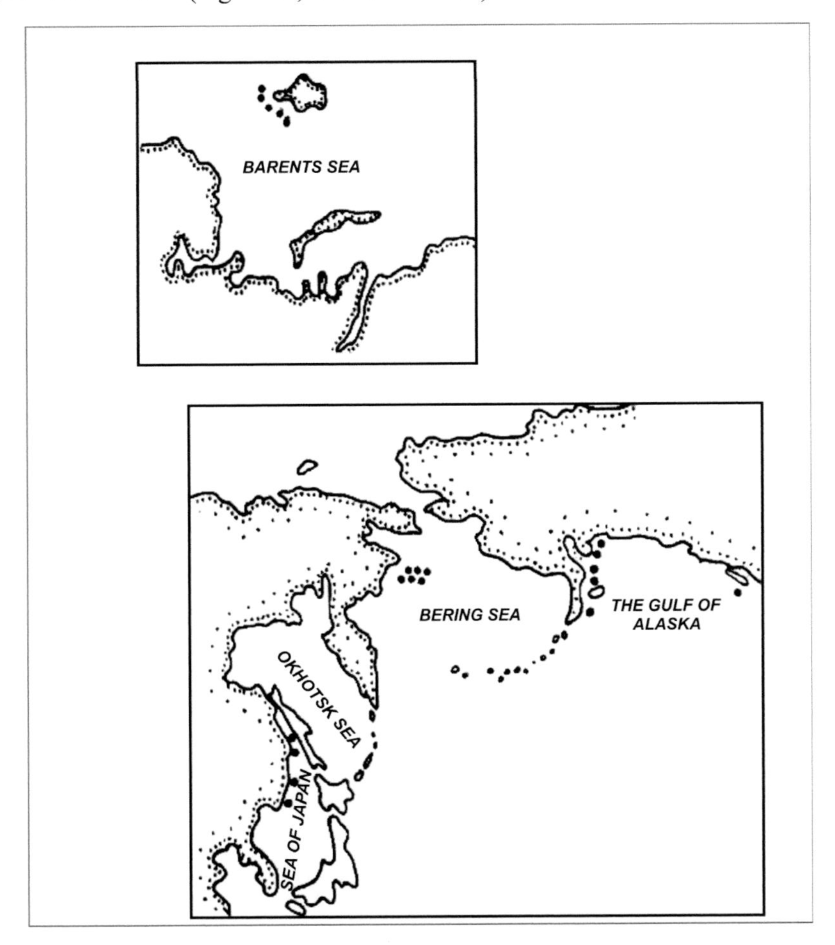

Fig. 8.4.1. Map of the area studied (from Kartavtsev 1994).

The most important result of the research was genetic homogeneity within a sea (or large gulf) and heterogeneity among them and in total (Fig. 8.4.2).

In accordance with the pattern of heterogeneity, there is a little if any genetic differentiation in terms of F_{st} within a sea basin and much more at the inter-basin level and in the total data set (Fig. 8.4.3). Ordination based on similarity of allele frequencies showed that samples from the same basin form much closer groups than samples belonging to different seas (Fig. 8.4.4).

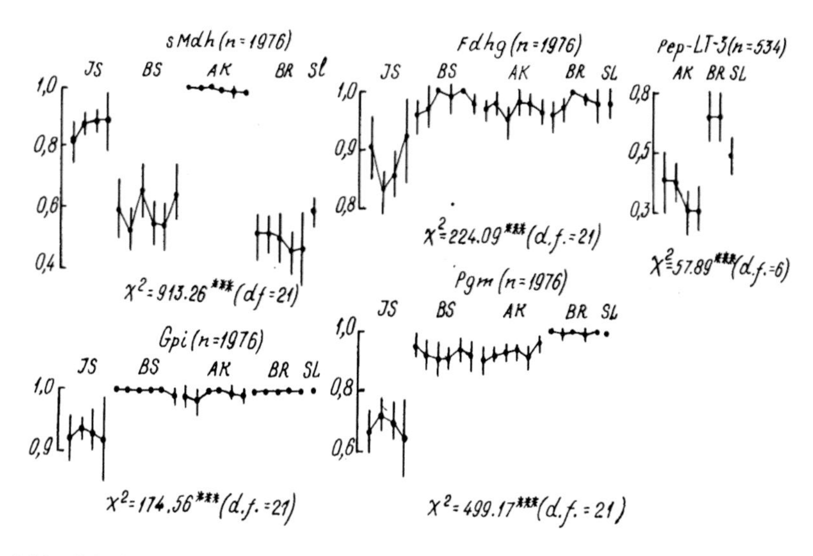

Fig. 8.4.2. Allele frequency variation at six allozyme loci in cohorts of pink shrimp, *Pandalus borealis*. Basins: JS, Sea of Japan, BS, Bering Sea, BR, Barents Sea, AK, Gulf of Alaska, SL, Saint Lawrence Bay. Asterisks show statistically significant heterogeneity: *** $p < 0.001$ (from Kartavtsev 1994).

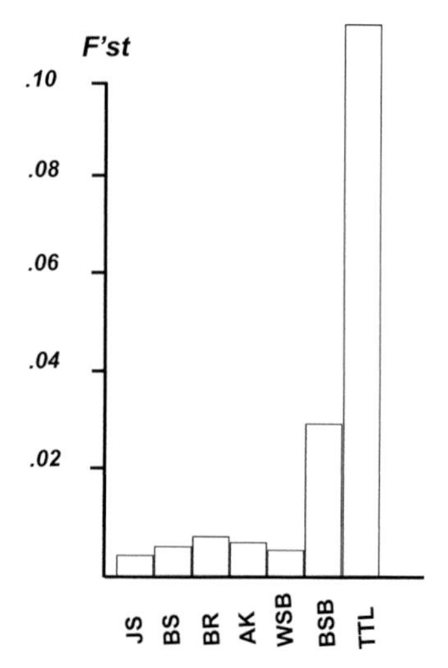

Fig. 8.4.3. Plot of F'st distribution within and among basins investigated in pink shrimp. JS, Sea of Japan, BS, Bering Sea, BR, Barents Sea, AK, Gulf of Alaska. WSB, mean within sea basin comparisons, BSB, mean between sea basins comparisons, TTL, total value for the whole area.

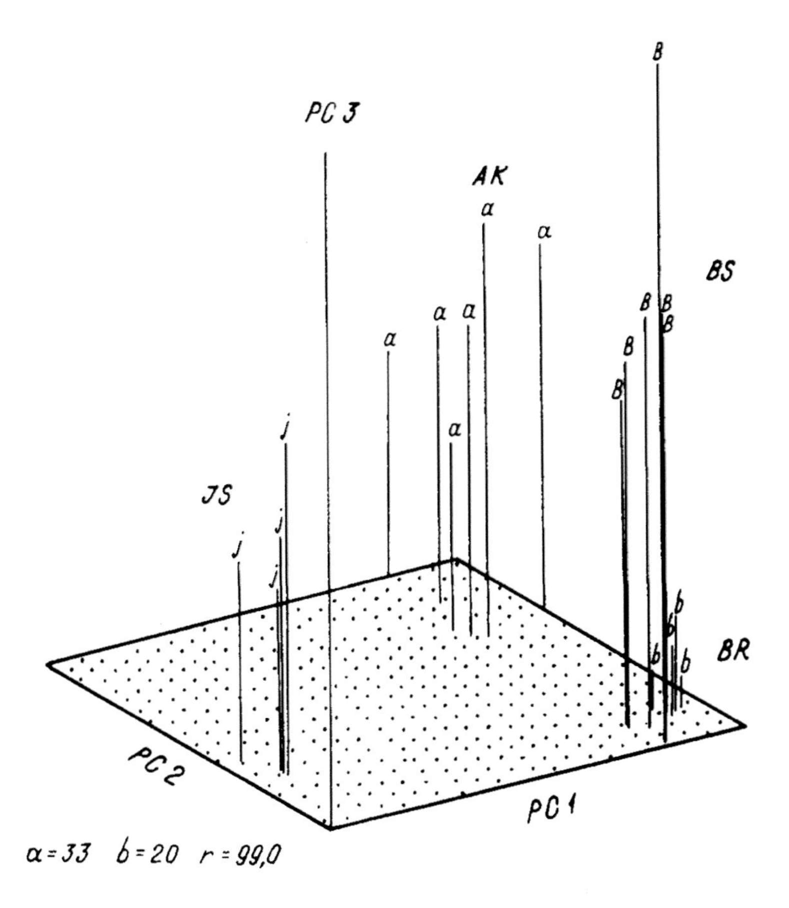

Fig. 8.4.4. Three-dimension distribution of principal components (PC) scores, taken from allele frequencies at allozyme loci.
JS—Sea of Japan, BS—Bering Sea, BR—Barents Sea, AK—Gulf of Alaska.

All populations of separate seas are unambiguously (100%) discriminated on the basis of allele frequencies at four allozyme loci tested, which separately illustrated for least different northern basins (Table 8.4.1).

Table 8.4.1. Matrix of sample classification of the pink shrimp, *Pandalus borealis*, for three northern basins, based on discriminant analysis of allele frequencies at four allozyme loci.

Group	Percentage of exact classifications	Number of Samples Classified		
		BS	BR	AK
BS	100	6	0	0
BR	100	0	5	0
AK	100	0	0	6
Total	100	6	5	6

Note. BS, Bering Sea, BR, Barents Sea, AK, Gulf of Alaska.

In general, genetic differentiation in pink shrimp agrees well with such data for other marine crustaceans and is small in comparison with freshwater or landlocked species with restricted gene flow (Fig. 8.4.5).

Morphometric research in general agrees with the population genetic data (Fig. 8.4.6).

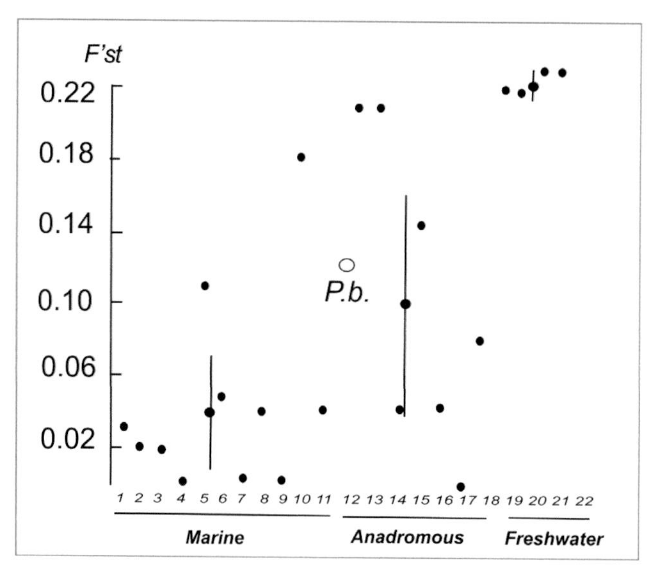

Fig. 8.4.5. Plot of $F_{'st}$ distribution among the higher crustaceans for three groups: marine (M), anadromous (A) and freshwater (F). The circle and the letters P.b. show the $F_{'st}$ score for the pink shrimp. Dots with vertical lines show mean $F_{'st}$ values with confidence intervals (95%) for the three groups M, A, F. Literature data from Chow and Fujio (1987) were recalculated in the scale $F_{'st}$ by the author (from Kartavtsev 1994).

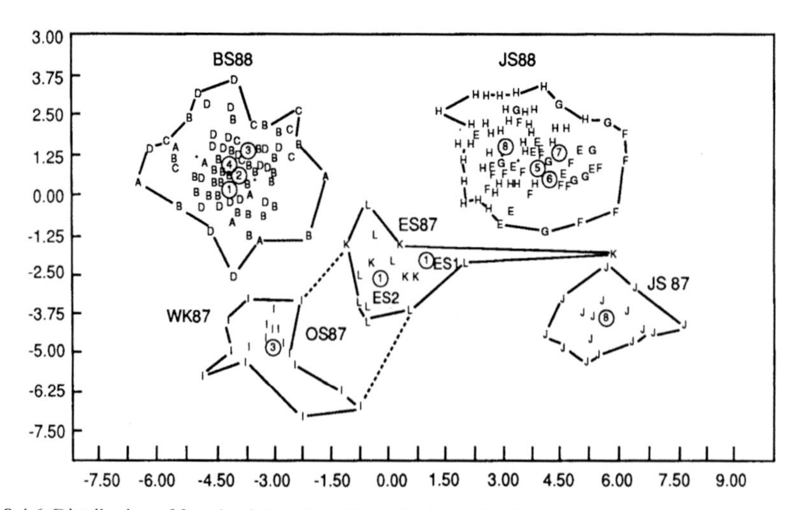

Fig. 8.4.6. Distribution of female pink shrimp, *Pandalus borealis*, along two canonical variables (CV1 and CV2) as exemplified by the for-sample, for-individual discriminative procedure based on the combined morphological traits and indices complex. The notations and abbreviations are the same as above (from Kartavtsev et al. 1993 with editor's permission).

8.5 TRAINING COURSE, #8

8.1 Make estimates of standard and minimal Nei's (Nei 1972) genetic distances on the pacific mussel data (SPECSTAT, File Edspec.dt1), and calculate as well the F_{st} and F'_{st} values using the program package SPECSTAT. Make calculations of F_{st} the program packages BIOSYS and GENEPOP.

What information does F_{st} (F'_{st}) provide?

8.2 Construct the UPGM dendrograms using the software BIOSYS for two different Nei's distance measures (D_n and D_m). Use the package example file for constructing distances and making dendrograms on their basis. Are the trees different?

8.3 Make numerical simulations using the POPULUS program package for the case where N_e, s and p parameters change.

8.4 Let us consider the following new terms.

F_{st} (G_{st}) is the standardized variance in allele frequency. This measure is used to estimate comparative levels of interpopulation or intrapopulation differentiation.

Population is a reproductive unit of organisms, which share a common gene pool. **Local populations** or **local stocks** are usually represented by subdivided populations, which consist of **demes** or **subpopulations**.

9

ANALYSIS OF GENETIC DIVERSITY WITHIN A SPECIES: THEORY AND EXPERIENCE

MAIN GOALS

SUMMARY

1. The main parameters that determine genetic dynamics of populations are (1) ***p*, allele frequency**, (2) N_e, **effective size of population**, (3) ***m*, coefficient of migration**, defining gene flow and (4) ***s*, coefficient of natural selection**.
2. Research on **intraspecific genetic diversity** requires a **complex approach**, taking into consideration the diversity of factors that cause variability and differentiation within a species.

9.1 MAIN GENETIC PARAMETERS OF POPULATION VARIABILITY

In the previous chapter we pointed out four major genetic parameters in a description of population structure within a species in a certain generation or at the current time. However, in general, over the whole set of generations in the evolutionary history of the population or species, mutations can also play a significant role. Their importance and evolutionary consequences were considered in Chapter 6. Let us remind ourselves once again which four main genetic parameters define the diversity among populations of a single species: (i) ***p***, *allele frequency*, (ii) N_e, *genetically effective size of population*, (iii) ***m***, *coefficient of migration*, (iv) ***s***, *coefficient of natural selection*. In Chapter 8, they were just introduced. Now let us look at them in more detail and consider some theoretical background for these parameters.

Allele Frequency, *p*

Empirically, under codominance or incomplete dominance, ***p*** can be estimated by a direct count of genotypes, as we already discussed (Chapter 4):

$$p_{A1} = (2D\,A_1A_1 + H\,A_1A_2)/2n = (D\,A_1A_1 + H/2\,A_1A_2)/n, \qquad (9.1)$$

where p_{A1} is frequency of A_1, D is the number in the sample of the genotype A_1A_1, H is the number in the sample of the genotype A_1A_2 and n is the general number of individuals in the sample. All other allele frequencies are obtained in the same manner. The frequency of the last allele need not be calculated because it is easily determined as in the two-allele case: $q_{A2} = 1 - p_{A1}$; p_{A1} is substituted here by the sum of all alleles except one. For the loci that are not linked with sex, the frequencies of alleles and gametes are equal. That is why in a large bisexual population with random mating there is a certain ratio between gamete frequencies and genotype frequencies known as the Hardy-Weinberg law (ratio):

$(p + q)^2 = p^2 + 2pq + q^2$ or in general notation for multiple alleles

$$(\Sigma\, p_k A_{ij})^2 = \Sigma\, p^2_k A_i A_i + \Sigma\, 2p_k(1 - p_k)\, A_i A_j, \qquad (9.2)$$

where, k = 1,2,3..j are ordered numbers of alleles. ***The equilibrium ratio between gametic and genotypic frequencies is established in a single generation of a random mating, and persists indefinitely, until it is disturbed by the action of natural selection, migration, population decrease or assortative mating that breaks the conditions of equilibrium.*** For the case of three alleles, the ratio may be written as: $(p + q + r)^2 = p^2 + q^2 + r^2 + 2pq + 2pr + 2qr$ and so on.

In a graphic mode, the relation between allelic (gametic) frequencies and genotypic frequencies may be shown as a parabola within a triangle (Fig. 9.1.1).

Allele frequency in principle may be considered as the probability of occurrence of an allele. Let us consider how to estimate an allele frequency in theory. Let ***p***, the frequency of the ***A*** allele, with $q = 1 - p$ for ***a***, be a random variate which follows the binomial distribution.

$$[p(A) + q(a)]^{2N}. \qquad (9.3)$$

Here ***N*** is the number of individuals. In reality, here and later ***N***, represents only reproducing individuals, and it is equal to, as we will soon see, N_e the effective size of population.

Let us denote $pA' = pi - p$ (where ***pi*** is allele frequency in sample ***i***) as the random deviation of the allele frequency from theoretical expectation (in a universe or total population), separate from deviation due to the action of systematic forces (selection,

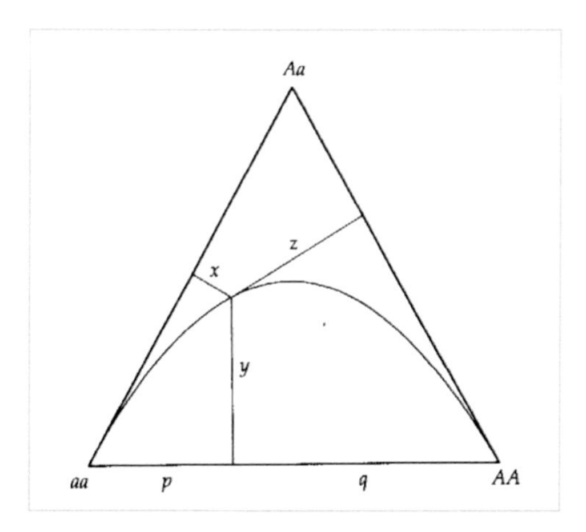

Fig. 9.1.1. De Finetti diagram showing the frequencies of the three genotypes by location in a triangular coordinate system.
The lengths of the line segments *x, y* and *z* indicate the frequencies of genotypes *AA, Aa* and *aa*, respectively. The vertices at the base represent fixation of the ***a*** allele (left) and the ***A*** allele (right). The top vertex represents a population with only heterozygotes and no homozygotes. The parabola that runs through the triangle is the set of points representing genotype frequencies in the Hardy-Weinberg equilibrium (from Hartl and Clark 1989 with permission).

migration, mutation). Thus, in a large randomly mating population (panmictic population), allele frequency variance for one generation is:

$$\sigma^2 pA' = [p\,(1 - p)]/2N. \quad \textbf{(9.4)}$$

Because sample estimates of variance is

$$\sigma^2 pAi = [p_i\,(1 - p_i)]/2n, \quad \textbf{(9.5)}$$

the equation $\sigma^2 \delta pA' = \sigma^2 pAi$ may be the null-hypothesis for testing whether a population corresponds to a large panmictic population with the H-W equilibrium. As a criterion of this equality of variances, it is convenient to use a chi-square (Chi^2) test of heterogeneity of allele frequencies (Workman and Niswander 1970) or its other versions in the case of many samples. In the case of wild populations, most of which, as we agreed in Chapter 8, are subdivided populations, the heterogeneity estimate should be more complex because the variance is more complex in this case.

$$\sigma^2 \delta'' = p\,(1 - p)\,F_{st}(t) + [(p\,(1 - p))/2N]\,(1 - F''_{st}). \quad \textbf{(9.6)}$$

We will return to this expression with explanations in Section 9.3, when we will consider models of the subdivided population.

Genetically Effective Size of Population, *N*

N_e is the number of individuals in an ideal reproducing population that has the same quality in a genetic drift sense as the real population of an organism. Consideration

of N_e, as a part of a reproducing population began with Wright (1931). Now there are at least 6 definitions of N_e (Kimura and Ohta 1971; Nei 1987; Altukhov 1989).

When generations are discrete (not overlapping), it is not difficult to estimate N_e in a panmictic population, if we have information on the numbers of males (m) and females (f) (Wright 1931). Discrete generations are not common in the animal world, however. Nevertheless, the supposition of discreteness is useful and acceptable for the preliminary estimation and explanation of what Ne is. We will not consider N_e for overlapping generations here; the relevant literature for them is available (Lande and Barrowclough 1987). Let us look at some definitions of Ne.

1. ***N_e under varying sex ratio.*** In Wright's paper cited above, N_e is defined as:

   $$N_e = (4N_m N_f)/(N_m + N_f), \tag{9.7}$$

   For a fixed number of parents, N, N_e will be maximal when $N_m = N_f$. If the sex ratio deviates significantly from 1:1, N_e depends largely on the less common sex.
2. ***N_e under varying number of offspring.*** Let k be the number of offspring of an individual over its whole life. Then N_e may depend on the input of males and females to the next generation, depending upon their fecundity, and may be expressed through N_m, N_f, mean and variance (V_k) for the k for each sex (Lande and Barrowclough 1987):

   $$Ne_m = (N_m \underline{k}_m - 1)/[\underline{k}_m + (V_{km}/\underline{k}_m) - 1],$$

   $$Ne_f = (N_f \underline{k}_f - 1)/[\underline{k}_f + (V_{kf}/\underline{k}_f) - 1].$$

   Under stable population number ($k = 2$)

   $$Ne = (4N - 2)/(V_k + 2). \tag{9.8}$$

   When the offspring number follows the Poisson Law ($k = V_k = 2$), N_e equals approximately the reproductive part of the population (N_b). However, in natural populations inequality usually holds, $V_k > k$. Consequently, in these populations $N_e < N_b$, and in general we can write: $N_e < N_b < N$, where N is the total size of the population, including non-mature individuals and others who could not take part in breeding.
3. ***N_e under fluctuating population size.*** When the size of the population has to fluctuate, the effective numbers in different generations, $N_e(1), N_e(2), \ldots, N_e(t)$, can be combined to yield a long-term effective population size using the formula

   $$N_e = t/\sum[1/N_e(i)]. \tag{9.9}$$

   In this equation, the harmonic mean determines the effective number in each generation. It is valid only for fluctuations occurring on a short time scale, during which gene frequencies remain nearly constant (Wright 1969, p. 214; Crow and Kimura 1970, p. 360). The harmonic mean for N_e in the above equation (9.9), therefore, should not be used in long-term management programs for effective size regulation.

4. ***Variance*** N_e. This definition of N_e was introduced by Crow (Crow 1954; Kimura and Crow 1963). In an ideal population, which comprises N_b mating individuals, the variance of allele frequency V_p, due to random sampling of gametes from one generation to another, is given by ratio 9.4. That is why it is natural to define N_e as

$$N_e = p\,(1-p)/2V\delta p. \tag{9.10}$$

There is also a definition of N_e in terms of inbreeding. We will omit it because we have not introduced inbreeding yet. Some practical outputs of estimates of N_e will be seen in Section 9.4. Now I'd like to present a simple scheme that illustrates ratios among N_e, N_b and N (Fig. 9.1.2).

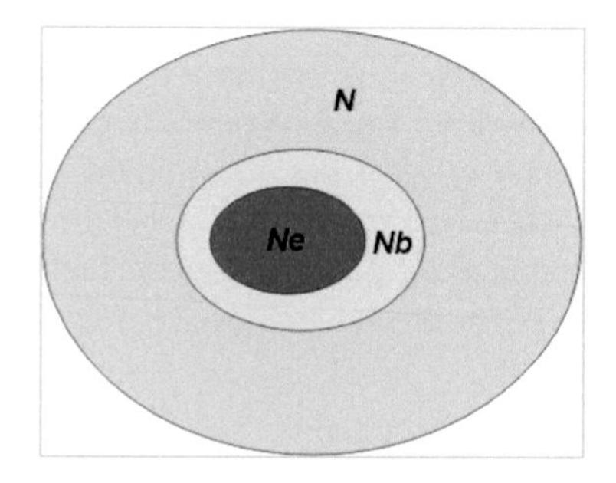

Fig. 9.1.2. Differences among population sizes.
N is the total number of individuals in a population: mature and not mature, N_b is the mature part of a population: ***spawning*** mates, N_e is the genetically effective size or a part of the spanning group: N_b is affected by sex ratio, fecundity variance and inter-generation abundance.

It is useful not to forget that N_e is a genetic abstraction, as evident from the definition given in the beginning of this section. In particular, from equation 9.7, it is easy to calculate that under an equal sex ratio N_e can reach the value of $4N$, which is not possible in practice.

How N_e may influence the variability of allele frequencies under random segregation in populations of different size is illustrated in Fig. 9.1.3 with numerical simulation details explained elsewhere (Kartavtsev 1996).

Gene Flow and the Coefficient of Migration, *m*

New alleles may come to a population through mutations but they also can enter a population as gametes brought by immigrants from other populations. This process of entering of alien or "foreign" genes we may call gene exchange, and if this process is constant and immigrants take part in reproduction we will call it a **gene flow**. *The effective size of a deme is often so small that mutations are likely to be negligibly rare in occurrence, even over considerable periods of time, thus giving no impact on genetic differentiation in the current generation* (That is why we do not consider the role of mutations in population structure analysis in space for a certain year or generation). On the other hand, immigration of individuals from other populations, followed by crossbreeding, must nearly always be a very important factor in current differentiation

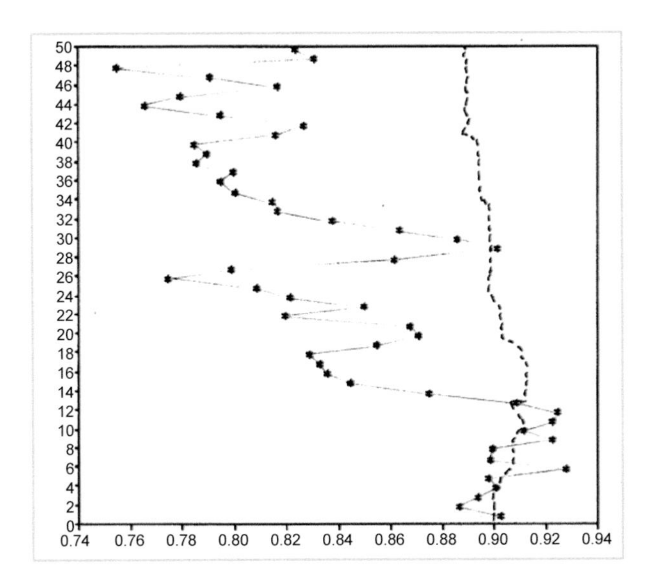

Fig. 9.1.3. ***Ne*** influence on gene frequency change (***q, X-axis***) through generations (***T, Y-axis***). Changes in a smaller population, $N_e = 118$, are shown with dots and a solid line; in a larger population, $N_e = 33182$, the variability is shown with a broken line. Initial allele frequency in both cases is $q = 0.9$.

and by extending in future, become their histories. Let $\boldsymbol{m}$ be the observed proportion of the population replaced per generation from outside and Q the observed frequency of a given gene in the immigrants in contrast with $\boldsymbol{q}$ for the same gene within the local deme (Wright 1931). The frequency of gene $\boldsymbol{a}$ in the initial (zero) generation is $\boldsymbol{q_0}$. The gene frequency in the first generation is $\boldsymbol{q_1}$, which may be defined as follows: $\boldsymbol{q_1 = q_0 (1 - m) + Qm = q_0 - q_0 m + Qm = q_0 - m (q_0 - Q)}$. Change in allele frequency for one generation is $\boldsymbol{Dq = q_1 - q_0}$. Substituting $\boldsymbol{q_1}$ here by the value in the last expression above, we have:

$$\boldsymbol{Dq = q_1 - q_0 = q_0 - m (q_0 - Q) - q_0 = - m (q_0 - Q).} \qquad \textbf{(9.11)}$$

Here $\boldsymbol{m}$ is a coefficient of migration or observed proportion of immigrants and $\boldsymbol{Dq}$ is the change in gene frequency for one generation. The migration may be defined in very different ways. Above we assumed that migration occurred only in one direction, for example, from a large mainland population to a small, semi-isolated island population. It is also suggested that the H-W equilibrium holds in both populations except that gene flow occurred. To understand the gene flow results better, let us see the numerical example, similar to that depicted by Mettler and Gregg (1972) (Fig. 9.1.4).

Migration affects not only the integrity of the gene pool but also the effective size of populations (Chapter 8, Fig. 8.3.13). Data in Fig. 8.3.13 correspond to high gene flow in marine invertebrates which is directly related to N_e (see Section 9.3, formula 9.22). This is easily understandable even on the intuitive level. Really, if three populations with N_{e1}, N_{e2} and N_{e3} have high rates of gene exchange, then this may mean the summation of their sizes is $N_e = N_{e1} + N_{e2} + N_{e3}$, which is naturally larger than the size of each separate population, when no migration exists.

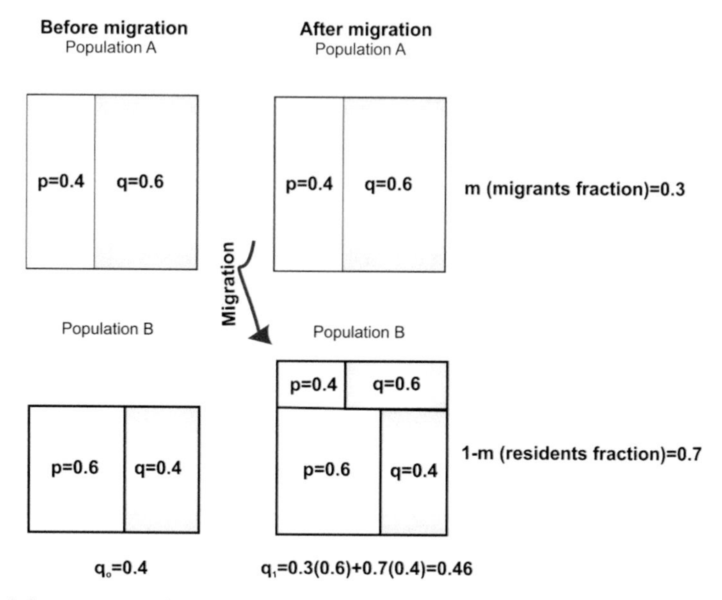

Fig. 9.1.4. Migration influence on gene frequency change.
Allele frequency (q) change in population B from $q_0 = 0.4$ (before migration) to $q_1 = 0.46$ (after migration).

Coefficient of Natural Selection, *s*

Relative levels of fitness, say for viability of genotypes (W), may be estimated from the ratio of the genotypes *AA, Aa* and *aa* in the offspring from the crosses of heterozygous ($Aa \times Aa$) parents by expression:

$$2n_{AA}/n_{Aa},\ 1,\ 2n_{aa}/n_{Aa}, \tag{9.12}$$

where n is number of genotypes.

If $W_{Aa} = 1$, then

$$W_{AA} = 1 - s_1 \text{ and } W_{aa} = 1 - s_2, \tag{9.13}$$

where s_1 is a coefficient of selection against AA, and s_2 is a coefficient of selection against aa. After one cycle of selection, mean fitness (at this single locus!) in a population becomes (see Table 9.1.5):

$$W = W_0 p^2 + W_1 2pq + W_2 q^2. \tag{9.14}$$

In other words, in our numerical example, as a result of selection, frequencies of both homozygotes decrease and the relative frequency of heterozygotes increase (Table 9.1.5). This change will reach a stable or a steady state when an allele frequency will come to an equilibrium value (p^ or q^) that holds as long as Ws are not changed. The POPULUS package defaults to t = 200 generations, q^a = 0.3 (Populus > Evolutionary Simulations > Selection > Autosomal Selection: Defaults).

Table 9.1.5. Results of one-cycle of selection (from Mettler and Gregg 1972).

Genotype	**AA**	**Aa**	**aa**	**Total**
Frequency before selection (F_0)	p^2	2pq	q^2	1
Fitness	W_0	W_1	W_2	-
Proportional impact	W_0p^2	$W_1 2pq$	W_2q^2	$\underline{W}$
Frequency after selection (F_1)	W_0p^2 ------- $\underline{W}$	$W_1 2pq$ ------- $\underline{W}$	W_2q^2 ------- $\underline{W}$	1

9.2 QUANTITATIVE MEASURES OF DIFFERENTIATION

Wahlund Effect

Some kind of quantification of the amount of differentiation in a structured population is desirable. In the above text, we defined this by F_{st} and G_{st} statistics, which will be further developed in the next subsection. However, a similar approach may be realized differently. Let us, as a basis for comparison, take a large, randomly mating population and consider a locus at which there is no selection. If the locus has two segregating alleles, *A* and *a* with respective frequencies *p* and *q*, the expected proportions of the genotypes *AA, Aa* and *aa* will be the Hardy-Weinberg proportions p^2, $2pq$ and q^2. If, instead, the population is subdivided and allele frequencies vary among subpopulations, then, as was first pointed out by Wahlund (1928), the proportion of homozygotes in the total population will be greater than the Hardy-Weinberg expectations. Let p_i, be the frequency of *A* in subpopulation *i* ($i = 1, ..., m$), and assume that in each subpopulation genotypes occur in the Hardy-Weinberg proportions given by the local allele frequencies. The frequency of *AA* in subpopulation *i* is then p^2_i, and if sub-populations have equal size, the proportion of *AA* in the entire population is

$$(\sum p^2_i)/m = \bar{p}^2 - V_p, \quad \mathbf{(9.15)}$$

where $\bar{p} = \sum p_i/m$ is the average frequency of *A* in the total population and V_p is the variance *of* p_i over subpopulations. Thus, V_p is equal to the excess of homozygotes and consequently, deficit of heterozygotes over the H-W expectation. Let us support this conclusion with a numerical example (Fig. 9.2.1).

Wright (1921) introduced a parameter *F*, called the fixation index, to characterize the genotypic distribution at a two-allele locus. For our case, the proportions of *AA, Aa* and *aa* in the total population can be written, respectively, as $p^2 + Fpq$, $2pq(1–F)$ and $q^2 + Fpq$; and comparing with (9.15) we see that $F = V_p/\bar{p}\,\bar{q}$. Wright (1943; 1951; 1965) developed this concept further by introducing the three *F*-statistics, F_{is}, F_{it} and F_{st}, whereby the overall deviation from Hardy-Weinberg proportions can be split up into deviation caused by subpopulation differentiation, deviation from the local Hardy-Weinberg proportions and in the total sample.

For subpopulation differentiation, the relevant *F*-statistic is F_{st}, which was defined by Wright as the correlation between random gametes, drawn from the same subpopulation, relative to the total. If the total is taken to be the currently existing

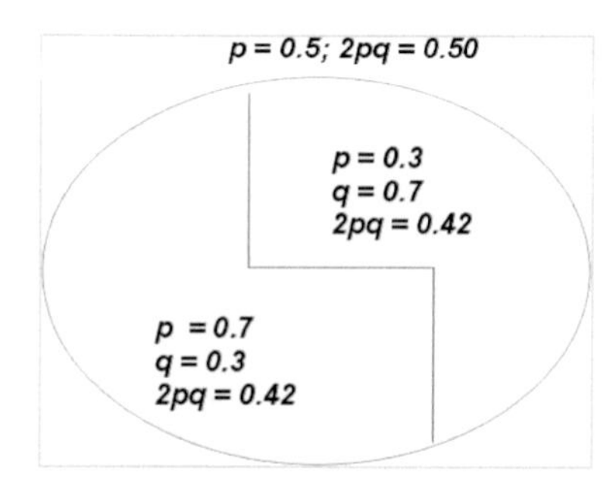

Fig. 9.2.1. Numerical example of Wahlund effect.
In a subdivided population, there is an excess of homozygotes and deficit of heterozygotes in comparison with H-W expectations in a single panmictic population. In a panmictic population with equal frequencies of alleles, as in the case of the numerical example, heterozygosity is $2pq = 0.50$, however in the subdivided population exemplified $2pq = 0.42$.

population, this definition leads to $F_{st} = V_p/\bar{p}\,\bar{q}$, which actually is the standardized variance of allele frequencies. Note, however, that F_{st} is given by $V_p/\bar{p}\,\bar{q}$ regardless of whether subpopulations are in Hardy-Weinberg proportions. A convenient formula for F_{st} can be obtained if we introduce $H_s = H_i/n$, where $H_i = 1 - (p^2_i + q^2_i)$ is the Hardy-Weinberg expectation of heterozygosity in subpopulation i, and $H_t = 1 - (\bar{p}^2 + \bar{q}^2)$. A simple calculation yields (Nei 1973)

$$F_{st} = V_p/\bar{p}\,\bar{q} = (H_t - H_s)/H_t = 1 - (H_s/H_t). \tag{9.16}$$

Gene Diversity Analysis

Nei (1973; 1977b) extended F_{st} by providing a measure of differentiation, called $\boldsymbol{G_s t}$, based on allele frequencies at several multiallelic loci. For a given locus, let $\boldsymbol{p_{ik}}$ be the frequency of allele $\boldsymbol{k}$ in subpopulation $\boldsymbol{i = 1,2..m}$ and define h_k, Hs and Ht as follows: $h_k = 1 - \sum^k_{i=1} p^2_i$, $H_s = \sum^m_{i=1} h_k/m$, $H_t = 1 - \sum^m_{i=1} \bar{p}^2_k$. If we accept as well that observed heterozygosity is $H_o = 1/n\ h_{obs}$, where h_{obs} is the observed number of heterozygotes at a given locus, and n is the number of individuals in a sample, we will define F-statistics for the multiple allele case (Nei 1973; 1977b; 1987), which gives easy averages for many alleles and many loci. For the single-locus case, F-statistics notations are as follows:

$$F_{is} = (H_s - H_o)/H_s,\ F_{it} = (H_t - H_o)/H_t,\ F_{st} = (H_t - H_s)/H_t. \tag{9.17}$$

F-statistics for several loci are obtained by averaging single locus values, with weighting as defined by Nei (1977b; see also details of averaging in Kartavtsev and Zaslavskaya 1982). To discriminate between Wright's F_{st} and the last F_{st}, Nei (1977b) suggested using a different notation, $\boldsymbol{G_{st}}$. In Nei's notation F_{st} is the measure of among-group gene differentiation. It is always positive, even if no actual heterogeneity exists, and that must be keep in mind when real populations are analyzed. Due to sampling errors, spurious diversity may be detected. F_{is} is the index of intra-population deviation from H-W expectation (and may be positive, when a heterozygote deficit exists relative to the H-W expectation, or negative, when an excess occurs). F_{it} is the index of total population deviation from the H-W expectation (which may be positive or negative but must always be greater than F_{is}).

Hierarchical Structure

It may be shown that $H_t = H_s + D_{st}$ (Nei 1973). Here $D_{st} = D_{kl}/s^2$ is the mean among-group diversity at the locus l; and D_{kl} is the minimal genetic distance among s subdivisions or subpopulations. With these notations another definition of G_{st} is appropriate. The relative magnitude of gene differentiation among subpopulations may be measured by

$$\boldsymbol{G_{st} = D_{st}/H_t}. \tag{9.18}$$

This varies from 0 to 1 and is called the *coefficient of gene differentiation.* Equation (9.18) can easily be extended to the case where each subpopulation is further subdivided into a number of colonies. In this case, H_s may be decomposed into the gene diversities within and between colonies (H_c and D_{cs}, respectively). Therefore,

$$\boldsymbol{H_t = H_c + D_{cs} + D_{st}}. \tag{9.19}$$

This sort of analysis can be continued to any degree of hierarchical subdivision. The relative degree of gene differentiation attributable to the colonies within subpopulations can be measured by $G_{cs(t)} = D_{cs}/H_t$. It can also be shown that $(1 - G_{cs})(1 - G_{st})H_t = H_c$, where $G_{cs} = D_{cs}/H_s$. For further discussion of this problem, see explanations in Chakraborty et al. (1982) and Chakraborty and Leimar (1987).

9.3 MODELS OF POPULATION STRUCTURE

Model of Panmictic Population

As was explained in Chapter 8 and above, the simplest model of population structure is the model of a panmictic population. The parameters of this model are $\boldsymbol{p}$ and $\boldsymbol{N_e}$. In the beginning of any population genetic analysis this model, with the absence of any genetic differentiation (heterogeneity) and with the H-W equilibrium, should be tested first. As formulated in a general notation for the multiple alleles (9.2), it is $\boldsymbol{(\Sigma p_k A_{ij})^2 = \Sigma p^2_k A_i A_i + \Sigma 2p_k(1 - p_k) A_i A_j}$, where $\boldsymbol{A_{ij}}$ denote allele $\boldsymbol{i}$ or $\boldsymbol{j}$, $\boldsymbol{p_k}$ its frequency, $\boldsymbol{k = 1,2,3..j}$ is ordering numbers of alleles and $\boldsymbol{\Sigma p_k = 1}$.

Models of Subdivided Population

Before we will consider the details, let us look at the schematic representation of the model (Fig. 9.3.1).

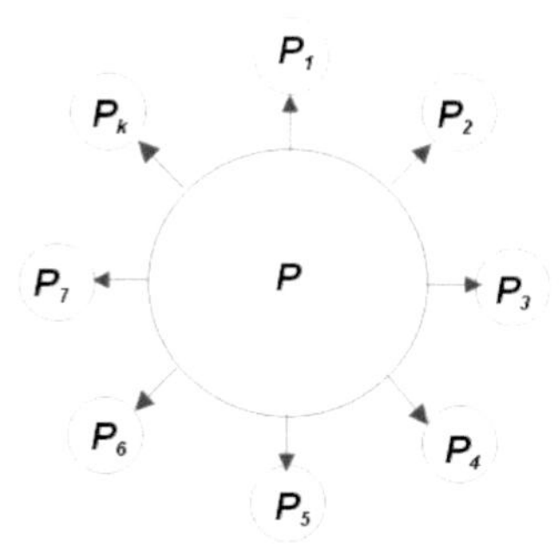

Fig. 9.3.1. One variant of the Island Model of population structure.
P is the mean frequency of alleles among $\boldsymbol{K}$ ($\boldsymbol{K} \rightarrow \infty$) demes ($p_1$ to p_k), which are "islands of an archipelago"; P equals as well the average allele frequency in a "mainland" population. Each deme gets genes from the "mainland" with the rate $\boldsymbol{m}$ (arrows).

Island Model

There are two variants of this model. The scheme for one of them is given above. In the second variant the "islands of the archipelago" are equal in size and all exchange genes with the same probability of $\boldsymbol{m}$ immigrants; "continent" is not considered in this variant. Gene pool of the migrants is formed by a joint contribution from all the demes. Distance among the demes does not influence the gene exchange. In their genetic consequences, these two variants of the model are identical.

We first will not take into account natural selection, i.e., only parameters $\boldsymbol{p}$, N_e and $\boldsymbol{m}$ are acting. In this case only genetic drift will govern differentiation among demes, and an equilibrium level of differentiation, determined by the balance of drift with migration, is given by equation (9.20), which is similar to (9.6) but reduced because here we consider the demes of equal size (and large enough to hold the H-W ratio).

$$\sigma^2_\delta{}'' = p\,(1 - p)\,F_{st}(t), \tag{9.20}$$

where $F_{st}(t) = (1 - m)^2[1/2N + (1 - 1/2N)\,F_{st}(t - 1)]$. **(9.21)**

If we express $\boldsymbol{F}$ in terms of $\boldsymbol{m}$ and vice versa in (9.21), we obtain well-known equations (Li 1978, p. 474):

$$(1 - m)^2 = 2N\,F_{st}\,/((2N - 1)\,F_{st} + 1), \tag{9.22}$$

$$F_{st} = (1 - m)^2/(2N - (2N - 1)(1 - m)^2). \tag{9.23}$$

These ratios indicate the negative relationship between migration and drift (N_e size) in determining the level of differentiation among groups. If the migration rate is small ($m \ll 1$), from (9.23) and (9.20) a simpler equation may be derived (Wright 1931; Kolmogorov 1935):

$$\sigma^2_\delta{}' = p\,(1 - p)/(4Nm + 1). \tag{9.24}$$

From this it is clear that $F_{st} = 1/(4Nm + 1)$, as was noted in the previous chapter. This makes clear that in any given generation neither ***m*** nor ***Ne*** alone, but instead their product, ***Nm***, determines the level of differentiation; i.e., the product *Nm* (= the absolute number of immigrants) is a key factor defining the level of differentiation. Here it is essential to recall that migration is in terms of gene flow; i.e., only those migrants that leave offspring are important!

Under the interaction of genetic drift (N_e change) and migration, the change in allele frequency over time has a complex function with a particular probability density distribution (Wright 1938; 1969). This is too complicated for this course to introduce. The relation between ***Nm*** and F_{st} at a steady state, when equilibrium between genetic drift and migration has been achieved, is shown in the next graph (Fig. 9.3.2).

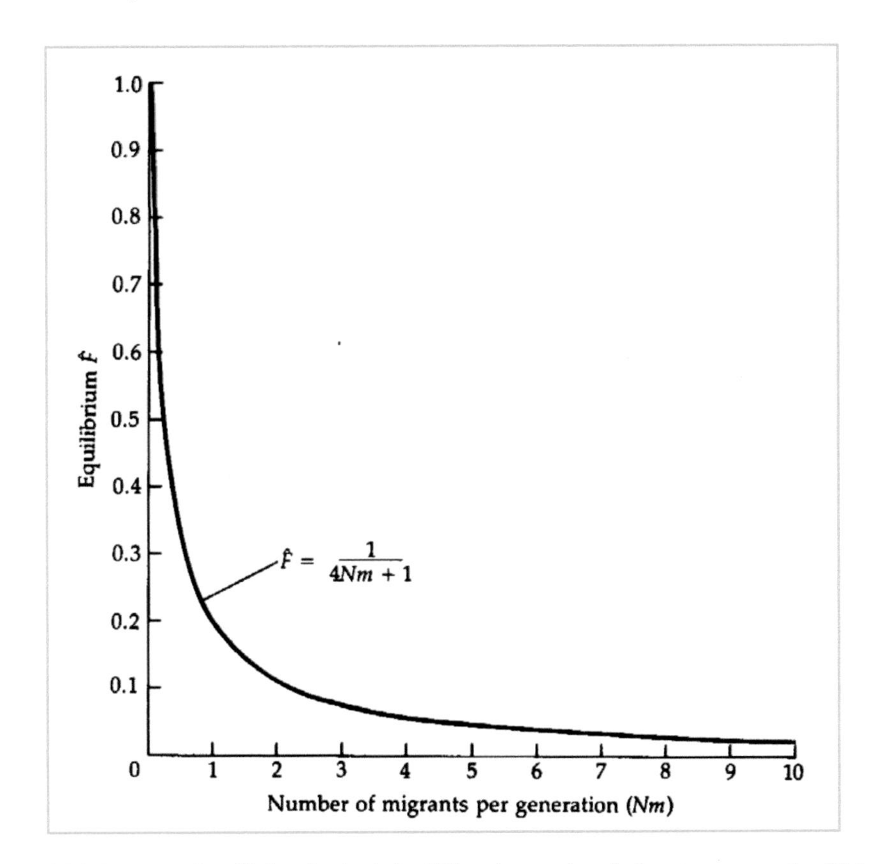

Fig. 9.3.2. Decrease of equilibrium fixation index (*F^*) against number of migrants per generation (*Nm*). Note that only a few migrants are necessary to reduce *F^* and thus a population differentiation, to a very small level. *F* = 0.50, *Nm* = 0.25 (one migrant every fourth generation), *F* = 0.40, *Nm* = 0.5 (one migrant every second generation), *F* = 0.33, *Nm* = 1 (one migrant every generation), *F* = 0.20, *Nm* = 2 (two migrants every generation), *F* = 0.11 (from Hartl and Clark 1989 used with permission from Sinauer Associates, Inc.).

The island model suggests that if a series of intraspecies divisions is involved, there must be an increase of genetic differentiation from lowest to highest levels of subdivision.

Stepping-stone Models

One-, two- and three-dimensional stepping-stone (stairs) models were developed by Kimura (1953), Kimura and Weiss (1964) and Weiss and Kimura (1965). We will not go into the details of these models. However, some general features will be considered (Fig. 9.3.3), and it may be concluded that an inverse relationship must exist between geographic and genetic distance in these cases.

In the theory of subdivided populations *there is clear-cut prediction of a relationship between degree of genetic differentiation and distance among demes* (Fig. 9.3.4), which is supported by the empirical research (Workman et al. 1973; Fig. 9.3.5, rearranged data from cited paper). Also, the general equilibrium ratio for the neutral case stays true: $F_{st} = 1/[4Nm + 1]$.

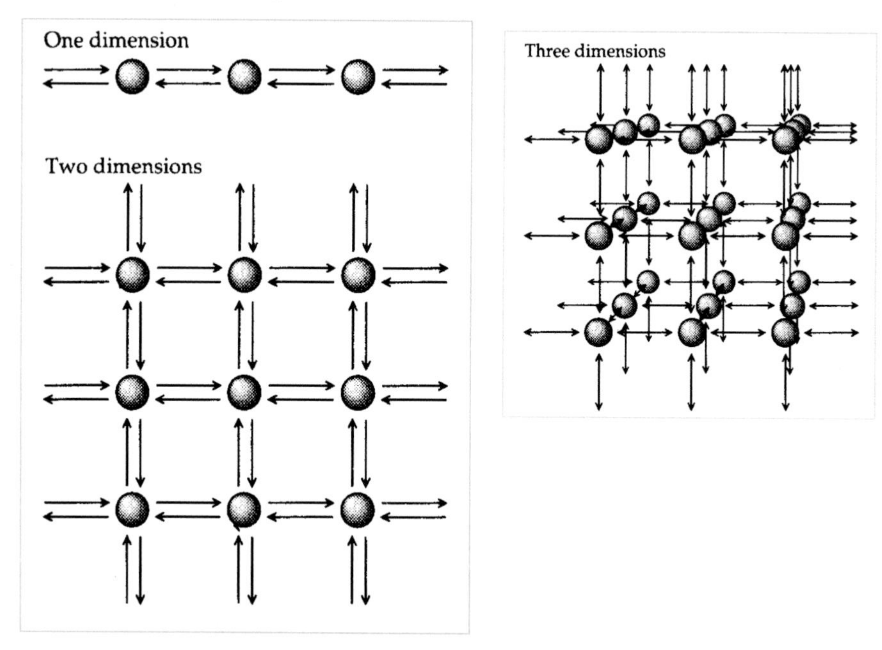

Fig. 9.3.3. Pattern of a migration in one-, two- and three-dimensional stepping-stone models. Each arrow represents a migration rate of $m/2$ in one dimension, $m/4$ in two dimensions and $m/6$ in each direction in three dimensions. Edge effects can be eliminated by wrapping the one-dimensional array into a circle, the two-dimensional array into the surface of a sphere and the three-dimensional array into a torus (from Hartl and Clark 1989 used with permission from Sinauer Associates, Inc.).

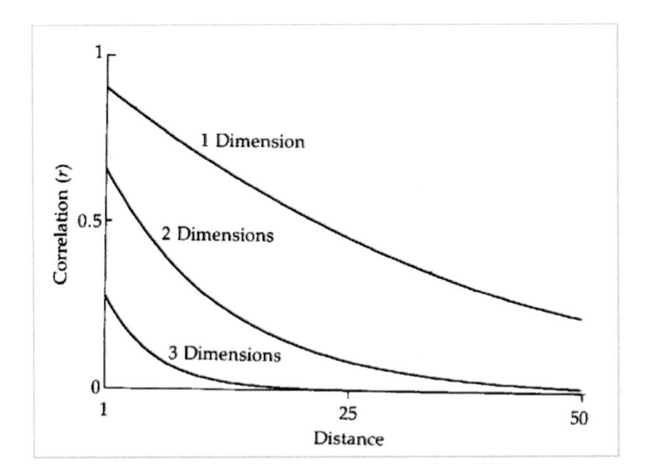

Fig 9.3.4. The decline in the correlation of allele frequencies with distance under the one-, two- and three-dimensional stepping-stone model (from Kimura and Weiss 1964; adopted from Hartl and Clark 1989 and used with permission from Sinauer Associates, Inc.).

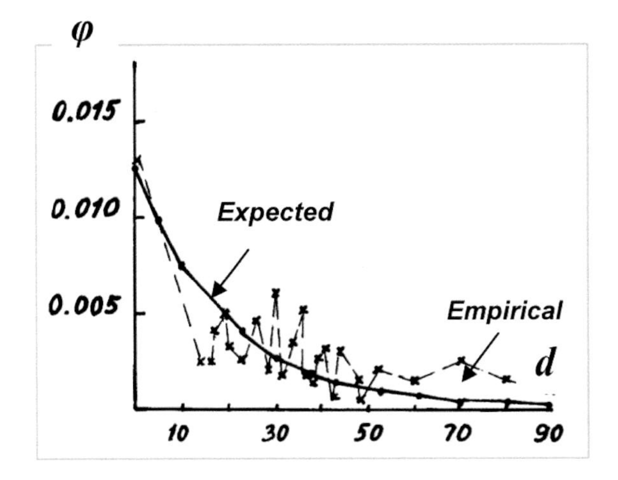

Fig. 9.3.5. Association between geographic distances (***d, km***) and genetic relatedness (***φ***). ***φ(d)*** = ***a*e – bd***, where a is coefficient of local similarity, *b* is function of wide-scale migration.

9.4 EMPIRICAL ESTIMATES OF PARAMETERS

Not many values of the necessary parameters have been estimated in natural populations. Let us consider available data on N_e (Table 9.4.1), G_{st} (Fig. 9.4.2), ***m*** (Table 9.4.2) and ***W*** (Table 9.4.3).

The data in Table 9.4.1 illustrate: (i) a reasonably small N_e size relative to total abundance of populations, (ii) the difference between vertebrates and invertebrates, especially marine invertebrates and (iii) larger N_e for pink salmon than for sockeye salmon (Table 9.4.1). Last point is relevant to the discussion in Chapter 8 on the sources of population structuring.

Table 9.4.1. Genetically effective size (N_e) in natural animal populations.

Species	N_e	**Source**
Vertebrates		
Oncorhynchus nerka	200	Altukhov et al. 1975
Same	174	Marriott 1964
O. gorbuscha	695	Kartavtsev 1995
Porcelio scaber	19–180	Brererton 1962
Chondrus bidens	50	Altukhov and Livshits 1978
Sceloporus olivaceus	250	Kerster 1964
Uta stansburiana	14	Tinkle 1965
Platycercus eximius	31–83	Brererton 1962
Mus musculus	10	Anderson 1970
Michigan populations	10–75	Rasmussen 1964
Arizona populations	30–130	Rasmussen 1964
Ovis canadensis	98	Geist 1971
Homo sapiens, Parma Valley, aboriginal	214–266	Cavalli-Sforza et al. 1964
Homo sapiens, Asiatic Mongoloids	45–218	Rychkov 1973; Rychkov and Sheremetyeva 1976
Homo sapiens, Indian Tribes, S. America	288–14400	Neel and Rothman 1978
Mean (averaged on minimal values)	**142**	
Mean (averaged on maximal values)	**1123**	
Land invertebrates		
Cepaea nemoralis	236–8440	Lamotte 1951
Same	190–12000	Greenwood 1975; 1976
Aedes aegipti	500–1000	Tabachnik and Powell 1978
Drosophila pseudoobscura	500–1000	Dobzhansky 1970b
D. subobscura	400	Begon 1977
Euphydrias edita	10–3700	Ehrlich 1965
Dacus olea	4000	Nei and Tajima 1981
Mean (averaged on minimal values)	**834**	
Mean (averaged on maximal values)	**4362**	
Marine invertebrates		
Mytilus trossulus	33182	Kartavtsev and Nikiforov 1993
Pandalus kessleri	730800	Sytnikov et al. 1997
Mean	**381991**	

In general genetic diversity, say in G_{st} terms, differs among species investigated, even those with close genetic proximity, as it exemplified by salmons (see Chapter 8, Fig. 8.3.10) and by data for the gastropod genus *Littorina* (Fig. 9.4.2). For *Littorina* in general averages are: $G_{st} = 0.047 \pm 0.010$, and $G_{st} = 0.154 \pm 0.028$, respectively for planktotrophic and lecitotrophic gastropods (Ward 1990; my recalculation for the averages and SE).

The difference of average G_{st} for these two groups is significant: $t_d = 3.6$, d.f. = 15, $p < 0.005$. Obtained differences in G_{st} levels in planktonic and non-planktonic *Littorina* show that species biology, which may influence migration rate, is most important for population genetic differentiation. Let us look at some direct data on migration rate.

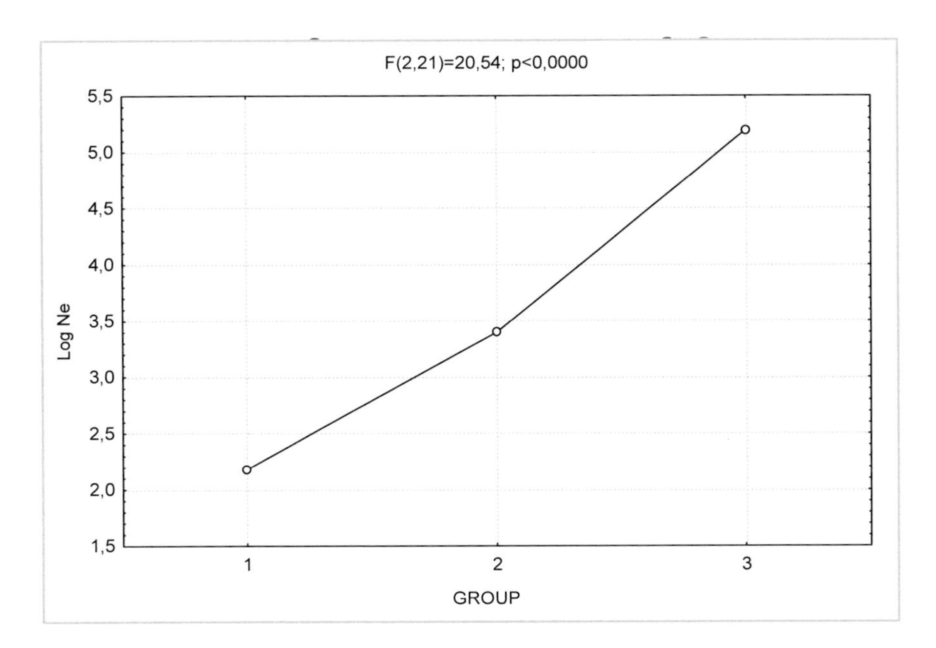

Fig. 9.4.1. Variance analysis of heterogeneity of N_e values in groups of animal taxa.
(1) Vertebrates, (2) Land invertebrates, (3) Marine invertebrates.

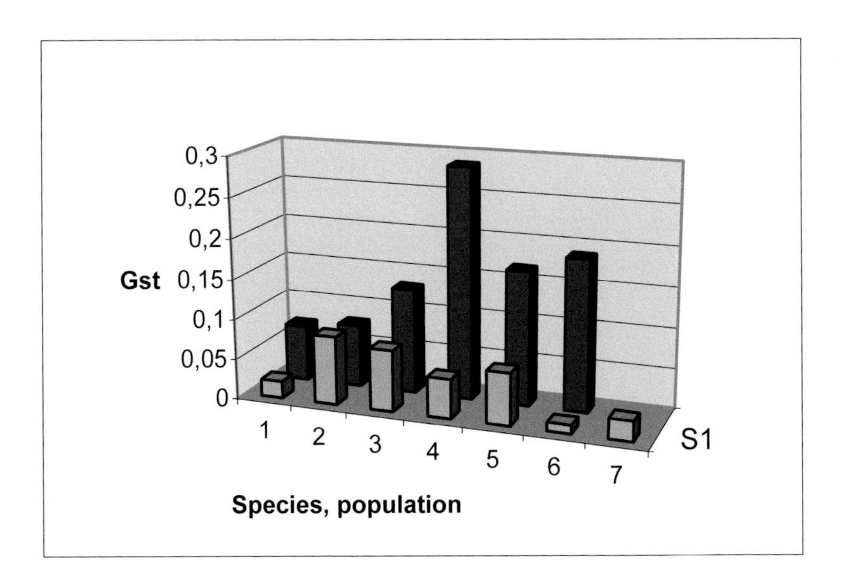

Fig. 9.4.2. Gene diversity among local populations of various *Littorina* species.
Front columns are populations of species that have pelagic larvae; back columns are populations of species that have no pelagic stage in a life circle. Pelagic: (1) *L. littorea*, (2) *L. scutulata*, (3) *L. plena*, (4) *L. angustior*, (5) *L. lineolata*, (6) *L. ziczac*, (7) *L. angulifera*; Non-pelagic: (1) *L. saxatilis*, (2–4) *L. saxatilis* (other populations), (5) *L. arcane*, (6) *L. nigrolineata* (Drawn on data from Ward 1990).

Table 9.4.2. Coefficients of migration (***m***) in salmon.

Species	m (%)	Source
O. nerka	3	Vernon 1957
same	<=3	Hartmann and Releigh 1964
same	2	Altukhov 1989
same	2	Iliin et al. 1983
same	0,001–0,4	Brannon 1982
O. gorbuscha	5*	Bams 1976
same	9[2]	Bams 1976
same	0–16*	Lane et al. 1990
same	0,1–11[1]	Rukhlov and Lubayeva 1979
O. kisutch	15–27[2]	Donaldson and Allen 1958
S. salar	5	Verspoor et al. 1991

Note. Values of ***m*** with an asterisk are taken from indirect sources. (1) This is an estimate for the hatchery stocks. (2) This is a return rate of transplanted fish.

Table 9.4.3. Estimates of relative fitness (W_i) of genotypes at allozyme loci in pacific salmon.

Species, Source	Values at a locus			Reference
	Homozygote 1	Heterozygote	Homozygote 2	
O. nerka, Spots	0.90	***LDH**** 1.00	0.86	1
Springs	0.98	1.00	0.97	1
Spring run	0.94	1.00	0.91	1
Lake	-	-	-	1
Stock, Total	0.98	1.00	0.97	1
Spots	0.94	***PGM**** 1.00	0.80	1
Springs	0.91	1.00	0.72	1
Summer run	0.92	1.00	0.74	1
Lake	0.99	1.00	0.97	1
Stock, Total	0.94	1.00	0.79	1
O. gorbuscha, parents-larvae	0.63	***MDH**** 1.00	-	2
larvae	0.66	-	-	3
-”-	0.58	***GPDH**** 1.00	-	2
-”-	0.61	-	-	3
-”-	0.71	***6PGD**** 1.00	0.43	2
-”-	0.58	-	-	3
-”-	0.64	***PGM**** 0.78	1.00	2
-”-	0.86	-	-	3

Note. W_i values belong to most common genotypes. Reference: (1) Altukhov 1989, (2) Salmenkova 1989, (3) Zhivotovsky et al. 1987.

Migration rate varies considerably among salmon species (Table 9.4.2).

The list in the Table 9.4.2 illustrates: (i) a reasonably small $\boldsymbol{m}$ in natural populations in comparison to hatchery populations. These data also show smaller $\boldsymbol{m}$ for sockeye than for pink salmon, and again they support the previous discussion on population structure in salmon species.

There are few data on the fitness estimates in wild populations. The fitness values are listed for the only two salmon species, and illustrate: (i) a reasonably small $\boldsymbol{W}_i$ differences among genotypes and (ii) small absolute scores of $\boldsymbol{W}_i$ (Table 9.4.3).

9.5 TRAINING COURSE, #9

9.1 Difference between F_{st} and F'_{st}. Distance measures that are most useful for population analysis: D_m, and R_s indices.

9.2 Joint effect of a migration and selection. Software "POPULUS" -> Evolutionary Simulations -> Differentiation Models -> Selection, Gene Flow and Clines. Test Two Different Modes with Two Numerical Sets of Parameters.

9.3 Selection in a Single Population. Software "POPULUS" -> Evolutionary Simulations -> Selection -> Autosomal Selection. Use defaults first and then change parameters.

9.4 Learn the few terms below.

Island Model of population structure is one of the models of a subdivided population and is most popular among researchers.

$\mathbf{F}_{\mathbf{st}}$ (and = $\mathbf{G}_{\mathbf{st}}$) is the standardized allele frequency variance, and is the most important measure of genetic differentiation of the populations within species. $\mathbf{G}_{\mathbf{st}}$ ($\mathbf{F}_{\mathbf{st}}$) is used as a tool for hierarchical population analysis. $\mathbf{G}_{\mathbf{st}}$ is also called the coefficient of gene differentiation.

10

NATURAL HYBRIDIZATION AND GENETIC INTROGRESSION IN WILD SPECIES

MAIN GOALS

10.1 What are Hybrids and Hybridization?
10.2 Methods of Detecting Hybridization
10.3 Genetic Changes in Hybrid Populations
10.4 Empirical Investigations of Hybrid Zones
10.5 TRAINING COURSE, #10

SUMMARY

1. The introduction of biochemical genetic and molecular genetic methods in **hybridization research** provided a real opportunity to investigate the phenomenon of hybridization in nature, including estimation of **hybrid frequencies** in populations and species.
2. **Hybridization** is a dynamic process. It includes mixing and reorganization of genomes at the individual level, as well as mixing and reorganization of gene pools at the population level. Hybridization is a process that generates genetic disequilibrium from former genetic equilibrium by mixing and interbreeding of two previously isolated gene pools. This process leads to a condition preventing management of fish and shellfish stocks as discrete population units or species.
3. **Genetic mixing** and the consequent creation of long-term **genetic disequilibrium** may have different impacts, both positive and negative. Positive impacts include the increase of genetic variance and building a special role for the **hybrid zone** (tension zone), which serves as a "field of battle" for species integrity. **Hybrids**, if advantageous, may be directly useful for aquaculture propagation or as a

source for breeding programs. Numerous negative impacts are created by (i) disequilibria, decreasing fitness in hybrid deme (taxa) and (ii) introgression of deleterious (or just non-adapted) genes into native populations or species, and as a result causing their degradation and substitution by other species (which as a rule are worse, both for humans and ecosystems).

10.1 WHAT ARE HYBRIDS AND HYBRIDIZATION?

Let us first answer the questions in the title of this section. A ***hybrid*** *is a mixture, an offspring of the cross(es) between genetically different organisms. Hybrids may be treated as well as individuals having a mixed ancestry.* In a certain sense, a heterozygote at one or more loci is a hybrid individual. So, the cross P_1: $A_1A_1\ B_2B_2 \times A_2A_2\ B_3B_3$ gives the hybrid $F_1 = F_H$: $A_1A_2\ B_2B_3$. ***Hybridization*** *is a process through which hybrids occur.* We should recognize the difference between the simple intra-deme crosses and crosses between different lines, demes and species. True hybrids are normally thought to be the offspring of more distant parents. The meaning of distant here is quite conditional, depending on the particular organism and its normal system of breeding. Beyond F_1 hybrids, other types of hybrids may develop: $F_1 \times F_1 = F_2$, $F_1 \times P_1 = F_b$ and so on. Hybridization may be either natural or artificial. In this chapter we will deal mostly with the former. Hybrids need not be exactly intermediate to the parental types, but depending on the complexity of the cross can be closer to one of them (Fig. 10.1.1). We will look at more details on this point later. It is accepted normally that fitness (***W***) is lower in hybrids in comparison to parental forms (Fig. 10.1.1), but this rule does not always hold. Sometimes the reverse is true: hybrid vigor can occur. However, these cases are beyond our theme, and belong largely to the artificial crosses between inbred

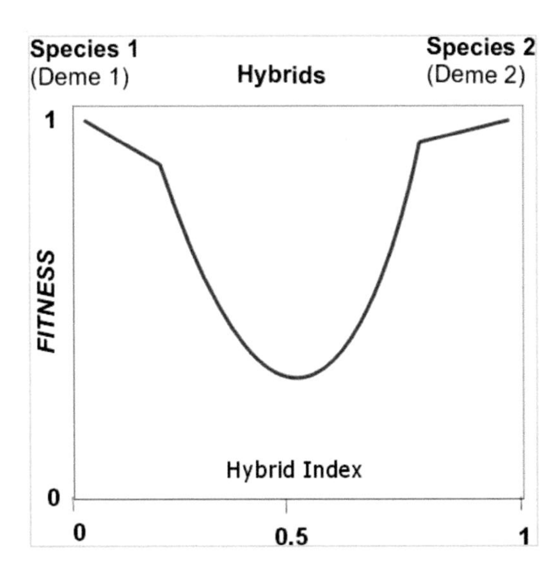

Fig. 10.1.1. A diagram showing the hybrids' origin and a gene pool change as a mixture between two parental species (demes)—1 and 2.

lines or brood stock kinds. The ability and propensity of taxonomically distinct fishes to interbreed and produce viable hybrid offspring are now well established. Schwartz (1972; 1981) compiled nearly four thousand references dealing with the natural and artificial hybridization of fishes. Much earlier Hubbs (1955) reviewed the literature on natural hybridization among North American fishes.

Natural hybridization is believed to be more common in fishes than in other groups of vertebrates. Common hybridization applies also to shellfish. One of the most important causes of more frequent hybridization in these groups may be weak chromosomal sex determination in comparison with more advanced organisms, such as mammals. In many fish species sex, determination depends upon many factors and sex chromosomes may be either absent or their effects negligible (Kirpichnikov 1979). Several other peculiarities of the biology of fish and shellfish species may account for this distinction, e.g., external fertilization, weak ethological isolating mechanisms, unequal abundance of the two parental species, competition for limited spawning habitat and susceptibility to secondary contact between recently evolved forms (Campton 1987). These peculiarities are affected to varying levels by local habitat.

Natural and man-induced changes in environmental conditions are often cited as causes of hybridization in fishes. For example, hybridization is relatively common among temperate freshwater fishes in areas where geological and climatic events since the Pleistocene have drastically altered aquatic environments. However, hybridization appears to be rare in marine and tropical fishes that inhabit more stable environments. Man-caused habitat changes in North America and other regions have also been correlated with hybridization between both previously allopatric and naturally sympatric pairs of species (Hubbs et al. 1953; Nelson 1966; 1973; Stevenson and Buchanon 1973). For salmon, such examples were summarized by Simon and Nobble (1968) and Altukhov and Salmenkova (1987).

Before we step into the section devoted to the methods of analysis, we need to define terms precisely. First, we will refer to gene flow as the process marked by selectively neutral (or nearly neutral) alleles and accept that **reproductive isolation** between biological species means absence of any gene flow (in the sense that F_1, and especially F_2 or F_b are infertile and unviable or low in fertility and viability); this is the strict BSC requirement. Second, the **hybrid zone** is an area in nature where hybrids occur between supposed two or more parental forms. Usually, in such an area, a cline arises. Third, **cline** is a gradual or abrupt allele frequency change in a mode: Species 1 (Deme 1) → Hybrids → Species 2 (Deme 2) and is maintained by a balance between dispersal and selection against hybrids (Modified from Barton and Hewitt 1985).

10.2 METHODS OF DETECTING HYBRIDIZATION

Detection of natural hybridization between fishes, as well as between other wild nature species, is often complicated due to several uncertainties. For example, two closely related taxa, or conspecific populations, may inhabit the same area without interbreeding yet hybridization may be suggested because of phenotypic overlap. Conversely, the hybridization may occur, but the hybrids themselves may never breed for one or several reasons, e.g., they could be infertile. If hybrids backcross with one or both parental species, introgression may occur or a hybrid population may originate. We'll consider examples later.

If the presence of hybridization has been established, one may want to distinguish individuals of mixed ancestry from those of the two parental taxa. However, as will be pointed out later, it is often a complicated task to distinguish true hybrids if hybridization has proceeded past the F_1. Below we will consider data on detecting hybridization between two taxa or populations obtained by different techniques.

Morphological Techniques

In such studies, meristic and metric morphological traits are counted and measured on the supposed hybrid individuals as well as on individuals representing the two potential parental taxa or populations. There are specific procedures for comparing morphological traits in general and in the case of hybridization research in particular (Bailey 1967; Hubbs and Lagler 1970). The fish are considered interspecific hybrids if their measured traits are intermediate on the average to the values for the two parental groups.

A problem that occurs is because morphological traits used to distinguish different taxa and especially populations are not always intermediate in the hybrids (Simon and Noble 1968; Ross and Cavender 1981; Leary et al. 1983). Hybrids frequently express a mixture of morphological traits, in which they may be similar to the parental forms in different combinations. In some cases the hybrid may be morphologically intermediate for nearly all traits, but for a specific character it may be identical to one of the two parental forms. A few pictures of trout and char help to show the difficulties with discrimination of hybrids by morphology (Fig. 10.2.1).

Specifically for this technique a statistic, called the **hybrid index**, can be used to measure the average morphological similarity of an individual fish to each of the two populations or taxa. The index (I) is calculated separately for each trait as $I = 100* [(u–X)/(Y–X)]$, where u is the value of the trait for the individual being evaluated, and X and Y are the mean values of the trait for population/taxa X and Y. An individual fish with a value for the trait equal to X or Y will have an index value equal to 0 or 100, respectively. An index value of 50 indicates the exact intermediary for the trait analysed.

There is a lot of weakness in the I index and in one-dimensional approaches in general. Instead multi-variate statistical methods (Neff and Smith 1979; Afifi and Azen 1982) are widely applied. Their major advance is the possibility to perform unambiguous classification of an individual to one or another group. A common approach is to derive the discriminant function using known individuals from the two parental species, and then to calculate the discriminant function scores for each of the supposed hybrid individuals.

Discriminant function analysis (DA) is one of multivariate statistical methods that are frequently used to analyze morphological data in studies of natural hybridization. DA derives weighted linear functions of the measured traits so that two or more groups of individuals are maximally separated or discriminated in multivariate space. Factor Analysis (FA) in the mode of principal components or principal axes is a second

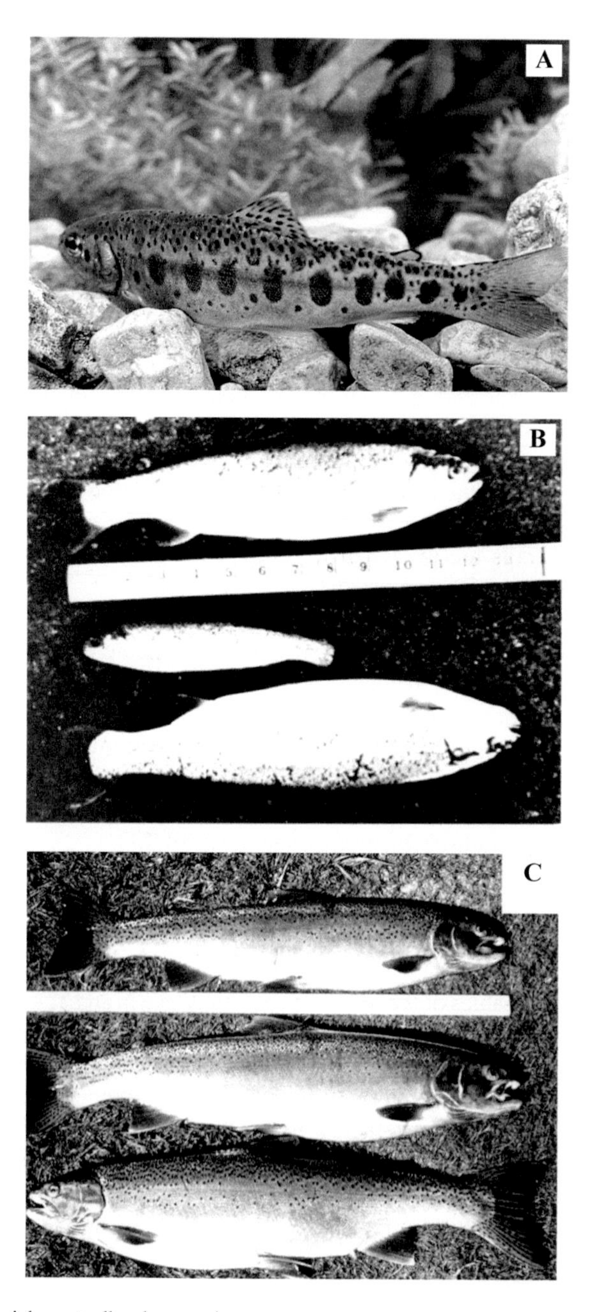

Fig. 10.2.1. The rainbow-steelhead comparison.
(A) On the colour slide is rainbow trout *Oncorhynchus mykiss* in a river (Image from FishBase. Entered: Lorenzoni, Massimo, 19.12.98). (B) The fish at the bottom is a rainbow trout *O. mykiss* from a cultivated strain; centre, a steelhead *O. mykiss* (*S. gairdneri*). At the top is a rainbow-steelhead hybrid. It is faster than the steelhead in growth, but with the steelhead's disposition to seek salt water. (C) Specimens of rainbow-steelhead hybrids that have returned from the sea (B and C from Hines 1976, Figs. 25 and 29; Neal O. Hines papers and items, 1942–1986 are open to all users).

multivariate statistical method that is often used in morphometric studies. Unlike DA, FA does not require the *a priori* identification of a group of individuals. By projecting these individuals (with the least-squares or other technique) from their positions in the n-dimensional score space, several principal component or axes can be calculated and visualized (Afifi and Azen 1982). A set of such projections for individuals collected from areas with and without hybrids can objectively summarize the morphological evidence for natural hybridization without making any *a priori* assumptions. Dowling and Moore (1984), in a study of natural hybridization between two cyprinid fishes, performed a similar analysis by plotting histograms of the first principal component scores for individual fish from each of the several locations (Fig. 10.2.2).

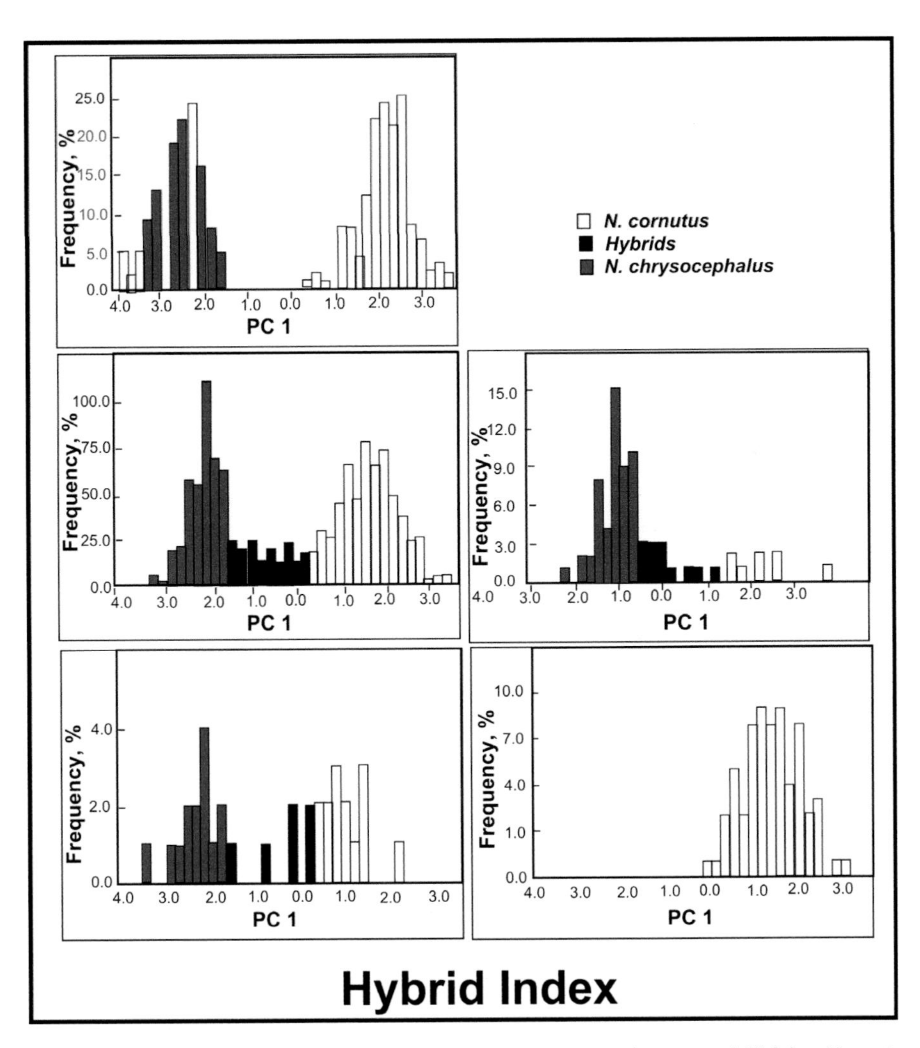

Fig. 10.2.2. Frequency distributions of first principal component scores for two cyprinid fishes, *Notropis cornutus* and *N. chrysocephalus*, and their supposed hybrids from each of the five geographic locations (from Dowling and Moore 1984; adopted from Campton 1987, Fig. 7.1).

Detecting natural hybridization using morphological characters suffers from many problems. In general, morphological data can provide only preliminary evidence for natural hybridization or introgression because hybrids are usually assumed *a priori* to be morphologically intermediate to the parental forms. However, frequently this does not hold.

Also, even under multivariate approaches the extent of hybridization or introgression cannot be determined from morphological data because of intergradation of variation and because F_1 hybrids may not be individually distinguishable from F_2 or backcrosses. As previously mentioned, F_1 hybrids often express a mosaic of morphological characters that separately resemble one or the other parental taxa. Detection of introgression may also be extremely difficult with such traits because introgressed populations could be morphologically identical to one of the parental taxa (Greenfield and Greenfield 1972; Busack and Gall 1981).

Because of these problems, we need more direct genetic methods for detecting natural hybridization.

Karyotyping

Only a few studies have used karyotypic methods or chromosomal comparisons to investigate natural hybridization in fishes (Setzer 1970; Greenfield and Greenfield 1972; Busack et al. 1980). Most attempts to stain differentially and thereby identify specific chromosomes in fishes by banding patterns have not been very successful. Some notable exceptions are reports on Q banding and staining of nuclear organizers in chromosomes of salmonid fishes (Phillips and Zajicek 1982; Delaney and Bloom 1984; Phillips et al. 1985; Frolov 2000). However, karyotypic analyses with new staining techniques can provide in future an objective and independent method for verifying the assumptions associated with the morphological detection of natural fish hybrids (Greenfield et al. 1973; Vasiliev 1985), and may supplement other genetic techniques.

Protein Gene Markers

Detecting natural hybridization and introgression by analysis of protein variability is relatively straightforward when the two parental taxa or forms are fixed for different alleles at two or more loci. Individuals of the two parental taxa will each be homozygous for different alleles, whereas F_1 hybrids will be heterozygous for those alleles at all diagnostic loci (Leary et al. 1983; Whitmore 1983). If hybridization has proceeded past the F_1 stage however, hybrid descendants will express a broad mixture of recombinant types, including the two parental types (see Avise and Van Den Avyle 1984). Consequently, an individual with a composite genotype that is identical to the genotype for one of the parental species could be a hybrid descendant if F_1 hybrids have reproduced among themselves or backcrossed with the parental taxa. The presence of recombinant genotypes can therefore be accepted as evidence of hybrid fertility and second-generation hybridization (F_2 or F_b); this holds true if there is the complete fixation of different alleles between taxa, giving unambiguous interpretation. If the two taxa are fixed for different alleles at only one locus, it is not always possible to distinguish F_1 or F_2 hybrids, such as when both the F_1 and the F_2 occur frequently. All individuals in such a population will express either the F_1 hybrid phenotype or one of the parents. However, when the F_1 frequency is low, then the probability of finding F_2 instead F_1 is relatively small even in single locus case (the frequency of F_2 is approximately one-tenth that of the F_1). The following example supports this point. Let A_1 and A_2 be alleles of parental forms (P_1 and P_2) that are fixed, and their frequencies along with F_1 frequency in a sample n = 100 from the population equal to: $P_1 = 0.01$, $P_2 = 0.99$, and F = 0.01. Gametic frequencies in hybrids equal in this case $p_{A1} = p_{A2} = 0.005$. Correspondingly, frequencies of heterozygotes in matings, under the assumption of random mating, will be: $F_1 \times F_1 = 0.005*0.005*2 = 0.00004$, $F_1 \times P_1 = 0.005 \times 0.99 = 0.005$ and $F_1 \times P_2 = 0.005 \times 0.01 = 0.00005$. Thus, the frequency of most frequent heterozygotes in the F_2 is approximately only one-tenth that of the F_1. In the situation when F_1s are frequent, the presence of heterozygous individuals in a mixture of both parental species can provide the only evidence that the two species have interbred; the extent of hybridization will be unclear. Many cases of natural hybridization have been documented on the basis of only a few heterozygous individuals (Soloman and Child 1978; Beland et al. 1981). That is why hybridization must be investigated quantitatively. One of the main questions is whether the proportion of fish with intermediate genotypic combinations is significantly greater than one could expect from the random mating within one or both taxa.

This question was considered on anadromous steelhead trout (*Saimo gairdneri*) and coastal cutthroat trout (*S. clarki clarki*) in the wild (Campton and Utter 1985). These two taxa have different common alleles at four electrophoretically detectable loci, but one or both species are polymorphic for both alleles at each locus. The investigation had revealed a surprisingly high number of individuals in two streams with intermediate genotypic combinations (Campton and Utter 1985). To quantify the likelihood of hybridization, a hybrid index (I_H) was suggested, measuring the relative probability that the composite genotype for each individual arose by random mating within each of the two taxa (Campton and Utter 1985).

$$\boldsymbol{I_H = 1.0 - [\log_{10}(p_x)/(\log_{10}(p_x) + \log_{10}(p_y))]}, \quad \textbf{(10.1)}$$

where $p_x = \Pi_{i=1}^{L} k_i \Pi_{j=1}^{Ai} (X_{ij})^{m_{ij}}$ and $p_y = \Pi_{i=1}^{L} k_i \Pi_{j=1}^{Ai} (Y_{ij})^{m_{ij}}$.

Here X_{ij} and Y_{ij} are the average frequencies of the *j-th* allele at the *i-th* locus for species (population) X and Y, respectively; m_{ij} is the number of alleles of the *j-th* type observed at the *i-th* locus for the individual being evaluated; A_i, is the total number of known alleles at the *i-th* locus for the two species combined; k_i, is the binomial sampling coefficient (e.g., $k_i = 2$ for Aa, $k_i = 1$ for AA or aa) associated with the genotype of the individual at the *i-th* locus; and L is the number of diagnostic loci used to distinguish the two species. The quantities p_x and p_y are the conditional probabilities that the composite genotype at all loci for an individual could have arisen by random mating within species X and species Y, respectively (it is assumed that sampled allele frequencies equal the averages for the two species and linkage equilibrium among all loci has been reached). This index can vary within the range 0–1, and will be close to one of these two values when individuals have a very high relative probability of belonging to species X or Y, respectively.

In the study of hybridization between steelhead trout and cutthroat trout (Campton and Utter 1985), the histograms of hybrid index scores clearly showed the presence of both species in the area of sympatry and the existence of a third, intermediate group at the separate location (Fig. 10.2.3). These authors interpreted this genotypically intermediate group of fish as natural hybrids.

The power to detect natural hybridization will increase as the number of distinguishing loci increases (Campton 1987). This is easily demonstrated by the simple example (Table 10.2.1). If the frequencies of alternate alleles (e.g., A_1 and A_2) at a locus are 0.8 and 0.2 in one population and 0.2 and 0.8 in the second population, the probability of an individual being heterozygous in each population is simply $2pq = 0.32$. If the two populations interbreed, then the expected frequency of heterozygotes among the F_1 hybrids is 0.68, which is only 2.13 times as great as the expected frequency of heterozygotes in each of the parental populations. As the number of distinguishing loci increases, however, this ratio also increases. For example, with

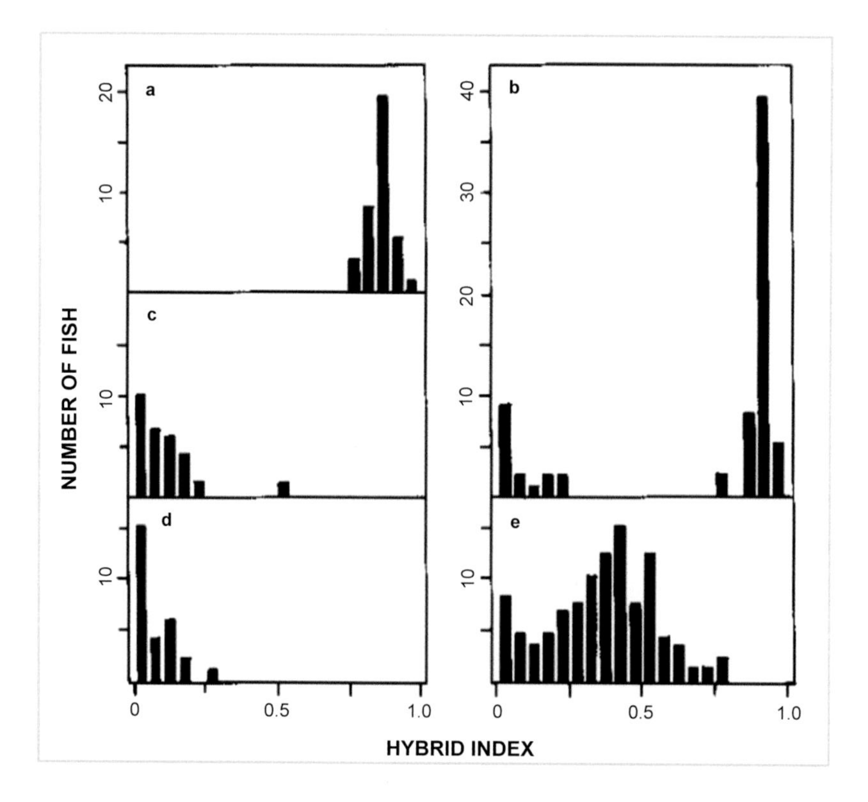

Fig. 10.2.3. Hybrid index scores for two sympatric species of trout, steelhead trout (anadromous *Saimo gairdneri*) and coastal cutthroat trout (*Saimo clarki clarki*), and their supposed hybrids from three sample sites within a small stream.

Individuals with values of the index close to 1.0 or 0.0 expressed composite electrophoretic phenotypes that have a high relative probability of occurring in *S. gairdneri* or *S. clarki clarki*, respectively: (a) Site 1, age 0 + fish (*gairdneri* only); (b) site 2, age 0+ fish (*clarki* and *gairdneri*, no hybrids); (c) site 2, age 1+ fish (*clarki* + 1 unknown or hybrid fish); (d) site 3, age 1+ fish (*clarki* only); (e) site 3, age 0+ fish (*clarki* + a large number of suspected hybrids) (from Campton and Utter 1985 with permission).

Table 10.2.1. Probability of an individual being heterozygous simultaneously at one to six distinguishing loci in each of two parental populations (P) and among F_1 hybrids (P_h), where the frequencies of alternate alleles at each locus are 0.8 and 0.2 in one population and 0.2 and 0.8 in the other. The probabilities were calculated as $P = [2(0.8)(0.2)]^L$ for the parental populations and as $Ph = [(0.8)(0.8) + (0.2)(0.2)]^L$ for the F1 hybrids, where L = number of loci. These expressions assume gametic equilibrium between all loci and Hardy-Weinberg genotypic proportions in each parental population (from Campton 1987).

Population	**Number of distinguishing loci**					
	1	**2**	**3**	**4**	**5**	**6**
Parental (P)	0.3200	0.1024	0.0328	0.0105	0.0034	0.0011
F_1 hybrid (P_h)	0.6800	0.4624	0.3144	0.2138	0.1454	0.0989
P_h /P	2.13	4.52	9.60	20.4	43.3	92.1

four distinguishing loci, the frequency of four-locus heterozygotes among F_1 hybrids is more than 20 times their expected frequency in each of the parental populations (Table 10.2.1). This ratio increases to 92.1 with six distinguishing loci.

However, even with four distinguishing loci, the expected frequency of four-locus heterozygotes within each of the parental populations in Table 10.2.1 is greater than 1%. Consequently, as previously pointed out, individuals cannot be unambiguously classified as hybrids if the two parental populations, or species, are not fixed for alternate alleles at one or more loci.

An advantage of the index method is that values can be calculated for each individual fish even when data are missing at one or more loci. The major disadvantage of the method is that it requires the *a priori* establishment of allele frequency profiles for each of the two parental taxa or populations. To overcome possible difficulties, we could evaluate the likelihood of hybridization by a series of two-dimensional plots of principal components.

Mitochondrial and Nuclear DNA as Genotype Markers

The use of restriction endonucleases for determining the nucleotide site variation of mitochondrial DNA (mtDNA) or nuclear DNA (nDNA) provides other sets of molecular markers and the possibility to investigate polymorphism in natural populations (see Chapter 13). The mtDNA is maternally inherited in most vertebrate and many invertebrate taxa, which provides an important tool in hybridization research. The maternal species can be identified in F_1 hybrids if the two parental populations or taxa are characterized by different mtDNAs.

Mitochondrial DNA genotypes, when used with nDNA markers or enzyme genotypes, can detect the introgression of mtDNA from one species into the nuclear background of another species, provided that hybrids and their offspring are fertile. Such introgressive hybridization requires the successive back-crossing of female hybrid descendants with males of the paternal species or other taxa. This introgression will be independent of the segregation and recombination events occurring within the nuclear genome if natural selection is not acting to maintain a nuclear-cytoplasmic compatibility (Takahata and Slatkin 1984; Nei 1987). Numerous examples of introgression of mtDNA (see below) lead us to think that such selection, if present, is not frequent in nature. Thus, hybridization events can be detected when individuals of a particular species, as identified by other methods, possess the alien mtDNA type similar to closely related species (taxa). Such interspecific transfers of mtDNA have been reported for congeneric species of *Drosophila, Mus* and *Rana* (Powell 1983; Ferris et al. 1983; Spolsky and Uzzell 1984; Yonekawa et al. 1981).

Nuclear DNA also provides strong evidence of hybrids in the marine realm (Fig. 10.2.4).

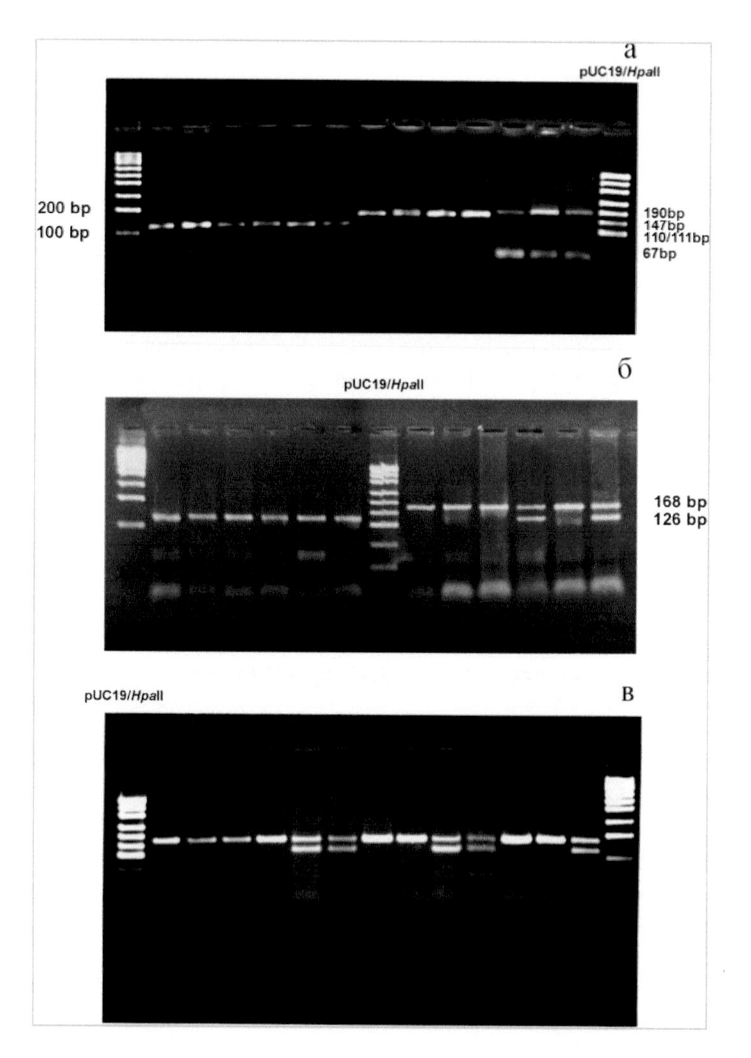

Fig. 10.2.4. Results of analysis of genotypes of three species belonging to the genus *Mytilus* by means of PCR of the nonrepetitive region of the adhesive protein of bissus.
From left to right: Ml, a 100-kb standard marker, M2, pUC19 DNA digested by the Hpall restriction endonuclease, (a): 1–3, *M. galloprovincialis* (France); 4–6, *M. galloprovincialis* (Ostrovok Fal'shivyi, Peter the Great Bay); 7–10, *M. trossulus* (Ostrovok Fal'shivyi); 11–13, *M. edulis* (Helgoland Island, Germany), (b): 1–6, *M. galloprovincialis* (Ostrovok Fal'shivyi); 7–12, *M. trossulus* Priboinaya Bay, the Vostok Bay (lanes 10 and 12, hybrid forms), (c): 1–6, *M. trossulus* (Priboinaya Bay); 7–13, the collector of the Vostok Biological Station (lanes 5, 6, 9, 10 and 13, hybrid forms) (from Skurihina and Kartavtsev et al. 2001).

The use of mtDNA for investigating natural hybridization in fishes began in the 1980s (Avise and Saunders 1984; Avise et al. 1984). Avise and Saunders (1984) used mtDNA in combination with allozyme markers to investigate the hybridization among nine species of sunfish (genus *Lepomis*) that inhabited two locations in the southeastern United States.

The major outcomes of the research can be summarized in four points:

1. Hybridization occurred at a relatively low frequency, but involved five of the nine species examined.
2. No mtDNA or allozyme evidence for introgression between species of *Lepomis* was detected. All hybrids appeared to be only F_1 offspring.
3. Every detected hybrid represented a cross between the most abundant and most rare species.
4. In six of seven possible hybrid crosses, the female parent was from the rare species, as determined by mtDNA genotype. This result was attributed to intense competition among males for mating partners and the general promiscuity of females.

The examples of mtDNA analysis may be easily enumerated. We will consider some of them in Section 10.4.

10.3 GENETIC CHANGES IN HYBRID POPULATIONS

Detecting natural hybridization requires the sampling of individuals from wild populations and their further experimental and statistical analyses. Until sampling and these analyses have been done, the researcher does not know, as a rule, whether hybridization has occurred. The simplest situation is a single randomly mating population, without past hybridization or mixture of two taxa, and with a few F_1 hybrids. Other and perhaps more complex situations are also possible in nature. If hybridization and introgression are detected, then it is necessary to evaluate the direction of crosses and how frequently these events take place. In this section, we will meet some theoretical models of population genetics that can be useful to interpret cases of hybridization.

Deviations from Hardy-Weinberg Proportions

The first phase of the analysis of allele frequency data is to test the null-hypothesis of one population (see Chapter 9). A single, randomly mating population is expected to fit the H-W proportions at each locus. On the other hand, a mixture of two, non-interbreeding populations with different allele frequencies is expected to yield a deficiency of heterozygotes compared with expected proportions. Thus, detection of the Wahlund effect is the main tool to detect hybridization in a mono-locus case (see Chapter 9). There are few specific formulae for the case of two subpopulations that form hybrids (Campton 1987).

However, there are complications in it. In particular, as was stressed in the two previous chapters, dynamics of allele frequencies usually have a complex nature, with four main factors acting, including natural selection. Consequently, an apparent Wahlund effect may be caused not only by subdivision of two (or more) reproductive units, but also possibly by selection. For example, that might be the case with the hybridization between two nominal species of *Notropis* (Cyprinidae) described above (Dowling and Moore 1984). In that research, a consistent deficiency of heterozygotes was observed. The authors (Dowling and Moore 1984) argue that these observations warrant recognition of *N. cornutus* and *N. chrysocephalus* as distinct species and not as subspecies, because "hybridization between subspecies should result in genotypic frequencies in H-W equilibrium, while hybridization between species should result in a marked deficiency of heterozygotes, either because of assortative mating (premating isolation) or selection against hybrids (post-mating isolation)". In a subsequent study, the authors were able to attribute the deficiency in heterozygotes to natural selection against hybrids and not necessarily to assortative mating (Dowling and Moore 1985). We know (Chapters 8–9) that not only subspecies, but also different demes give deficit of heterozygotes due to the Wahlund effect, so some further interpreting is required here.

Linkage Disequilibrium

Linkage disequilibrium (or gametic disequilibrium) refers to the nonrandom association of alleles (gametes) between different loci of a diploid genome. A vast literature describes the conditions necessary to generate and maintain gametic disequilibrium (Hedrick et al. 1978; Zhivotovsky 1984). The subdivision of a population into several subpopulations, as well as the interbreeding of two or more previously isolated taxa or populations, are ways in which gametic disequilibrium can be generated (Lewontin 1974; Ohta 1982; Zhivotovsky 1984). A theoretical assessment of gametic disequilibrium is therefore essential in understanding the dynamics of hybridization and introgression. A population is in gametic equilibrium if the frequency of gametes carrying alleles from two (or more) loci is equal to the product of the respective allele frequencies. Let us consider two diallelic loci. If p_1 and p_2 are the frequencies of two alleles (A_1 and A_2) at one locus and q_1 and q_2 are the frequencies of two alleles (B_1 and B_2) at a second locus, then gametic equilibrium is achieved in the population if $X_{11} = p_1q_1$, $X_{12} = p_1q_2$, $X_{21} = p_2q_1$ and $X_{22} = p_2q_2$, where X_{ij} is the frequency of gametes in the population carrying alleles A_i and B_j. If alleles are not randomly associated between loci (i.e., $X_{ij} \neq p_iq_j$), then one can show that

$$X_{11} = p_1q_1 + D \; X_{22} = p_1q_2 - D \; X_{21} = p_2q_1 - D \; X_{22} = p_2q_2 + D. \quad (10.2)$$

Here D measures the amount of interlocus association between alleles within gametes (Hedrick 1983). D is termed the gametic (linkage) disequilibrium coefficient, and is defined as $D = X_{11} - p_1q_1$ for a two-locus system (Lewontin and Kojima 1960). By substituting $X_{11} + X_{12}$ for p_1 and $X_{11} + X_{21}$ for q_1, in the preceding definition of D, it may be derived

$$D = (X_{11} X_{22}) - (X_{12} X_{21}). \quad \textbf{(10.3)}$$

Thus, the amount of linkage disequilibrium between two loci is simply the difference between the products of the gametic frequencies in cis-position (i.e., frequencies of A_1B_1 and A_2B_2) and the products of the gametic frequencies in trans-position (i.e., frequencies of A_1B_2 and A_2B_1). In other words, D measures deviation of non-allelic gene frequencies from random expectations. Gametic disequilibria can also be defined in terms of more loci, but these expressions are more complicated (Bennett 1954; Crow and Kimura 1970, p. 50), and are far from the scope of the chapter. The range of possible values for D depends upon the population allele frequencies at the two loci under consideration. From expression (10.3) we may obtain that D reach a maximum value of 0.25 if $X_{11} = X_{22} = 0.5$ ($X_{12} = X_{21} = 0$) and a minimum value of $D = -0.25$, if $X_{12} = X_{21} = 0.5$ ($X_{11} = X_{22} = 0$). These maximum and minimum values of D are achieved only if the frequencies of both alleles at each locus are equal, i.e., when $p_1 = q_1 = p_2 = q_2 = 0.5$. That is why the absolute value of D may be difficult to interpret, especially when several D values are being compared. A more appropriate statistic has been proposed: $D' = D/Dmax$, which may be used as a measure of gametic disequilibrium (Lewontin 1964; 1974). Here $Dmax$ is the maximum, absolute value of D for a particular set of allele frequencies (Hedrick 1983). The value of D' ranges from –1.0 to +1.0, and it provides a relative measure of the nonrandom association among alleles between loci. Other measures of disequilibrium have also been defined (Weir 1979). A good measure of gametic disequilibrium is R, the correlation coefficient in which D is normalized by covariation between alleles at different loci:

$$R = \sqrt{[D/(p_1 q_1 * p_2 q_2)]}. \quad \textbf{(10.4)}$$

In assigning the value 1 to A_1 and B_1 and 0 to A_2 and B_2, R can also achieve values in the range –1.0 to +1.0, but this happens only when allele frequencies at the two loci are equal ($p_1 = p_2$ and $q_1 = q_2$). More convenient for calculations is the variant of D estimate proposed by Weir (1979), which has the advantage of using observed genotype frequencies and also has criteria of statistical significance of disequilibrium (Zhivotovsky 1991, p. 120).

Contrary to the case for a single locus, the equilibrium for two loci is not established in a single generation of random mating. Instead, it is reached gradually in time, so that after t generations,

$$D_t = (1 - r)^t * D_0. \quad \textbf{(10.5)}$$

Here r is the coefficient of recombination and D_0 is the disequilibrium in generation zero.

For tightly linked loci (i.e., $r << 1/2$), the disequilibrium can persist almost indefinitely, especially in small populations where recombination will be counterbalanced by genetic drift.

In the above sections the theoretical aspects of linkage disequilibrium were considered. However, the evaluation of this disequilibrium in real natural populations will be based on the estimation from samples of individuals (not gametes). So far, the gametic genotypes usually cannot be observed directly, but must be inferred from the diploid genotypes. For example, $A_1A_1B_1B_2$ individuals can be interpreted as representing the union of A_1B_1 and A_1B_2 gametes. In the same way, gametic genotypes can be inferred for all individuals that are homozygous at one or both loci. In contrast, gametic genotypes cannot be inferred from double heterozygotes because the gametic phase for these individuals is usually not known. Normally, distinguishing between A_1B_1/A_2B_2 and A_1B_2/A_2B_1 is possible only in planned crosses. That is why gametic disequilibrium cannot be exactly calculated for natural populations of diploid organisms. Nevertheless, several methods have been proposed for the estimation of gametic disequilibrium in natural populations when double heterozygotes of both types cannot be distinguished. The maximum likelihood method of estimation (MLE) (Hill 1974) and Burrow's composite method (Weir 1979) are most familiar. In the second method D is portioned in two components ($D = DW + DB$) and reasonably good estimates are achieved in a step-by-step approach. This technique was originally developed earlier by Cockerham and Weir (1977).

An example that illustrates the use of the MLE and the composite method of linkage disequilibrium is given in Table 10.3.1.

Table 10.3.1. Estimates of linkage disequilibrium using Hill's (1974) maximum likelihood method and Burrows's composite measure (Weir 1979) for suspected hybrids of steelhead trout (*Saimo gairdneri*) and coastal cutthroat trout (*S. clarki clarki*). The statistic R is the estimated correlation between alleles, and X^2 is chi-square the goodness of fit statistic ($X^2 = NR^2$) with d.f. = 1 for testing $Ho: D = 0$ (MLE) or $D + DB = 0$ (Weir 1979). The data and estimates are for the same fish as shown in Fig. 9.2.2e in Campton and Utter (1985).

Loci	Maximum Likelihood			Composite Method		
	D	R	X^2	$D + D_B$	R	X^2
GLD-1, ME-4	0.105	0.427	16.79***	0.026	0.239	5.26*
GLD-1, SDH-1	0.037	0.161	2.29	0.033	0.229	4.69*
ME-4, SDH-1	0.049	0.218	3.93*	0.062	0.248	5.09*

*$P < 0.05$, ***$P < 0.001$

Estimating Admixture Proportions

Hybrid populations that have progressed past the F_1 stage with negligible decrease in fitness (viability or fertility) are often described as genetic or population admixtures. Admixture refers to the production of new genotypic combinations through recombination, and it was first used in this context to describe the interbreeding of human races following secondary contact in the western hemisphere (Glass and Li 1953; Glass 1955). Many human populations in North and South America actually represent genetic admixtures of two or more previously isolated human races (Cavalli-Sforza and Bodmer 1971). If hybridization has been detected, and if it is possible to conclude that a single, randomly mating population has resulted (Campton and Johnston 1985), then the necessity rises to estimate the proportional contributions of the parental taxa or populations to the genetic admixture. If hybridization has occurred for several generations in one direction, such that the hybrid population is formed in each generation from a fraction m of one parental population (say population 2) and a fraction $1 - m$ of the hybrid population from the previous generation, then the expected frequency of the allele in the hybrid (admixture) population at generation $t[p_h(t)]$ is given by (Campton 1987):

$$P_h(t) = (1 - m)^t p_1 + [(1-(1 - m)^t] p_2. \quad \textbf{(10.6)}$$

If t is known, one can estimate m, the mean admixture rate per generation, as a ratio:

$$m = 1 - [(P_h(t) - p_2)/(p_1 - p_2)]^{1/t}. \quad \textbf{(10.7)}$$

This ratio can be viewed as a simple one-way migration model, where $P_h(0) = p_1$. The principal scheme of allele frequency change, along with a numerical example, was explained in Chapters 8–9. Equation (10.7) originated from Glass and Li (1953) to describe the dynamics of racial intermixture in the United States, where the case of the introgression of genes from Caucasians to the American Black population predominates (Glass 1955; Roberts 1955; Saldanha 1957). Equation (10.7) is also relevant to hybridization problems in animals including marine organisms, because of the common practice of creating hatchery stocks (for many Pacific salmon, for instance), which are derived from two or more native local populations and later repeatedly interbred with another population. Continuous stocking of an exotic species into the native habitat of a second species raises the same problem at the interspecific level if hybrids are fertile (Busack and Gall 1981).

Equation (10.7) (and 14 in Campton 1987) provides estimates of admixture proportions for two populations based on data for only a single allele or locus. Data for many loci are available, however, and more than two populations could be contributing to the genetic admixture. Elston (1971) described the formal maximum likelihood estimators and least squares of admixture proportions for these generalized situations. The details of these procedures are beyond the scope of this chapter, but interested readers are welcome to consult Elston's (1971) presentation (Fig. 10.3.1).

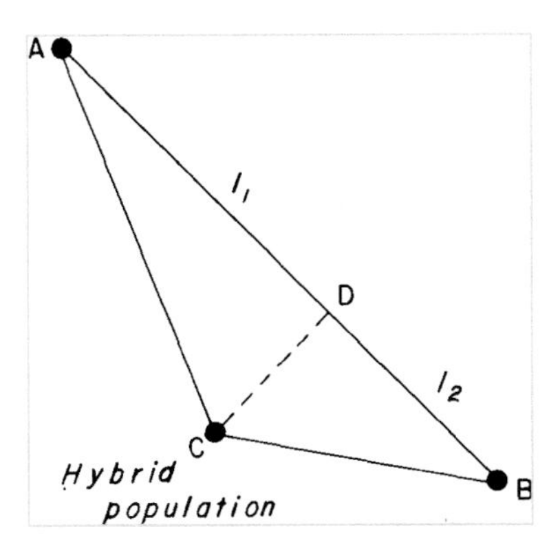

Fig. 10.3.1. Graphical representation of the relative positions of two hypothetical parental populations and a hybrid population in the n-dimensional sample space of estimated allele frequencies.
The hybrid population is not collinear with parental populations because of founder effects, genetic drift, sampling error, and possibly natural selection. As a result, estimates of admixture proportions obtained by equation 14 (Campton 1987) will vary among loci. The least squares estimate projects the hybrid population to point *D*. If I_1 and I_2 are the lengths of line segments AD and BD, respectively, then the least squares estimates of I_1 and I_2 (the admixture proportions of populations 1 and 2, respectively) are $I_2/(I_1 + I_2)$ and $I_1/(I_1 + I_2)$ respectively (from Thompson 1973 with permission).

10.4 EMPIRICAL INVESTIGATIONS OF HYBRID ZONES

Molecular genetic markers, including allozyme loci, are especially useful for studying large hybrid zones. Based on allozymes, a wide hybrid zone between two morphologically distinct subspecies of bluegill sunfish, *Lepomis macrochirus macrochirus* and *L. m. purpurescens*, was described (Avise and Smith 1974). These two subspecies have different and nearly fixed alleles at two allozyme loci in the zones of their allopatry in the southeastern United States. Also there is an area of genetic intermixing, where the ranges of these two subspecies overlap (Fig. 10.4.1).

A similar zone of overlap and hybridization in the southeastern United States occurs between the northern and Florida subspecies of largemouth bass, *Micropterus salmoides salmoides* and *M. s. floridanus* (Phillipp et al. 1983). The geographic distributions of allele frequencies in the hybrid zones for these subspecies are basically similar to Fig. 10.4.1. Similarly on the basis of biochemical genetic data the apparent introgression between native and introduced rainbow trout in the Yakima River of

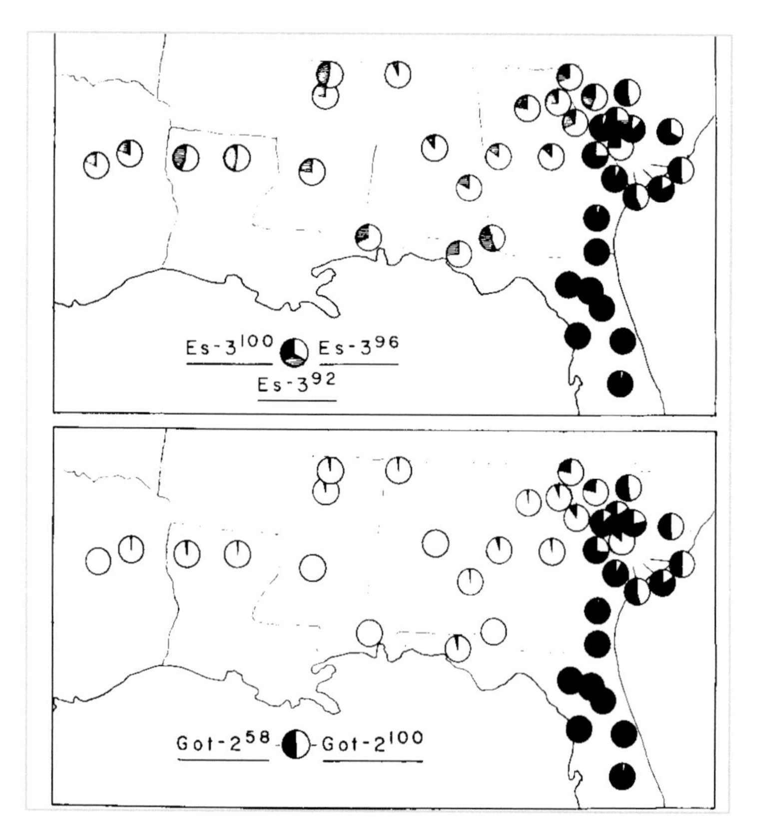

Fig. 10.4.1. Geographic distributions of allele frequencies at two allozyme loci (**Es-3*** and **Got-2***) for two subspecies of bluegill sunflsh, *Lepomis macrochirus macrochirus* and *L. m. purpurescens*, in the southeastern United States.
On the map circles with mixed colours denote hybrids occurrence (from Avise and Smith 1974 with permission from Wiley & Sons publisher).

Washington State was detected by Campton and Johnston (1985) as shown in Fig. 10.4.2. Avise and Smith (1974) suggested that terrestrial islands or refuges formed during the Pleistocene glaciations in the southeastern United States, which isolated the freshwater fauna of this region prior to secondary contacts in historically recent times. Presence of such hybrid zones is very significant for research in evolutionary genetics, and has been discussed repeatedly (Remington 1968; Woodruff 1973; Moore 1977; Barton 1979; 1983; Barton and Hewitt 1985).

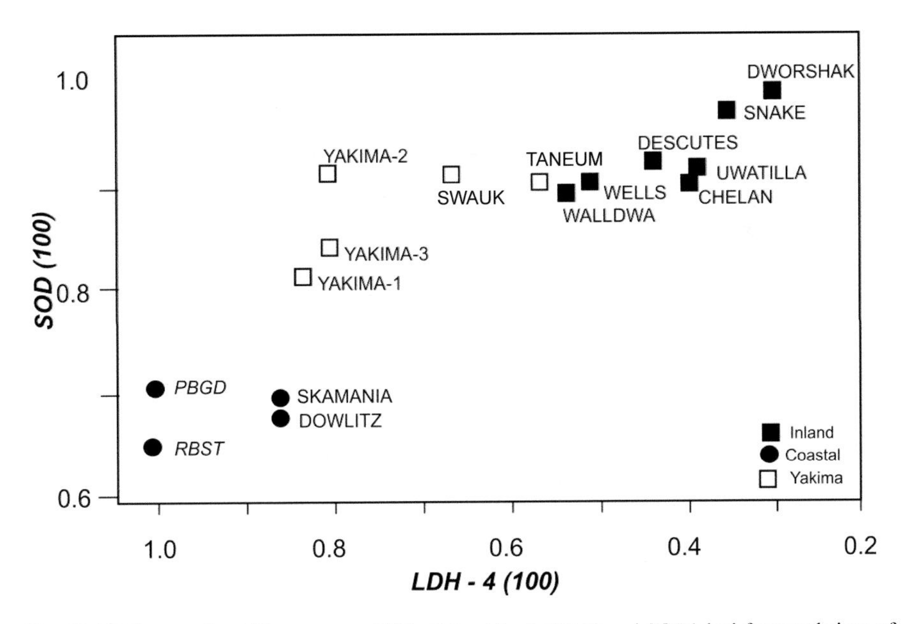

Fig. 10.4.2. Frequencies of the common (100) allele at the **LDH-4*** and **SOD*** loci for populations of rainbow trout (*Salmo gairdneri*) in Washington state.
The Yakima River is a major tributary of the Columbia River east of the Cascade Mountains, but rainbow trout inhabiting this river (sites 1, 2 and 3) and two tributary creeks (Swauk and Taneum) expressed allele frequencies intermediate to those for other inland populations and those for introduced coastal populations. RBGD and RBST represent two hatchery populations of nonanadromous rainbow trout (from Campton and Johnston 1985 with publisher permission).

Profound results of investigation of the ground crickets *Allonemobius fasciatus* and *A. socius* provided strong evidence for both the Wahlund effect and disequilibria at many loci of the eastern United States (Howard and Waring 1991) (Fig. 10.4.3, Tables 10.4.1–10.4.2). The Illinois transect is represented by populations 1–17, the mountain transect by populations 18–45, and the east coast transect by populations 46–70. Positive F_{is} scores are evidence for heterozygote deficit and a Wahlund effect (Howard and Waring 1991).

It was concluded that one of the theoretical predictions of the BSC was not supported by the data. Contrary to the prediction, the strength of reproductive isolation between the two species was as strong in Illinois as in the Appalachian Mountains. This result suggests that if reinforcement has occurred in the zone, the width of the zone has not been a major factor in the process.

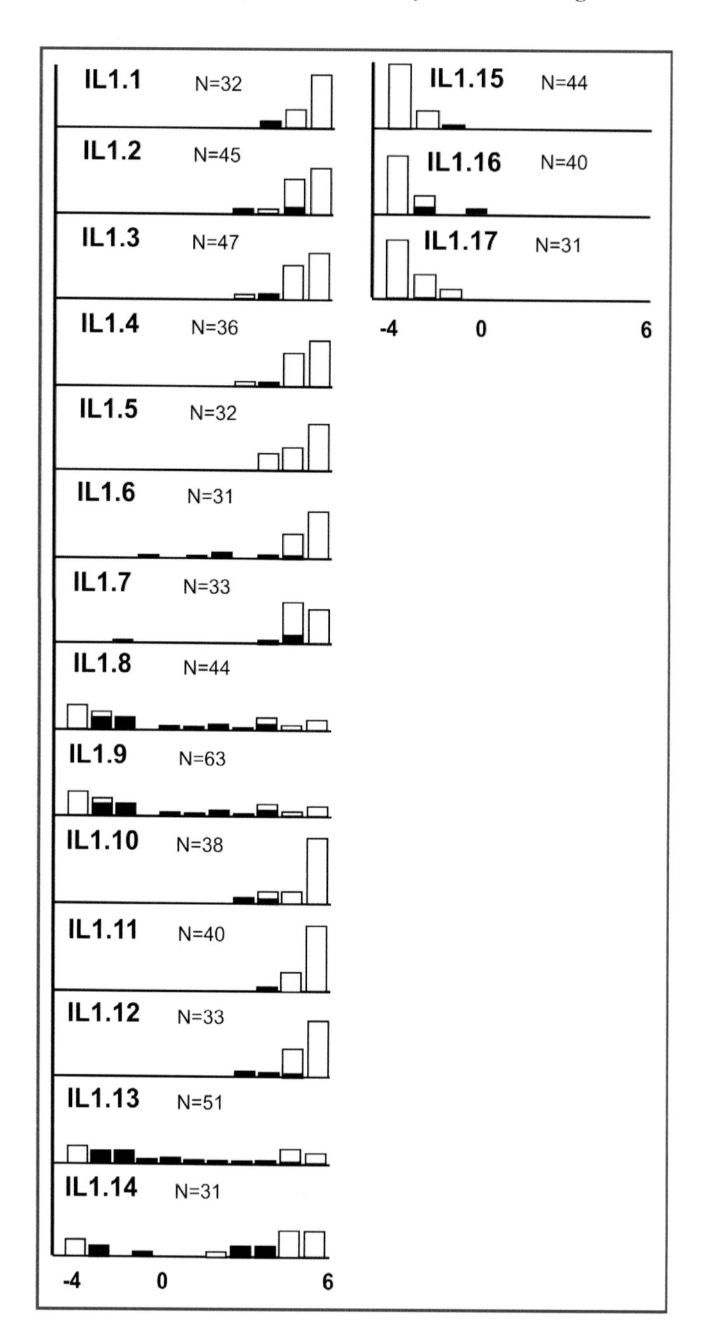

Fig. 10.4.3. Character index profiles for ground cricket populations *A. fasciatus* and *A. socius* collected along the Illinois transect (ILL1–ILL17).
The height of the bar gives the proportion of individuals in a population having that score. The solid part of the bar represents the proportion of individuals with that character-index score having a genotype indicative of mixed ancestry. Each graph represents a different population. The N is sample size (Modified from Howard and Warning 1991 with permission from Wiley & Sons publisher).

Table 10.4.1. The inbreeding coefficients (F_{is} values) of marker loci in mixed populations of ground crickets (from Howard and Waring 1991).

Locus/Site	Idh-1*	Hk*	Pep-3*	Got*
ILL8	0.399	0.763	0.573	0.385
ILL9	0.560	0.753	0.721	0.559
ILL 13	0.608	0.606	0.595	0.329
ILL 14	0.682	0.843	0.928	0.129
MTN26	0.646	0.812	0.546	0.189
MTN27	0.550	0.725	0.892	0.402
MTN33	0.332	0.714	0.514	0.260
MTN37	0.211	0.365	0.458	–0.113
MTN40	0.092	0.697	0.414	0.045
EC57	0.109	0.419	0.596	0.246
EC58	0.135	0.664	0.453	0.388

Remarkable results were obtained for "common mussel", *Mytilus edulis* species group, by Penney and Hart (1999), which investigated collections from Newfoundland both with morphology and allozymes (Figs. 10.4.4–10.4.5). In another paper for this area the mussel research made with molecular markers (Comesana et al. 1999).

It is notable: (i) Presence of hybrid individuals as seen from diagnostic ***Mpi**** locus, (ii) increased portion of recombinant genotypes for this locus in composite genotypes in edulis-like morphotype (E/T) as compared to trossulus-like morphotype (T/E; Fig. 10.4.6, see ***Mpi*** – E/T и T/E), which evidenced for T → E introgression direction prevailing, (iii) the possibility that hybrids may not only be F1 cannot be ruled out as it was proved by DNA results (Comensana et al. 1999).

Different results were obtained for this species group (*Mytilus* spp. group) from the area of Vancouver Island (Fig. 10.4.6).

The distribution of alien alleles at the two marker loci differed among sampling sites, suggesting differential introgression. The widespread incidence of alien alleles, combined with evidence of extensive hybridization between the native and introduced species, indicates that the alien alleles will probably persist in the gene pool of British Columbia mussels for some time (Heath et al. 1995).

Research in another geographic region, in England, also showed remarkable results for common mussels. (Fig. 10.4.7).

Using two DNA markers it was shown that *hybrid mussels* from Whitsand Bay UK carry alleles that are the products of intragenic recombination. The high frequency (10%) of these recombinant alleles within the hybrid population suggests that recombination is fairly frequent within this gene or that hybridization between *M. edulis* and *M. galloprovincialis* is substantial and has been occurring over considerable evolutionary time.

Table 10.4.2. Burrow's composite measure of linkage disequilibrium (Weir and Cockerham 1989) for mixed populations of ground cricket *A. fasciatus* and *A. socius estimated at allozyme loci.* The numbers in the row beside the estimates of linkage disequilibrium (in bold) represent X^2 values and test the hypothesis $D = 0$ (d.f. = 1) (from Howard and Waring 1991).

Site	***Idh-1*/Hk****	**Idh-1*/Pep-3***	**Idh-1*/Got***	**Hk*/Pep-3***	**Hk*/Got***	**Pep-3*/Got***
ILLS	***0.229*** *15.208***	***0.229*** *16.912***	***0.238*** *22.644***	***0.286*** *21.458***	***0.219*** *18.140***	***0.206*** *15.799***
ILL9	***0.323*** *100.968***	***0.270*** *56.571***	***0.271*** *62.075***	***0.317*** *65.853**	***0.309*** *65.407***	***0.295*** *68.521***
ILL 13	***0.239*** *18.080***	***0.265*** *22.207***	***0.198*** *24.600***	***0.289*** *26.421***	***0.218*** *19.493***	***0.223*** *37.843***
ILL 14	***0.308*** *15.204***	***0.341*** *17.374***	***0.182*** *9.778***	***0.320*** *14.872***	***0.185*** *9.396***	***0.211*** *9.903***
MTN26	***0.350*** *122.048***	***0.320*** *101.375***	***0.146*** *15.094***	***0.377*** *168.916***	***0.164*** *16.138***	***0.166*** *17.389***
MTN27	***0.324*** *132.606***	***0.334*** *112.741***	***0.235*** *68.255***	***0.350*** *108.934***	***0.223*** *59.123***	***0.232*** *46.449***
MTN33	***0.234*** *16.987***	***0.213*** *16.241***	***0.149*** *11.303***	***0.297*** *23.606***	***0.218*** *18.279***	***0.198*** *17.505***
MTN37	***0.164*** *21.811***	***0.145*** *17.513***	***0.038*** *2.481 NS*	***0.198*** *36.796***	***0.026*** *1.117NS*	***0.020*** *0.692 NS*
MTN40	***0.067*** *3.970**	***0.044*** *2.570 NS*	***0.016*** *0.535 NS*	***0.042*** *2.659 NS*	***0.012*** *0.317 NS*	***0.006*** *0.103NS*
EC57	***0.025*** *9.612***	***0.053*** *2.575 NS*	***0.092*** *5.991**	***0.051*** *2.695 NS*	***0.099*** *6.568**	***0.007*** *0.027 NS*
EC58	***0.178*** *33.994***	***0.035*** *3.270 NS*	***0.110*** *18.269***	***0.065*** *11.526***	***0.156*** *30.450***	***0.057*** *11.253***

NS = not significant, * $p < 0.05$, ** $p < 0.005$.

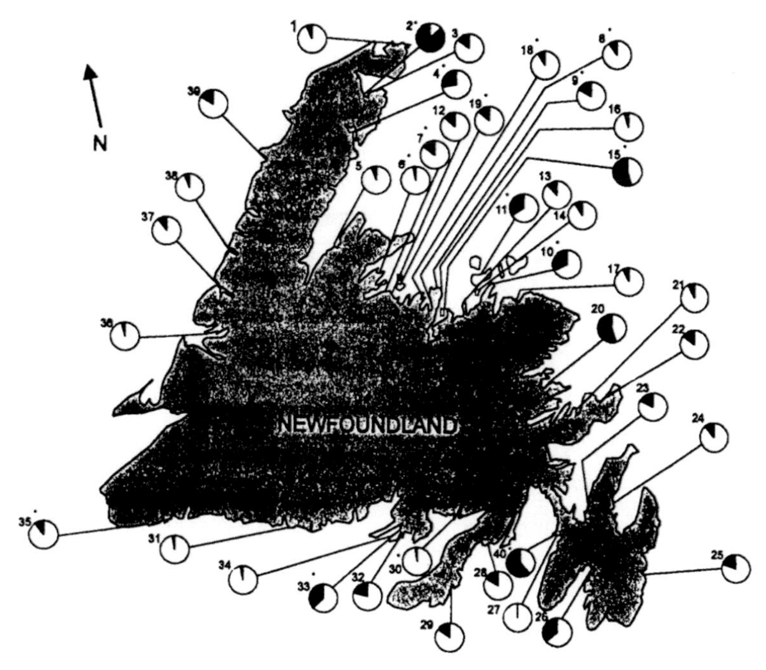

Fig. 10.4.4. Proportions of *Mytilus edulis* (white) and *M. trossulus* (black) in both wild and cultured populations in Newfoundland. Cultured populations are denoted by an asterisk (from Penney and Hart 1999 with publisher permission).

The cline in Fig. 10.4.7 may be interpreted as sign of natural selection. While it is not easy to obtain such evidence for selection in normal conditions, this association, together with others for *Mytilus ex. group edulis* hybrid zones (Skibinski et al. 1983; Gardner and Skibinsky 1988; Wilheim 1993), leads us to the conclusion of stronger natural selection, including selection against hybrids.

This section looked at only a few examples of hybrid zones. The aim was simply to stress that they are real in nature, and as real as are populations and species themselves. This fact is well documented by several genetic techniques.

It is clear that introgression exists, although even in a wide zone of *Mytilus* spp., for example, it may be quite restricted (Rawson et al. 1999) or be asymmetric (see above and Heath et al. 1995; Rawson and Hilbish 1998). If we accept, bearing in mind examples in Section 10.3–10.4, that the sexually reproducing species in marine and terrestrial realms are widely introgressed, then we should recognize that the orthodox biological species concept, as we define it earlier in Chapters 7–8, is inadequate. Happily, species do not know anything about concepts and most of them keep their integrity despite the failure of concept.

Genetic mixing and creation of long-term genetic disequilibrium may have different impacts: positive and negative. Positive impacts are due to the increase of genetic variance, and a special role of hybrid zones (tension zones) is as a "field of battle" for species integrity. Hybrids, if sometimes advantageous, may be used directly

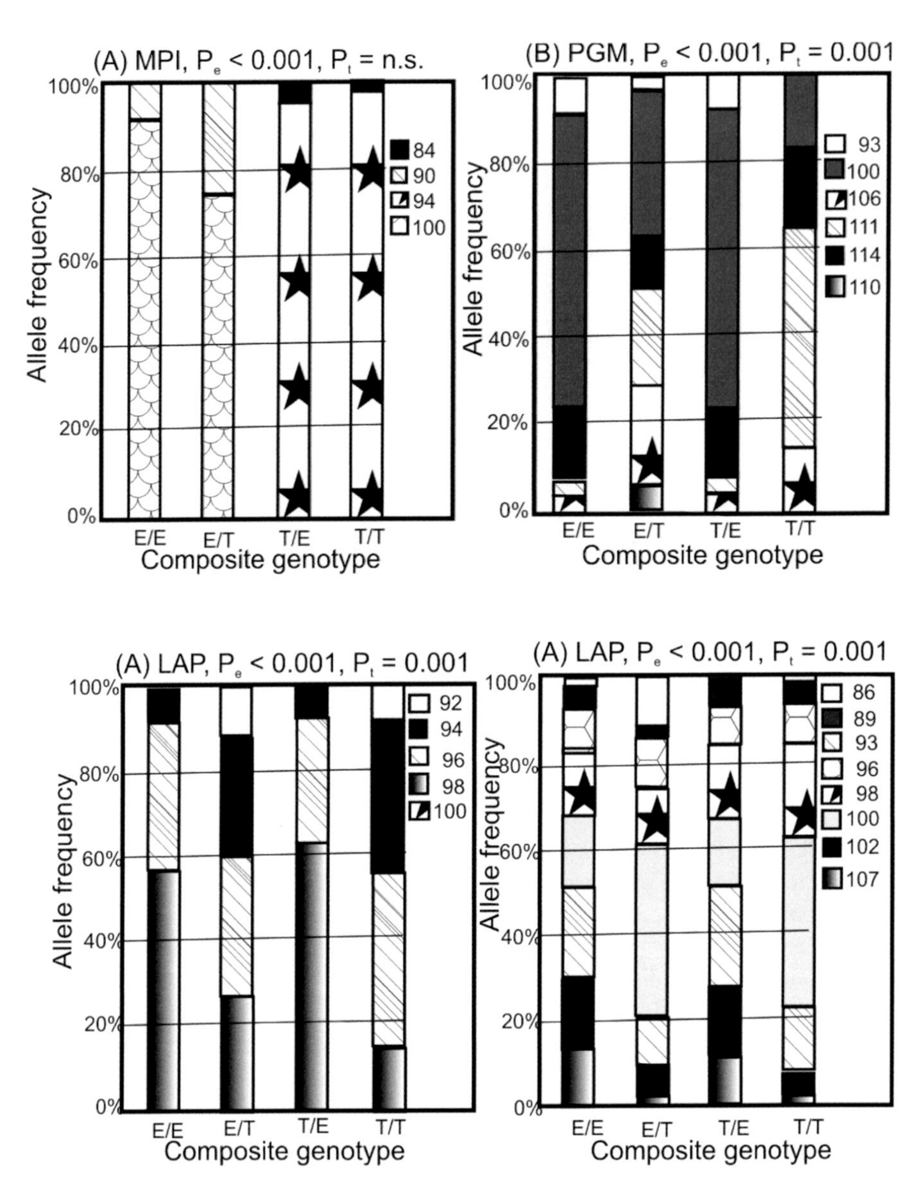

Fig. 10.4.5. Comparison of allele frequencies among the four composite genotype classes at each of the four loci (a) **Mpi***, (b) **Pgm***, (c) **Lap*** and (d) **Gpi***. Vertical bars denote individual alleles numbered as per the side legend. Alleles at very low frequency are pooled with the nearest adjacent allele at each locus. Probability values, Pe and Pt are for X^2 tests for intraspecific differences between allele frequencies (Pe = E/E vs. E/T; Pt = T/T vs. T/E) (Modified from Penney and Hart 1999 with publisher permission).

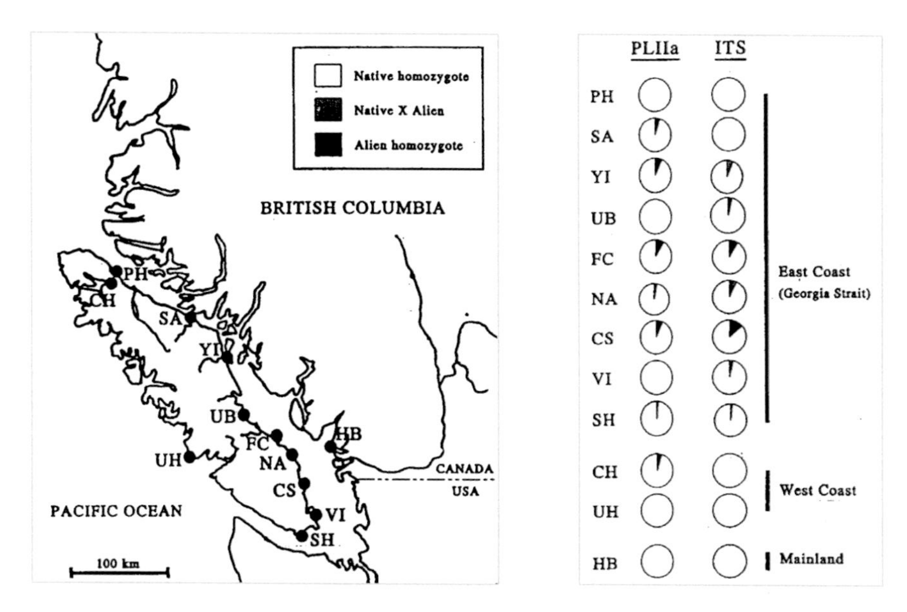

Fig. 10.4.6. Southwestern British Columbia and Vancouver Island showing the 12 blue mussel sampling sites and the presence of the alien (*M. edulis* and/or *M. galloprovincialis*) and native (*M. trossulus*) genotypes as determined by the two independent nuclear DNA markers (PLIIa and ITS) (from Heath et al. 1995 with permission).

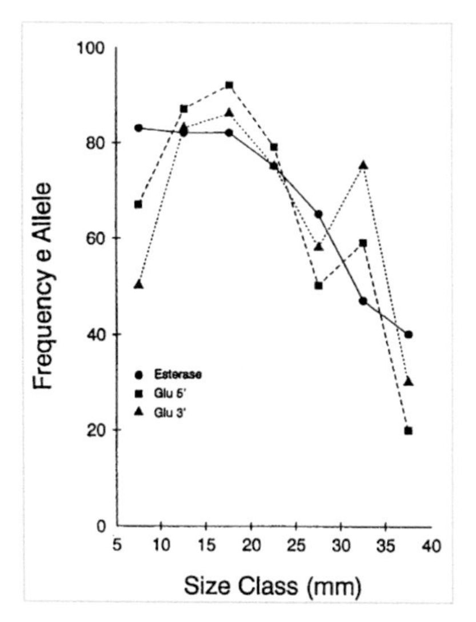

Fig. 10.4.7. Frequency of *Mytilus edulis* (e) alleles at the Esterase-D (Est), Glu-5' and Glu-3' DNA markers as a function of size for the Whitsand Bay, UK, hybrid zone population.
For estimating allele frequencies at the Glu-5' marker, *M. edulis*-specific 350 bp and 380 bp products were treated as separate alleles whereas the *M. galloprovincialis* 300 bp and 500 bp products were assumed to be produced by a single allele. Thus, mussels with 350 bp, 380 bp, or 350/380 bp banding patterns were scored as *M. edulis* homozygotes, those with 300 bp patterns were scored as *M. galloprovincialis* homozygotes and those with a 300 bp band in combination with either the 350 bp or 380 bp band were scored as heterozygotes (from Rawson et al. 1996 with permission from Wiley & Sons publisher).

for the aquaculture rearing or as a source for breeding programs. However, numerous negative impacts are created by: (i) disequilibria, decreasing fitness in hybrid demes (taxa), and (ii) introgression of deleterious (or weakly adapted) genes into native populations or species and as a result, their degradation and consequently substitution by other species (which as a rule are worse for humans and ecosystems).

10.5 TRAINING COURSE, #10

10.1 Software SPECSTAT. Consider specific usage of the NEI-DIST and DISTANCE Units (see Read.me File).

10.2 Software NTSYS-pc: acquaintance with programs and principles of their managing.

10.3 Consider and learn new terms below:

Hybrid is a mixture, an offspring of cross(es) between genetically different organisms. Hybrids may be treated as well as individuals having a mixed ancestry. In a certain sense, a heterozygote at one or more loci could be named a hybrid individual. However, normally the term hybrid is attributed to more distant cross, e.g., between different species.

Hybridization is the process through which hybrids occur.

Reproductive isolation between biological species in a pure sense means absence of any gene flow.

Hybrid zone is an area in nature where hybrids occur between supposed parental forms.

Cline is a gradual or abrupt allele frequency change in a mode: Species 1 (deme 1) → hybrids → Species 2 (deme 2) and that is maintained by balance between dispersal and selection against hybrids.

Gametic disequilibrium refers to the nonrandom association of alleles between loci.

11

QUANTITATIVE TRAITS: INHERITANCE AND EVOLUTION

MAIN GOALS

SUMMARY

1. **Continuous variation** is exhibited in crosses involving traits under **polygenic control**. Such traits are quantitative in nature and are inherited as a result of the cumulative impact of **additive alleles**.
2. **Polygenic characters** can be analysed using statistical methods, which include the **mean**, the **variance**, the **standard deviation** and the **standard error** of the mean. Such statistical analysis is descriptive, and can be used to make inferences about a population or to compare and contrast sets of data.
3. For many phenotypic characters, it is difficult to ascertain when variation depends on genetic or on environmental factors. **Heritability** is an estimate of the relative importance of **genetic factors** versus **non-genetic factors**, and is determined correspondingly as a ratio of the **genotypic variance** to **phenotypic variance** in populations, $\boldsymbol{h}^2_{B} = \sigma^2_{G}/\sigma^2_{PH}$. This ratio measures the heritability, which can be calculated for many characters, and is especially useful in selective breeding of commercially valuable plants and animals. The **coefficient of heritability** is known in two modes as **narrow and wide sense heritability**.

4. Loci bearing genes that control a quantitative trait are called **quantitative trait loci (QTLs)**. Location and distribution of QTLs within a genome can be ascertained by the application of molecular genetic markers.

11.1 QUANTITATIVE VARIATION AND ITS LAWS

As we remember from Chapter 3, a quantitative trait (QT) is the trait that changes in continuous mode, where no discrete classes can be defined. However, we can measure such a trait and find limits of its variation within a scale that is defined in a certain way. The QT's distribution can be explained by the theory of Gauss, by the **normal distribution**, like this (Fig. 11.1.1):

$$F(x) = j\ (x;\ \mu\sigma^2). \tag{11.1}$$

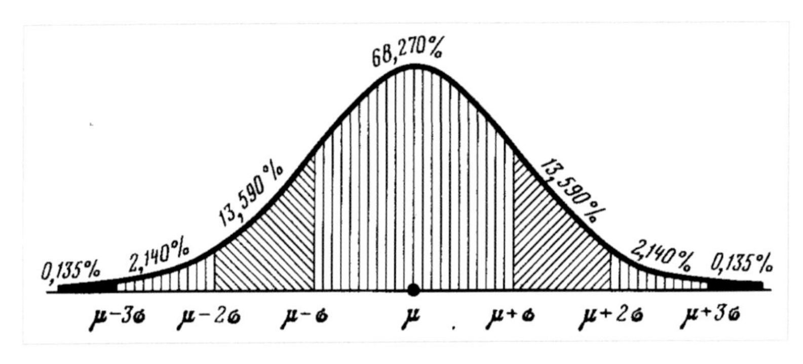

Fig. 11.1.1. Diagram representation of the shape of the normal distribution $F(x) = j\ (x;\ \mu s^2)$. Numbers above the curve denote a square (in %) or probability of putting a stochastic continuous variable in these limits or intervals (from Zhivotovsky 1991 with permission).

Thus, the normal distribution is a distribution function of density of probability with two parameters; μ and σ^2 are parameters of a function $F(x)$, called mathematical expectation and variance; these parameters can be evaluated through the estimates of mean and standard deviation.

Normal distribution is a classical law of mathematical statistics. This distribution is important for several reasons. (i) Firstly, it is a limiting case for many distributions. For example, from the calculation point of view it is difficult to define a binomial distribution for large samples. For this case, the normal approximation fits well. Thus, it is known that distribution of an estimate $p = k/N$ is approximated with good precision by the normal distribution with mean value that equals to μ and variance (σ^2), $\mu\ (1 - \mu)/N$. (ii) Secondly, sampling estimates of mean values for quantitative traits, even those that may deviate substantially from the normal distribution, are described more or less precisely by normal distributions with mean values that equal the mathematical expectation (general mean) and variance σ^2/N, where σ^2 is general variance of the trait in question. Figure 11.1.2 shows the distribution of shell height (M_HGT) in several samples of a gastropod from the Sea of Japan.

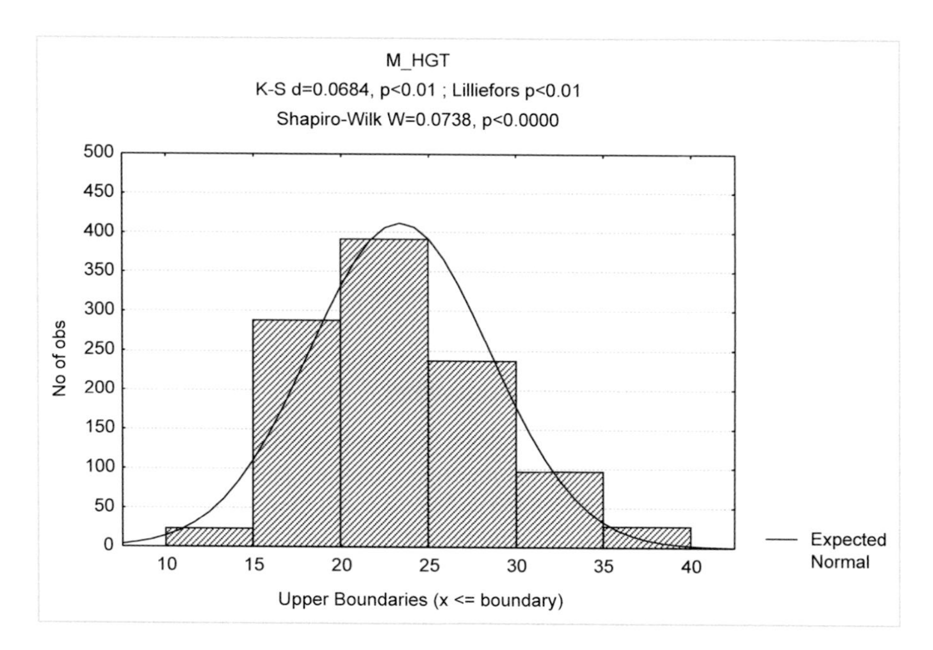

Fig. 11.1.2. Distribution of a gastropod mollusk *Nucella heyseana* shell height (M_HGT) in samples from Peter the Great Bay, Sea of Japan.
Histogram shows the empirical raw variation, and the curve presents the approximation by a normal distribution. There is visually a satisfactory fit of two distributions, although statistically significant deviation is observed for some of them at precise criteria (Based on data from Kartavtsev et al. 2006).

True values for μ and σ^2 are not known in a general totality. It is possible to judge their values only by taking a random sample from the total population, for which we can calculate mean ($\bar{x}$ or M), and mean square deviation (**mean square**, for short, ***MS*** or s^2_x) or standard deviation (s_x), which are estimators for the above two parameters. We will call them estimates to discriminate from parameter notations. However precisely the estimate is a numerical value, say 2.3 or 5.6, for either $\bar{x}$ or s_x.

Mean and Standard Deviation

The mean may be defined as

$$\bar{x} = (\sum x_i)/n = 1/n\ (\sum x_i), \qquad \textbf{(11.2)}$$

where x_i is a value of an individual measure of a trait in fish, shellfish or plant, and n is the number of measurements.

MS and standard deviation for a series of such measurements are

$$MS = \sum (x_i - \bar{x})^2/(n - 1).\ \ s_x = \sqrt{[\sum (x_i - \bar{x})^2/(n - 1)]}. \qquad \textbf{(11.3)}$$

Several other important statistics are used in genetic investigations of QTs. Most frequent is the coefficient of variation, ***Cv***.

$$Cv = (s_x/\bar{x}) * 100\%. \qquad (11.4)$$

This coefficient standardizes the variance estimates of different traits by dividing by the averages, thus making them more comparable. We will become acquainted with some other QT standardizations and transformations in the Training Course #11. A very important statistic is the normalized deviate (t): $t = (x_i - \bar{x})/s_x$. Firstly, this equation leads to prediction of x_i as follows: $(x_i - \bar{x}) = t*s_x$ or $x_i = \bar{x} + t*s_x$. Secondly, all traits after t-transformation have the mean equal to zero and the variance equal to one: $\bar{x} = 0$ and $s_x = 1$. This property is widely used in multi-dimensional statistical analyses because it makes all the variables comparable.

Confidence Limit and Confidence Probability

The estimate of a mean $\bar{x}$ (or M) is called a point estimate, because it has a certain numerical value for a certain sample, or may be presented as a point on a numerical axis. Because of randomness of the sample and its finite size this point will not coincide with the true μ. That is why it is useful to define a limit of the variable values in which the μ will definitely occur. How do we find this limit? We may know the certain function, say it is $F(x)$, t, χ^2 or F. The difficulty here is that we do not know the value of the parameter μ to calculate probabilities and define the limit. To avoid this difficulty, however, there is a simple way. We may substitute μ with a certain sample estimate, $\bar{x}$, and make inferences about the total population via sampling. The consequent distribution will be like a theoretical distribution, although it will be biased from μ to $\bar{x}$ (Fig. 11.1.3A). Inside this distribution, it is possible to find a limit of values I_1, I_2

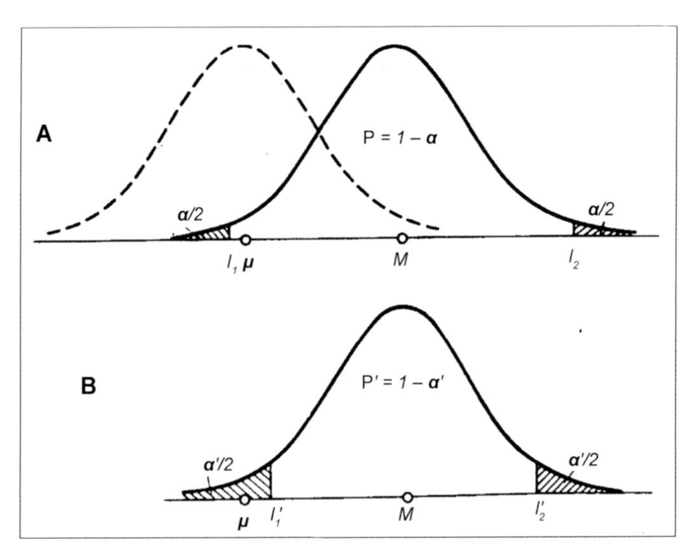

Fig. 11.1.3. Diagrammatic representation of a confidence limit and a confidence probability. (A) The distribution of the theoretical estimates' values is shown with the dotted line. The solid line is the same distribution in which as a "centre" is given the sample estimates of M (mean). I_1, I_2 is a confidence limit, which includes μ, a general parameter, with a confidence probability of P. (B) The same distribution, but the confidence limit I_1, I_2 shown with lower confidence probability, $P' < P$ ($\alpha' > \alpha$) and it no longer includes μ, a general parameter value (from Zhivotovsky 1991 with permission).

where the total squared values under the curve and outside this limit will be equal to α ($\alpha/2$ from both sides); square inside the limit is $P = 1 - \alpha$. A segment I_1, I_2 is called a **confidence limit** for estimating the parameter μ with **confidence probability** $P = 1 - \alpha$. The segment I_1, I_2 covers 100% of the confidence limit. Thus, for example, the 95% confidence limits corresponds to $P = 0.95$, 99% to $P = 0.99$, and 99.9% to $P = 0.999$.

Standard Error of Statistics

We also need to know the standard error (SE) of the mean ($s_{\bar{x}}$) and variance. ***SE*** of the mean is $s_{\bar{x}} = s_x/\sqrt{n}$. In this case, we deal with the estimate of the arithmetic mean, and the SE is necessary to compare one arithmetic mean with others. ***SE*** for the standard deviation is $s_x = Cv = s_x/\sqrt{2n}$. Similarly, SE is needed for comparing the s_x and Cv with other such estimates. The same rule holds true for other statistics.

Analysis of Quantitative Traits

The main purpose of the analysis is to understand whether there is a difference of a trait mean and variance between two or more lines, kinds, offspring generations, etc.? I will give to you other examples with STATISTICA software. With modern software you will be able to compare a lot of other qualities of the distributions of traits. You should understand how and why the comparisons are made. The simplest case is when two mean values or two standard deviations are compared. What we need first in this case is the SE estimates for them. Then, we can compare these two values, say two average values, on the basis of a null-hypothesis of the stochastic nature of the difference found. If the difference is greater then the expected stochastic error, we will reject the null-hypothesis of no difference, and may speak of two distinct populations (lines, etc.). The main tool for such a judgment is Student's **t-statistic.** For example, for samples 1 and 2, an observed t value for arithmetic means is: $t = (M_1 - M_2)/\sqrt{(s^2_{\bar{x}1} + s^2_{\bar{x}2})}$. As in the case of chi-square, the t observed is compared to t expected (t-table): d.f. = $n_1 + n_2 - 2$; M is mean, $s_{\bar{x}}$ is standard error and n_1, n_2 are sample sizes for each of the two samples; d.f. is degree of freedom. We have to remember that the above formula holds when samples are approximately equal in size and they are big enough. If these conditions do not hold, it is necessary to use different algorithms for calculations.

The principle of analysis of variance is simple too. We will see it below in Section 11.3.

11.2 THE MULTIPLE FACTOR HYPOTHESIS

At the beginning of the 20th century, geneticists noted that many characters in different species had similar patterns of inheritance, such as height and weight in humans, seed size in the broad bean, grains colour in wheat and kernel number and ear length in corn. In each case, offspring in the succeeding generation appeared to be a mixture of the parents' trait scores. The question of whether QT variation could be considered in Mendelian terms was not widely accepted in the beginning of 20th century. Some, like W. Batson and G. Yule, suggested that a large number of factors or genes could account for the observed patterns. This idea, called the **multiple-factor hypothesis,** supposed that large number of loci could affect the phenotype in a cumulative or quantitative way. However, other geneticists doubted that Mendel's discrete unit factors could account for the blending of parental phenotype in the QTs and such patterns of inheritance. It was well known that one can observe the intermediate scores in QTs in F_1 and segregation in F_2, which lead to wider variance in the next generation than in F_1. However, scientifically representative results obtained by East (1936) proved the Mendelian basis of QT inheritance (Fig. 11.2.1).

East's experiments demonstrated that although the variation in corolla length looks as continuous, experimental crosses resulted in the inheriting of distinct phenotypic classes, as caused by directional selection in the three independent F_3 categories. This principal observation comprised the basis for the multiple-factor hypothesis, which explains how traits can deviate considerably in their expression.

Additive Alleles: The Basis of Continuous Variation

The multiple-factor hypothesis, suggested by the observations of East and others, includes the following major points:

1. QTs, which exhibit continuous variability, can be quantified by measuring, weighing, counting and so on.
2. Two or more loci dispersed in the genome account for the hereditary influence on the QTs in an additive way. Because many genes can be involved, inheritance of this type is often called *polygenic.*
3. Each locus may be occupied by either an **additive allele**, which adds its impact to the phenotype, or by a **nonadditive allele**, which does not contribute cumulatively to the phenotype.
4. The total effect on the phenotype of each additive allele, while small, is approximately equivalent to all other additive alleles at other gene loci.
5. Together, the genes controlling a single character produce its substantial phenotypic variation.
6. Analysis of polygenic traits requires the investigation of large samples of progeny from a population of organisms.

The points summarized above centre on the concept that additive alleles at polygenic loci control quantitative traits. To illustrate this, let's examine Herman Nilsson-Ehle's experiments involving grain colour in wheat performed in the beginning of the 20th century. In one of the experiments, wheat with red grains was crossed with wheat with white grains (Fig. 11.2.2).

Inheritance of the grain colour occurred in relation to the "dose" of additive alleles, and follows the ratio 1:4:6:4:1. As we have seen in the cross presented above, two gene pairs are involved, the segregation of which give five F_2 phenotypic categories with expected ratios of 1:4:6:4:1. Therefore, as we see, there is a solid base to accept that continuous variation can be explained in a Mendelian fashion. There is no reason why three, four or more gene pairs cannot control various phenotypes. As greater numbers

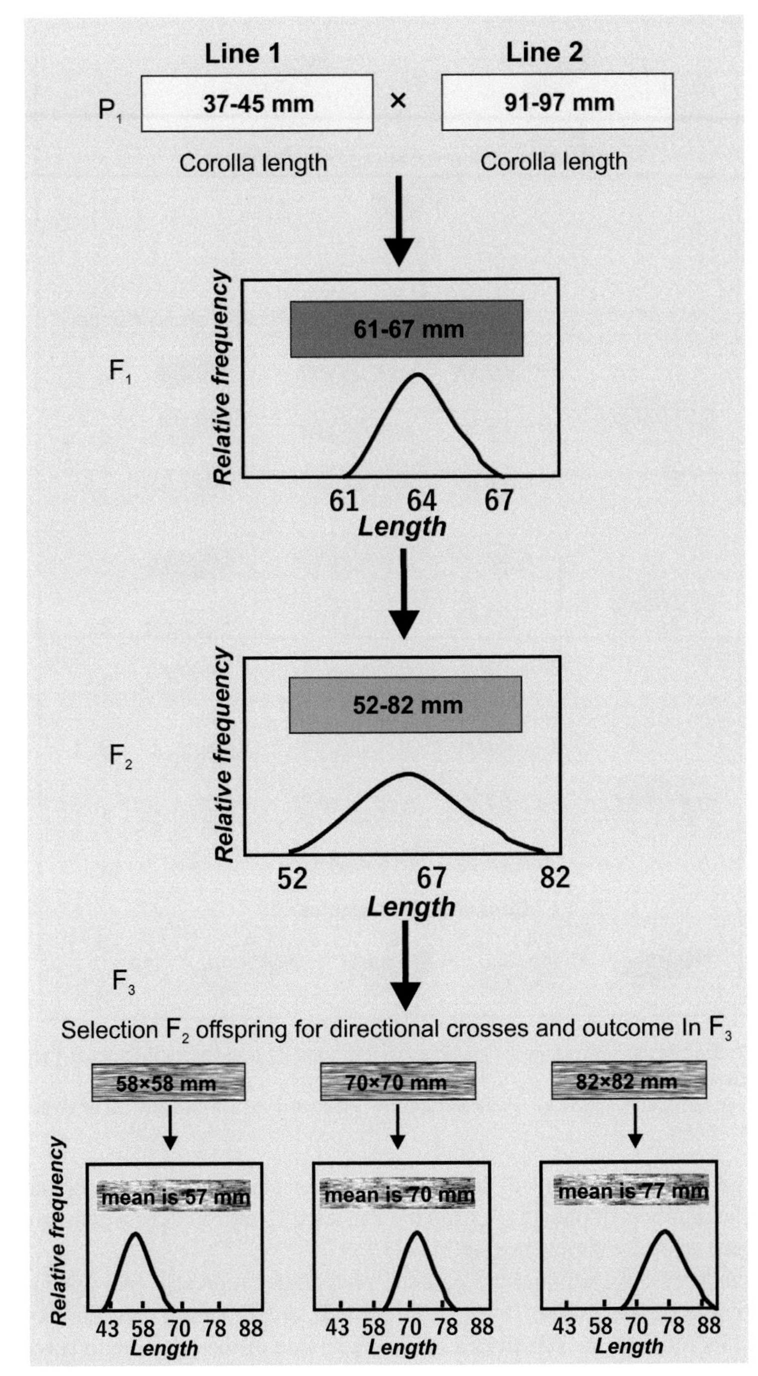

Fig. 11.2.1. The F_1, F_2 and selected F_3 results of East's cross between two strains of Nicotiana with different corolla lengths. Plants of line 1 vary from 37 to 43 mm, while plants of line 2 vary from 91 to 97 mm.

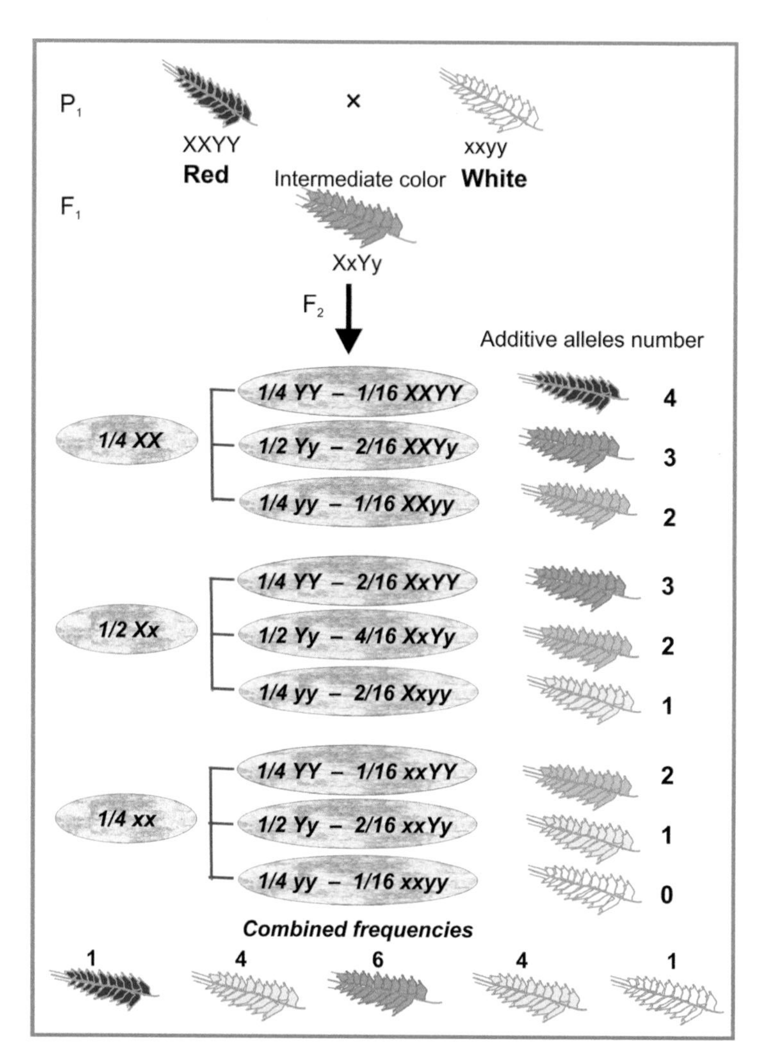

Fig. 11.2.2. Scheme that shows how the multiple-factor hypothesis accounts for the 1:4:6:4:1 phenotypic ratio of grain colour in wheat.
All alleles designated by uppercase letters are additive and contribute an equal amount of pigment to the grain or phenotype.

of gene pairs become involved, the number of classes increases, with more complex ratios. The numbers of phenotypes and the expected F_2 ratios of crosses involving up to five gene pairs are shown next in Fig. 11.2.3.

In our previous example, the parental phenotypes represent two extremes, the red and white grains. In Fig. 11.2.2, 1/16 of the F_2 are either red or white like the P_1 classes; this ratio can be substituted on the right side of the equation to solve for n: $\mathbf{1/4^n = 1/16, 4^n = 16}$ **or** $\mathbf{n = Log_4 16 = 2}$**.** In general, the **gene number for a quantitative trait** is $\mathbf{n = Log_4 X}$, where **X** is the denominator of the segregating ratio.

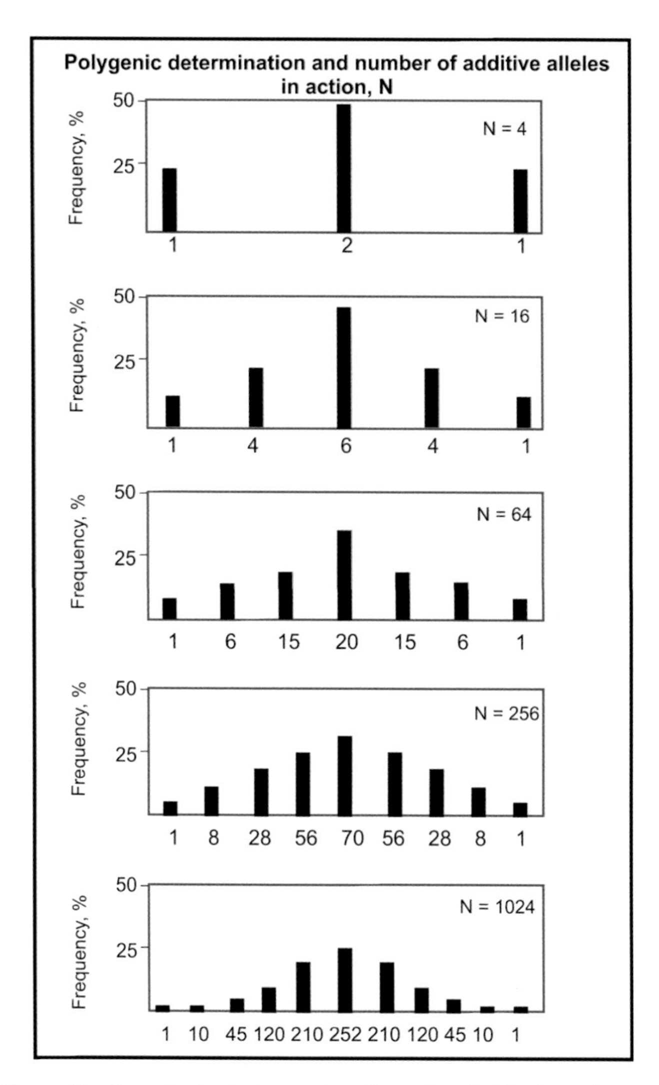

Fig. 11.2.3. The results of crossing two heterozygotes when polygenic inheritance is operative with 1–5 gene pairs.
Each histogram bar indicates a distinct phenotypic class from one extreme (left bar) to the other extreme (right bar). Each phenotype results from a different number of additive alleles.

The Significance of Polygenic Inheritance

Polygenic inheritance is an important concept. It serves as the genetic basis for a vast number of traits involved in animal breeding, agriculture and aquaculture. Such traits as height, weight, and body shape in animals, size and grain yield in crops, meat and milk production in cattle, egg production in chickens, fish and shellfish are all known to be under the polygenic control. It is important to note, that the genotype establishes the potential range within which a particular phenotype is expressed. However, environmental factors determine how much of the genotype will be realized. In the crosses described above, we have assumed an optimal environment, which minimizes a variation from different sources. To measure the effects of these sources is a special task in the genetics of QTs.

11.3 MAIN TASKS OF GENETIC RESEARCH ON QUANTITATIVE TRAITS

Three tasks are among the most important : 1. Analysis of the shape of the distribution, and description of within-sample and among-sample variability; 2. Determination of the heritability coefficient; 3. Detection of the number of controlling genes. We will not consider this last goal here because it is presented in brief at the end of the previous section, 11.2.

Analysis of the Distribution Shape and Variability

Defining the shape of the distribution is very important because it helps to apply exact theoretical approximation for a certain trait under investigation. Usually distributions of QTs are fit to the normal distribution or its peculiar cases like Student's or Fisher's distribution (**t** and **F** distributions). Thus, in the first step of the analysis, we must accept or reject the null-hypothesis (H_n) that the variation statistically significantly deviates from a normal distribution: $H_n >= P >= 0.95$ ($P <= 0.05$) or higher (Section 11.1). Test statistics to detect deviations may be of different types: Chi-square (X^2obs $< X^2$exp), **t**-, **F**-statistics, etc. The second step of the analysis is to accept or reject the H_n of homogeneity of samples analysed. This is done with analysis of variance if more than two lines or collections form the data set, or with the **t**-statistic in the case of two samples, as we have seen before. Again, we will solve some numerical tasks through the Training Course #11 with the STATISTICA software. Some students may think that it is too difficult to comprehend what the analysis of variance (ANOVA) is. Actually the idea of ANOVA in the majority of cases (in research on natural populations) is clear and easy. This idea is based on the following equation:

$$\boldsymbol{\sigma}^2_{\mathrm{y}} = \boldsymbol{\sigma}^2_{\mathrm{A}} + \boldsymbol{\sigma}^2_{\mathrm{Z}}. \qquad \textbf{(11.5)}$$

Here $\boldsymbol{\sigma}^2_{\mathrm{y}}$ is the total variance, $\boldsymbol{\sigma}^2_{\mathrm{A}}$ is the among-group variance and $\boldsymbol{\sigma}^2_{\mathrm{Z}}$ is the average within-group variance. In genetic analysis of QTs, the variances may be further portioned into components, depending on the complexity of the research and the scheme of the cross (we'll see this below). In the first step, geneticists prefer to present the total variance in the following way:

$$\boldsymbol{\sigma}^2_{\mathrm{PH}} = \boldsymbol{\sigma}^2_{\mathrm{G}} + \boldsymbol{\sigma}^2_{\mathrm{E}} + \boldsymbol{\sigma}^2_{\mathrm{GE}}. \qquad \textbf{(11.6)}$$

Where $\boldsymbol{\sigma}^2_{\mathrm{PH}}$ ($=\boldsymbol{\sigma}^2_{\mathrm{y}}$) is the total or phenotypic variance, $\boldsymbol{\sigma}^2_{\mathrm{G}}$ is genotypic variance, $\boldsymbol{\sigma}^2_{\mathrm{E}}$ is environmental variance and $\boldsymbol{s}^2_{\mathrm{GE}}$ is a covariation member of the equation (it is usually negligible and may be omitted).

Determining of a Heritability Coefficient

The estimation of a coefficient of heritability is one of the most important tasks in the genetics of QTs. In a general or in a wide sense, it is easy to define the coefficient of inheritance (h^2), which is simply derived from equation 11.6. Thus,

Wide-Sense Heritability (h^2_B) is

$$h^2_B = \sigma^2_G/\sigma^2_{PH}. \quad \textbf{(11.7)}$$

In other words, heritability of a QT is the proportion of genotypic variance relative to the total variance. The variance σ^2_G is complex in its nature: $\sigma^2_G = \sigma^2_A + \sigma^2_D \sigma^2_I$, where σ^2_A is **additive variance**, σ^2_D, and σ^2_I are the intra-locus and inter-locus variances' interaction (The first one is the dominance deviation or deviation from additive allele effects, and the second is the deviation caused by the effects of epistasis). Thus, in a narrow sense, h^2 could be defined as follows:

Narrow-Sense Heritability (h^2_A) is

$$h^2_A = \sigma^2_A/\sigma^2_{PH} = \sigma^2_A/(\sigma^2_A + \sigma^2_E + \sigma^2_D \sigma^2_I). \quad \textbf{(11.8)}$$

The effect of σ^2_I is negligible and can be omitted. How do we determine the genotypic variance σ^2_G and the rest of the variances? This is the fundamental question in the genetic analysis of QTs. There are special tools in genetics to apportion the phenotypic variance into components, and we'll see them below. Before that, one simple example will help to understand the essence of σ^2_G detection. Imagine that there is a **clone**, *a unity of genotypically identical individuals.* Within a clone, no variation can be attributed to genotype and all variability that occurs is environmental, i.e., equal to σ^2_E. If we performed the analysis of many clones, that differ genetically, we could also detect the total variance, σ^2_{PH}. Thus, we can calculate that $\sigma^2_G = \sigma^2_{PH} - \sigma^2_E$. Such logic may be applied further to different genetic levels: inbred lines, families and siblings (sibs, half-sibs, etc.). One useful way is to see the results of artificial selection.

11.4 HERITABILITY

Artificial Selection

The process of selecting a specific group of organisms from an initially heterogeneous population for breeding purposes is called **artificial selection.** A relatively high h^2 value predicts that selection will likely succeed in altering a population. As you can imagine based on our preceding discussion, measuring of the components necessary to calculate h^2 is a complex task. A simplified approach involves measuring the central tendencies (the change of means) of a trait from

1) a parental population exhibiting a normal distribution (*M*),
2) a "selected" segment of the parental population that expresses the most desirable quantitative phenotypes (*M1*) and
3) using the offspring (*M2*) resulting from interbreeding the selected *M1* group.

When this is accomplished, the following relationship of the three means and h^2 exists:

$M2 = M + h^2 (M1 - M)$. Solving this equation for h^2 gives us

$h^2 = (M2 - M1)/(M1 - M)$. The equation can be further simplified by defining $M2 - M$ as the **response** (R) to selection and $M1 - M$ as the **selection differential** (S), where h^2 reflects the ratio of the response observed to the total response possible. Thus,

$$h^2 = R/S. \qquad \textbf{(11.9)}$$

This is so-called realized heritability. The numerical example for calculation of such heritability is:

$h^2 = (13–20)/(10–20) = –7/–10 = 0.70.$

The longest-running artificial selection experiment known is still being conducted at the State Agricultural Laboratory in Illinois (USA). Since 1896, corn has been selected for both high and low oil content. After 76 generations, selection continues to increase oil yield and heritability remains reasonably high. Heritability measured in this experiment at several generations (9, 25, 52 and 76) was 0.32, 0.34, 0.11 and 0.12 (Klug and Cummings 2002, p. 112). Heritability for some commercially significant QTs is summarized in Table 11.4.1. Some results of the response to selection in salmon breeding are very impressive (Fig. 11.4.1).

Heritability coefficient may vary between 0 and 1 (0–100%). In the above table, there are many coefficients with values close to 0.4–0.5; this is moderate heritability. Quite frequently, most significant commercial traits are weakly inherited, $h^2 = 0.1–0.2$ and less.

Fig. 11.4.1. Trout selected for fertility by L. Donaldson (from Hines 1976 with permission). The increase is from 3000 eggs in wild fish (bottom) up to some 30000 eggs in "Susie's" body (top). Heritability was not determined.

Table 11.4.1. Heritability estimates (h^2, %) (from Allendorf et al. 1987 and Klug and Cummings 2002).

Organism, Trait	***h^2***
Human	
Stature	65
Systolic blood pressure	48
Cattle	
Body weight (adult)	65
Birth weight	51
Milk yield	20–44
Butterfat	40
Conception rate	3
Pig	
Back fat thickness	70
Weight gain per day	40
Chickens	
Body weight (at 32 weeks)	50–55
Egg weight (at 32 weeks)	50
Egg production	20
Egg hatchability	15
Sheep	
Fibre diameter	20–50
Body weight	20–40
Mice	
Tail length (at 6 weeks)	40–60
Body weight (at 6 weeks)	35–37
Litter size	15
Drosophila melanogaster	
Body size	40
Thorax length	25–50
Rainbow trout	
Body weight (juveniles)	12
Body weight (adults)	17
Body length (juveniles)	24
Body length (adults)	17
Meatiness	14
Atlantic salmon	
Body weight (juveniles)	8
Body weight (adults)	36
Body length (juveniles)	14
Body length (adults)	41
Meatiness	16
Common carp	
Body weight (juveniles)	15
Body weight (adults)	36

You should not think that all $\boldsymbol{h}^2$ in the table above were obtained as realized inheritance. Many other techniques are used. It is very common to find $\boldsymbol{h}^2$ from the correlation (or regression) among relatives:

$\boldsymbol{h}^2$ = r = b (both parents of the offspring known), $\boldsymbol{h}^2$ = 2r = 2b (siblings or one parent known), $\boldsymbol{h}^2$ = 4r = 4b (half-siblings). One example of $\boldsymbol{h}^2$ detection is given below for eel-pout (Fig. 11.4.2). The most precise and most difficult method is the scheme of obtaining $\boldsymbol{h}^2$ by subdividing the phenotypic variance with ANOVA. We'll try to learn this scheme in the Training Course #11.

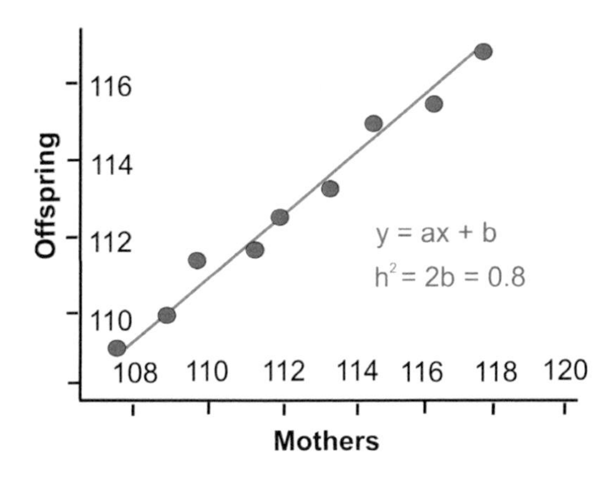

Fig. 11.4.2. Regression of vertebrae number in eel-pout Zoarces viviparus (Modified from Smith 1921).

11.5 MAPPING OF QUANTITATIVE TRAIT LOCI (QTL). QTL EVOLUTION

Because quantitative traits are influenced by multiple gene loci, the task to determine their location in the genome is more complicated. One of the important questions is whether they are linked on a single chromosome or located on several chromosomes? The initial approach is to localize (map) genes controlling QTs on a certain chromosome or chromosomes. In the context of this analysis, these genes are called **quantitative trait loci (QTLs)**. As we will see later, to map these genes there are some special methods.

As an example of such an approach, let us consider the resistance to the insecticide DDT in *Drosophila*. To obtain the loci controlling QT, two kinds of fly lines, selected for resistance to DDT and strains selected for sensitivity to DDT, were crossed with lines carrying dominant genes, which serve as markers on each of the four chromosomes of *Drosophila*. In the experimental crosses, offspring that contained many different combinations of marker chromosomes and chromosomes from either resistant or sensitive strains were obtained. After that, individuals with various chromosome combinations were tested against the resistance to DDT by their survival rate during

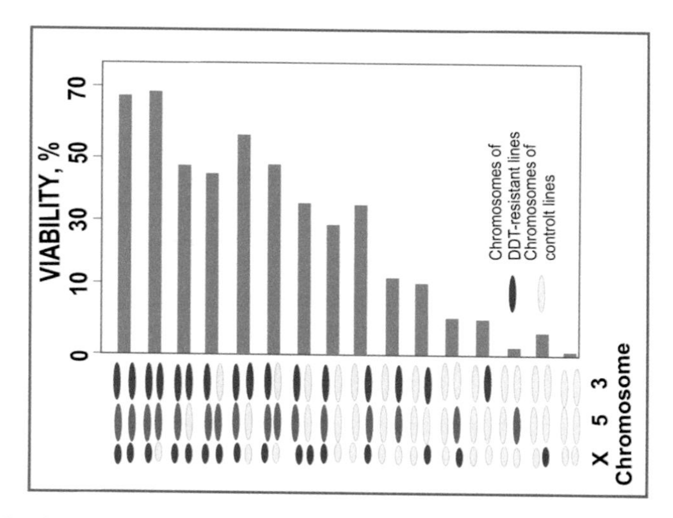

Fig. 11.5.1. Survival rates of *Drosophila* carrying combinations of chromosomes from DDT-resistant and DDT-sensitive (control) strains when exposed to DDT. The results indicate that DDT resistance is polygenic, with genes on each of the major chromosomes making a contribution. Chromosome 4 carries only a few *Drosophila* genes and was omitted from this analysis.

exposition to insecticide (Fig. 11.5.1). This experiment obviously shows that genes contributing to the resistance are present on most chromosomes in *Drosophila*. In other words, the loci bearing the genes that control DDT resistance are spread throughout the *Drosophila* genome as presented earlier (Klug and Cummings 2002, Figs. 6–8).

Now it is possible to map the position of QTLs more specifically along each chromosome in *Drosophila* and other species if we previously detect the presence of molecular markers on each chromosome. The locations of QTL are determined relative to the known positions of these markers. One group of these markers, called **restriction fragment length polymorphisms (RFLPs)**, represents specific nuclease cleavage sites. Other markers may be applied too.

This approach enumerates and maps the loci responsible for QTs. RFLP markers are now available for many organisms of agricultural importance, making possible systematic mapping of QTL. For example, hundreds of RFLP markers have been located in the tomato. They are located on all 12 chromosomes of this plant. Sometimes a marker gene and a phenotypic trait of interest are jointly expressed; in this case, it is so-called **cosegregation**. Significant cosegregation establishes the presence of a QTL near the RFLP or other marker on the chromosome. Whenever both a marker and a QTL responsible for the trait under investigation are linked on a single chromosome, they are more likely to demonstrate an association throughout breeding than if they are not linked. Obviously, this finding can be valuable for practical breeding. When numerous QTL are located, a genetic map is created for all the involved genes.

Correlated Response and Genetic Correlation

Many genes have pleiotropic effects on the phenotype. This means that each gene could potentially affect many traits in the organism; these effects may be either primary or indirect. This matter is quite complicated, as was introduced in Chapter 5 (Section 5.4). The alleles that are favourable for one quantitative trait may have deleterious effects on another quantitative trait. When these alleles are increased in frequency due to artificial selection (thereby improving the phenotypic score of the selected quantitative trait), the same alleles may bring deterioration to some other traits of the phenotypic performance. Pleiotropy is one cause of such a **correlated response**. A second possible cause of correlated responses is linkage disequilibrium; a favourable allele for one trait that increases in frequency under selection may correlatively impact an allele of another, linked gene that has a detrimental effect on an unselected trait. We will see also in Chapter 12 different allelic interactions within a locus, and we will consider their association in multiple-locus genotypes that may affect the QT.

Correlated responses are quite common in artificial selection, and often, but not always, result in deterioration in reproductive performance. In the case of Leghorn chickens, for example, 12 generations of selection for increased shank length reduced the egg hatchability by nearly half (Lerner 1958). Viability in the line of Donaldson's rainbow trout, which was mentioned above, out of very strict temperature regime is so low that it makes impossible its use in a case of a poly culture, where there is no necessity of hard temperature regulation. Strictly speaking, any breeding leads to the creation of a line (kind), which is fit to certain "selected" conditions. Outside these conditions, the high vigour of the organisms of the line will not be realized; moreover, the line may be even worse than unselected organisms under environmental fluctuations (Zuchenko 1980).

The complexity of breeding response is well illustrated by selection of weight gain in rats. An artificial selection experiment was carried out for 23 generations in favour of weight gain in rats, between the third and ninth week (Baker et al. 1975). Six strains were characterized for 17 skull traits in order to examine patterns of correlated responses (Atchley et al. 1982). The characters included 6 measured traits such as skull length, skull width, inter-orbital width, etc. The experiment consisted of two unselected control groups, two lines selected up for weight gain, and two lines selected down for weight gain, with all lines originating from a common stock. The direct response was remarkable for the magnitude of asymmetry: the males selected for greater weight gain were 3.46 standard deviations larger than the controls, while the males selected for less weight gain were 1.30 standard deviations smaller than the controls. Similarly, females selected for greater weight gain were 1.46 standard deviations larger, and females selected for less weight gain were 1.02 standard deviations smaller. Asymmetric response is generally attributed to opposing natural selection, and no more specific mechanism can be offered here. Correlated responses showed similar degrees of asymmetry. However, while the replicate lines showed statistical consistency in the magnitudes of *direct* responses, 35 of 51 comparisons between males in replicate lines were significantly different in their **correlated** responses. The irreproducibility of correlated response may be due to the greater sensitivity of genetic covariances to the perturbing effects of selection and random drift.

Atchley et al. (1982) performed multivariate analyses of the correlated characters, and obtained canonical variates (linear combinations of the characters that account for the greatest variance). In comparing the heritability of the individual characters to the heritability of the canonical variates, they found that the latter generally had higher narrow-sense heritability. This finding underscores the arbitrariness of the way characters are measured and gene action is inferred, because ***the pattern of genetic correlation suggests that the genetic variation affects suites of characters in coordinated fashion. Rather than imagining genes affecting skull length and jaw width, the data suggest that complex patterns of traits are the targets of gene action.*** Similar ideas were developed by Zhivotovsky (1984). The complexities of examining evolutionary change in multiple characters were apparent from the rat data, and it might seem that a complete description of multiple-trait selection would be hopelessly complex. Indeed, actually there is no theory yet formed for QT gene control. The Gaussian distribution is only an approximation. However, the theoretical problem becomes tractable by making certain assumptions. The most important is that the multidimensional distributions of phenotypes and of genetic effects are assumed to be multivariate normal. If the initial distribution is multivariate normal, and selection favours the reproduction of some phenotypes over others, then the distribution would be expected to depart from normal after selection. Although this is true, if the selection is sufficiently weak, then the phenotypic distribution will remain approximately normal, at least in the short term.

Evolution of Quantitative Traits: Mutation-Selection Balance

Directional selection, as has been seen in the previous sections, enormously changes traits and lines of organisms. Since Darwin's effort, it is commonly recognized that in nature, directional selection is the main cause of evolution. However, this is not the only form of selection. Stabilizing (purifying) selection is of great interest for the role it may play in the persistence of heritable genetic variation in multifactorial traits. Intuitively, it seems reasonable to suppose that observed levels of additive genetic variation might result from a balance between stabilizing selection, which tends to reduce genetic variation, and new mutations, which tend to increase it. Although reasonably simple when stated verbally, the models become complex if one attempts to produce an exact mathematic theory. Complications include the number of genes, the type of action of alleles and their interactions, linkage between genes, the type and intensity of selection and the influence of selection on other traits, which are related through pleiotropy or epistasis. Pleiotropy is the widespread tendency of genes to affect several traits simultaneously, sometimes including seemingly unrelated traits, which usually results from the fact that complex phenotypic traits are determined by the interactions of the products of many genes during development.

The number of genes affecting a trait is relevant to mutation-selection balance because, as Crow and Kimura (1970) showed, for a given genetic variance, the selection intensity per locus decreases as the number of loci increases. If the total mutation rate is fixed, the per-locus mutation rate must decrease as the number of loci increases. For a population in mutation-selection balance, the intensity of selection per locus is approximately independent of the number of loci (Turelli 1984). Complex as the situation is, the hypothesis that additive genetic variation might be maintained by stabilizing selection balanced against mutation has been supported by mathematical approximations in the models studied by Lande (1975; 1977; 1979; 1980).

On the other hand, the approximations appear to be quite sensitive to the particular genetic and mathematical assumptions that are made. The simplest models ignore dominance and epistasis, and they assume a large number of alleles per gene, the distribution of effects of which are approximated by a normal or Gaussian distribution. Under such simplifying assumptions, solutions may be found (Kimura 1965; Lande 1975; Turelli 1984). More details on this theme may be found in the book by Hartl and Clark (1989, p. 502).

11.6 TRAINING COURSE, #11

11.1 Find h^2 for tilapia (see Table 11.6.1).

11.2 Solve exercises on the inheritance of quantitative traits: #1–3 (Textbook, Klug and Cummings 2002, P. 117, Problems and Discussion Questions).

11.3 Solve one example to define mean, standard deviation and SE with STATISTICA software.

11.4 Principles of ANOVA for h^2 determination. Schemes of cross: (i) hierarchical and (ii) polyallelic. Finding variances in complete 2 × 2 diallelic scheme of the analysis.

11.5 Consider model of directional selection for QT in software POPULUS > Evolutionary Simulations > Quantitative Genetics > Directional Selection at Quantitative Character.

11.6 Let us consider several new terms:

- **Polygenic trait** is a trait that is controlled by several genes.
- **Heritability** is the ratio of genetic variance to the total or phenotypic variance.
- **Additive allele.** Each locus may be occupied by either an **additive allele,** which add an impact to the phenotype, or by a **nonadditive allele,** which does not contribute cumulatively to the phenotype.
- **Clone** is a unity of genotypically identical individuals.
- **Variance**

  Genetic variance is a component of variability at a QT that can be attributed to genotypic control. It has a complex nature.

Table 11.6.1. Determination of h^2_b in fish tilapia through selection by C.M. Thieu (Kirpitchnikov 1979).

SEX	*P*			*F1*			$h^2 = R/S$
	Small	**Mid**	**Large**	**Small**	**Mid**	**Large**	
Female	15.7	20.2	24.8	11.8	12.9	12.9	$h^2_{L/S} = 0.12$ $h^2_{L/M} = ?$ $h^2_{M/S} = ?$
Male	16.8	23.8	28.4	19.3	20.3	20.7	$h^2_{L/S} = 0.12$ $h^2_{L/M} = ?$ $h^2_{M/S} = ?$

Environmental variance is a component of variability at a QT that cannot be attributed to genotypic control.

- **Normal distribution** is a distribution of density of probability with two parameters, which are **mathematical expectation** (μ) and **variance** (σ^2). Estimates of these parameters under sampling procedures are **mean** and **standard deviation**.

Scheme of crosses. To perform ANOVA it is necessary make crosses by special schemes (Fig. 11.6.1).

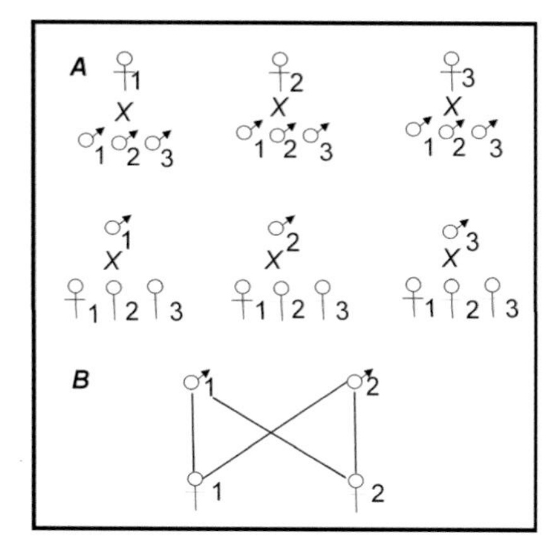

Fig. 11.6.1. Scheme of crosses: (A) Hierarchical complex (two variants) and (B) Diallelic cross.

12

GENETIC HOMEOSTASIS, HETEROSIS AND HETEROZYGOSITY

MAIN GOALS

12.1 Concept of Genetic Homeostasis
12.2 Heterosis and Heterozygosity
12.3 Empirical Results and Interpretation
12.4 TRAINING COURSE, #12

SUMMARY

1. **Genetic homeostasis** is a feature of a population to balance its genetic composition; it came to genetics from physiology. The concept of homeostasis is many-sided, and can be attributed particularly to ontogenesis.
2. The essence of **heterosis** and **heterozygosity** is different; however, the two notions overlap. A positive relationship between heterozygosity and heterosis may exist, but the former is not always a cause of the latter.
3. A relationship between heterozygosity and fitness can be explained both from the position of selective neutrality and from the position of selective significance of marker loci. The most recognized now are the **hypothesis of inbreeding depression**, which suggests a neutrality of gene markers, and the **balance hypothesis**, which is in principle an adaptive one. Neither of these hypotheses complies fully with empirical data. There is a need for a new model, which should be developed presumably on the basis of the balance hypothesis.
4. The new model should include a mechanism of balancing and relaxing selection, which acts bi-directionally during the life cycle, (1) as a regulator of the optimal heterozygosity in a population (support for a stable segregation load), and (2) as a regulator of a phenotypic homeostasis for **quantitative traits** and also accounting for identity vs. linkage disequilibria.

12.1 CONCEPT OF GENETIC HOMEOSTASIS

Cannon (1932) in the field of physiology developed the concept of physiological homeostasis. In general, it may be said, that the "***homeostasis*** *refers to the property of the organism to adjust itself to variable conditions, or to the self-regulatory mechanisms of the organism which permit it to stabilize itself in fluctuating inner and outer environments*" (Lerner 1970). Homeostasis may be attributed to different systems: to human society, to some properties of chemical substances (buffers), etc. So, in a certain sense we may talk of ecological homeostasis and **genetic homeostasis** (GH). The *GH can be defined as the property of the population to equilibrate its genetic composition and to resist sudden changes* (Lerner 1970). The GH includes not only group effects on the population level but also regulation at the individual level, including developmental homeostasis.

Three main premises should be accepted to discuss the GH (Lerner 1970; with a few additions).

1. The evolution of cross-fertilized organisms has effects on the genotypes that produce developmental patterns with a considerable degree of self-regulation. In embryological terms, the process of development of a specific form has become canalized, leading to a uniformity of phenotypic expression in individuals of a given population, in spite of the genetic variability between them.
2. Mendelian populations also possess self-equilibrating properties, tending to retain a genetic composition that produces a maximum average fitness in the particular environments in which these populations exist. This may extend to upper levels of a subdivided population too. The most likely mechanism for both types of homeostasis lies in the superiority with respect to fitness of the heterozygous over the homozygous genotypes: (a) ontogenetic self-regulation (*developmental homeostasis*) is based on the greater ability of the heterozygote to stay within the norms of canalized development; (b) self-regulation of populations (*genetic homeostasis*) is based on natural selection favouring intermediate rather than extreme phenotypes.

In this chapter, we will try to consider both the theory (or specifically, some models), and the empirical results on GH with respect to population level and the relationship of heterozygosity and fitness traits (FT). FT are major components of fitness, such as weight, height, fecundity, viability, etc., and that may direct GH.

The results of artificial selection have an important place in understanding the role of FT for GH. As depicted in Fig. 12.1.1, when selection is suspended the value of FT may return to the adaptive norm ($\underline{W}_1$; bottom, broken curve) or may shift to another condition norm, if a certain restriction level has been exceeded ($\underline{W}_2$; top curve).

Such results led to the idea that the genotype of the organisms has balanced combinations of genes. Due to selection it may be changed, but it tends to return to the adaptive norm ($\underline{W}_1$). If change is above some tolerance limit, however, than another adaptive norm ($\underline{W}_2$) may be achieved (Fig. 12.1.1, upper curves). This is a similar idea to the shifting balance concept, developed in genetic terms by Wright (1935; 1968).

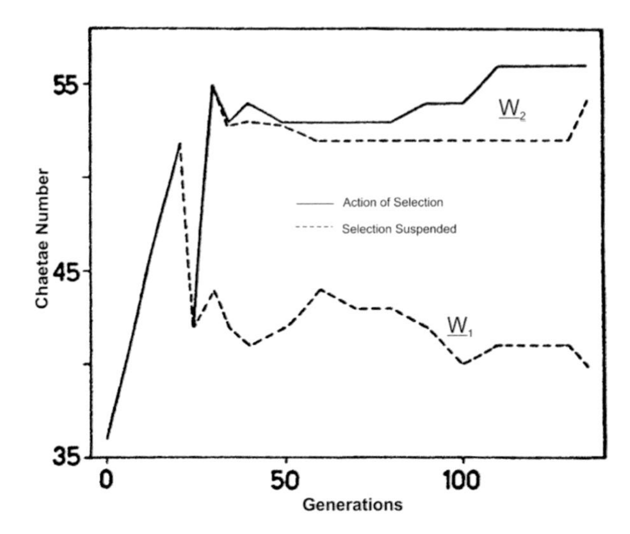

Fig. 12.1.1. Schematic representation of the results of selection for a high number of abdominal chaetae (FT) in *Drosophila melanogaster* (from Mather and Harrison 1949).

12.2 HETEROSIS AND HETEROZYGOSITY

The essence of heterosis and heterozygosity is different, although these two ideas overlap. The term ***heterosis*** was introduced by Shull (1914), for *hybrid vigour or increased "power" of F1 offspring as compared to both parent forms* without reference to a mechanism of this action. Vigour may occur in any quantitative trait (QT) value, like weight, length, fertility, etc. Formally it may be written as follows: $F_{P1} \times F_{P2} = F_H$; thus, that $F_{P1} < F_H > F_{P2}$. Here: F_{P1} and F_{P2} are QT scores in the parental lines or populations, F_H is the QT score in the hybrid offspring. Because the majority of QTs are connected with fitness in Darwin's sense, heterosis became a synonym for the increased fitness of hybrids. And because hybridization normally leads to an increased heterozygosity in F_1 (F_H), heterosis is sometimes equated to heterozygosity. Nowadays, at least four different definitions of heterosis are mentioned (Hershenson 1983; Strunnikov 1986) (Fig. 12.2.1).

Thus, we should clearly understand that a positive association between heterozygosity and heterosis can exist, heterozygosity is not always the cause of heterosis. Only in the variants 1 and 3 is heterozygosity the cause of heterosis (Fig. 12.2.1).

Symbolically, a relationship between the polymorphic gene *A* and quantitative expression of a trait under its control can be expressed like this:

1) $A_1A_1 = A_1A_2 = A_2A_2$: ***neutrality,***
2) $(A_1A_1 + A_2A_2)/2 > A_1A_2$; $A_1A_1 > A_1A_2 > A_2A_2$ or $A_2A_2 > A_1A_2 > A_1A_1$: ***additivity,***
3) $(A_1A_1 + A_2A_2)/2 < A_1A_2$; $A_1A_1 = A_1A_2 > A_2A_2$ or $A_2A_2 = A_1A_2 > A_1A_1$: ***dominance,***
4) $A_1A_1 < A_1A_2 > A_2A_2$: ***over-dominance.***

1. Compensation of a deleterious effect of recessive alleles.

2. Additive effect of favourable gene combinations.

3. Overdominance or mono-hybrid heterosis.

4. Super epistasis of viability modifiers.

1. Let A, B, C, D define a norm, while a, b, c, d are recessive markers and QTs are controlled by genes x, y, z. Then: P_1: aaBBccDD × P_2: AAbbCCdd, $F_1 = F_H =$ AaBbCcDd and $F_{P_1} < F_H > F_{P_2}$ or $x_1x_1y_1y_1z_1z_1 > x_0x_0y_0y_0z_0z_0$, where subscript 0 and 1 stands for P and $F_1 = F_H$.
2. Let $G_1W_1Y_1R_1$ and $G_2W_2Y_2R_2$ be the norm, $G_1Y_1R_2$ is a favourable genetic combination that increases QTs' value. Then in cross: P_1: $G_1G_1W_1W_2Y_1Y_2R_1R_1$ × P_2: $G_2G_2W_2W_2Y_1Y_2R_2R_2$, part of $F_1 = F_H = G_1G_2W_1W_2Y_1Y_2R_1R_2$ and $F_{P_1} < F_H > F_{P_2}$.
3. Let $A_1A_2A_3$ be the norm. Then if $A_1A_2, A_1A_3, A_2A_3 > A_1A_1, A_2A_2, A_3A_3$, $F_1 = F_H = F_{P_1} < F_H > F_{P_2}$ or $x_1x_1y_1y_1z_1z_1 > x_0x_0y_0y_0z_0z_0$, when subscript 0 and 1 stands for P and F_1.
4. Let $x_1x_1y_1y_1z_1z_1$ and $x_1x_1y_1y_1z_2z_2$ be sub-vital inbred lines. In them, the recessive genes are x, y control QTs and z is a modifier of their viability, such that it keeps their vitality at a sustainable level. If $z_1, z_2 > x_1, y$: (epistasis), then under the cross P_1: $x_1x_1y_1y_1z_1z_1$ (epistasis) × P_2: $x_1x_1y_1y_1z_2z_2$ (epistasis), super-epistasis is established: $F_1 = F_H = x_1x_1y_1y_1z_1z_2$ and $F_{P_1} < F_H > F_{P_2}$.

Fig. 12.2.1. Four schemes of heterosis: 1–4 (Kartavtsev 2009b).

The association between the heterozygosity for marker genes and the variability of an adaptively important trait can be theoretically explained in terms of both neutrality and selection theory. Currently, two concepts are most widely accepted: the inbreeding depression model (Mitton and Piers 1980; Chakraborty 1981; Zouros 1987); and the varying selection or general balance model, introduced via adaptive distance (Smouse 1986; Bush et al. 1987). These may be considered as basically neutral in respect of marker genes and selection respectively. Several attempts have been made to find analytical solutions for heterozygosity and the quantitative trait (QT = FT) relationship (Chakraborty 1981; Chakraborty and Ryman 1983; Bush et al. 1987; David 1997). Recent theoretical development (David 1997) basically supports the inbreeding depression model. Updates on the subject have been presented in two reviews (Chapman et al. 2009; Szulkin et al. 2010).

Several authors have analysed the expected relationships and presented numerical simulations for heterozygosity and QTs (Bezrukov 1989; David 1997; Kartavtsev 2005a). Mitton and Piers (1980) and Chakraborty (1981) concluded that the positive association between heterozygosity and QTs found in some studies of allozyme loci merely resulted from multiple-locus averaging, while the absence of correlation in other empirical studies was attributed to differences between genomic homozygosity and sampled heterozygosity or weak sampling. These and other authors believe that only inappropriate research design (Szulkin et al. 2010) or weak correlations between genomic and sample heterozygosities (Zouros 1987) prevent researchers from obtaining sufficient experimental evidence for these relationships. This is one of the most serious problems related to the interpretation of experimental results in favour of one of the two aforementioned alternative hypotheses. The analysis of the relationship between heterozygosity and QT is also related to many other problems (Mitton and Grant 1984; Zouros 1987; Zouros and Foltz 1987; Chapman et al. 2009). One of them is the heterogeneity of the compared sets of objects (populations, species, years of research, etc.) with respect to mean heterozygosity, which, according to the results of some studies (Mitton and Piers 1980; Waples 1991), inevitably causes pseudocorrelations or creates other statistical problems of the comparison (Chapman et al. 2009).

Chapman and coworkers (Chapman et al. 2009) highlight another aspect of the research, a selective neutrality of marker genes. In their survey they considered only microsatellite loci, which are widely assumed by most researchers to be selectively neutral; thus, when a suitable number of loci is used, they may be the best markers of a genome's heterozygosity level or inversely the level of inbreeding. Notwithstanding the numerous studies and reviews on various aspects of the relationship between heterozygosity and QTs (Mitton and Grant 1984; Zouros 1987; Zouros and Foltz 1987; Mitton and Lewis 1989; Chapman et al. 2009; Szulkin et al. 2010, etc.), researchers seem to lose sight of one important aspect of heterozygosity vs. QT relationships; namely, the existence of single-locus allelic effects on QT and the non trivial combination of these effects in a multiple-locus genotype. In the study by Bezrukov (1989) cited above, intralocus effects were mainly used for a classification of single-locus action. David (1997) considered to a certain degree interlocus effects. Kartavtsev (2005a) considered intralocus effects and their impact on multiple-locus averaging, but only limited numerical simulations were made, without considering QT variance at all.

Inbreeding Depression Hypothesis

This hypothesis suggests that the marker loci themselves are selectively neutral. They are only markers of genome heterozygosity (Mitton and Piers 1980; Chakraborty 1981). The lower the marker heterozygosity, the higher is genomic homozygous and inbreeding, i.e., consequently, there are more deleterious genes in this condition. Of principal significance here are dominance-and-recessiveness relationships.

We will consider this hypothesis by the example from Zouros (1987) (Table 12.2.1). Let QT (say weight of an animal at age X), be under the control of two loci, ***A*** and ***B***, each with two alleles. Genotypic scores for locus ***A*** are $A_1A_1 = 2$, $A_2A_2 = 6$, and $A_1A_2 = 5$. These are means for the three genotypes, weighted across their frequency in population. Genotypic scores for the three genotypes of locus ***B*** are 2, 8 and 7. There is no overdominance, neither for ***A*** ($2 < 5 < 6$) nor for ***B*** ($2 < 7 < 8$). Let us, for the sake of simplicity, obtain two-locus genotypic scores as averages of the single-locus scores (Table 12.2.1). In each cell of the table, slashes separate scores averaged as arithmetic, geometric and harmonic means.

In the second part of Table 12.2.1 (b) the 9 two-locus genotypes are combined in three groups (with 4, 4 and 1 scores in the group), in accordance with the degree of their heterozygosity ($H_o = 0$, 1 and 2). Average values were then calculated for each group. We see that the higher the heterozygosity, the greater is genotypic score (Table 12.2.1b). This holds true for different averaging modes. Thus, without overdominance, averaging single-locus data into multiple-locus sets gives a correlation, in this example a positive one. In the second example, for figures shown in brackets in the table, all these calculations were repeated with changed genotypic scores for heterozygotes (A_1A_2 changed from 5 to 3 and B_1B_2 from 7 to 3) (Table 12.2.1). Now the correlation between

Table 12.2.1. Numerical example of the averaging effect of genotypic values at multi-loci genotypes in accordance with certain heterozygosity rank (Zouros 1987).

(a) Obtaining of two-loci genotypic values from single-locus values

Genotypic Value		**A_1A_1 2(2)**	**A_2A_2 6(6)**	**A_1A_2 5(3)**
B_1B_1	**2 (2)**	2/2/2 (2/2/2)	4/3.46/3 (4/3.46/3)	3.5/3.16/2.86 (2.5/2.45/2.4)
B_2B_2	**8 (8)**	5/4/3.2 (5/4/3.2)	7/6.93/6.86 (7/6.93/6.86)	6.5/6.32/6.15 (5.5/4.9/4.36)
B_1B_2	**7 (3)**	4.5/3.74/3.11 (2.5/2.45/2.4)	6.5/6.48/6.46 (4.5/4.24/4)	6/5.91/5.83 (3/3/3)

(b) Mean values for two-locus genotypes in accord with their rank of heterozygosity

Model	**Heterozygosity**		
	0	**1**	**2**
1. Arithmetic	4.5 (4.5)	5.25 (3.75)	6.0 (3)
2. Geometric	4.1 (4.1)	4.92 (3.51)	5.91 (3)
3. Harmonic	3.76 (3.76)	4.64 (3.29)	5.83 (3)

Note. The second set of scores (second example) is shown in brackets.

H_0 and QT is negative (Table 12.2.1b). In the first example, the situation of intralocus interaction (3) or dominance is considered. In the second example, a case close to the situation (2) or the single-locus additivity is presented. The numerical example for case (2) of additivity is not supported by theoretical predictions for that case (Bezrukov 1989, Fig. 1), but is supported for the case (3) of dominance (Chakraborty and Ryman 1983), as well as for case (4) of overdominance (Bezrukov 1989, Fig. 1) (see discussion below). However, in accord with my consideration (Kartavtsev 2005a; 2009b), the theory predicts both the negative H_0 & QT association when additive effects exist, the positive H_0 & QT association under single-locus dominance and overdominance. Bezrukov (1989, Fig. 1) showed absence of correlation under additivity, but here the negative correlation is predicted (Kartavtsev 2005a; 2009b; see also Fig. 12.3.5). Thus, for general judgment in support of the inbreeding depression hypothesis, it is not necessary to use averaging or the concept of multi-locus overdominance (see the above and the next paragraphs). We only need to know whether dominance prevails in nature in comparison with the additive mode or with other types of gene action. However, I am not sure that the fact of dominance prevailing is really established. Contrary to this, additive gene action in all quantitative genetics and breeding practice is widely accepted, and only for viability and fertility traits is the role of nonadditive effects recognized as considerable (Kirpichnikov 1987). However, Zouros' (1987) opinion is different, and he supports his conclusion with reasonable evidence.

In connection to the above discussion, two questions were raised (Mitton and Piers 1980; Chakraborty 1981; Zouros and Foltz 1987). (1) Do the marker loci investigated represent a sub-sample of all the genome's many hundreds or even thousands of loci? (2) If so, do they represent a random sample in terms of the association of H_0 & QT and in terms of dependence of variability of QT on H_0 rank? These authors' conclusion was negative. That is, if the number of loci investigated is small, say less than 20 (as in the majority of experimental research), the rank of the individual's heterozygosity is practically not correlated with the sampled heterozygosity, and is independent of the QTs. In the logic of the authors cited, this is the main reason why in the majority of studies, correlations between H_o & QT were not found. I may disagree here with them. The reason may be simply the absence of prevalence of dominance, as we noted earlier.

Balance Hypothesis or Hypothesis of Overdominance

In a case of overdominance, theory predicts straightforward and positive proportional association of H_o & QT, and a number of formal notations are available on the subject (Fig. 12.2.2, Fig. 12.3.6–Fig. 12.3.7). This hypothesis predicts that marker loci themselves or closely linked markers influence the association of H_o & QT.

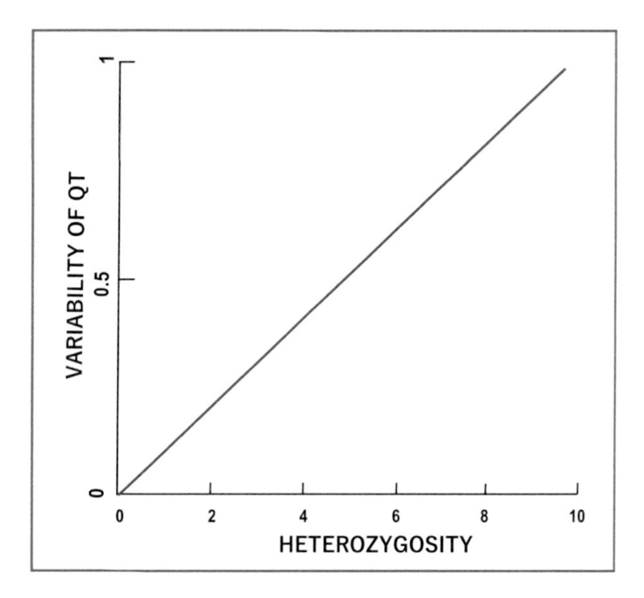

Fig. 12.2.2. Dependence of mean QT scores from the number of heterozygous loci: the overdominance case. On the X-axis is H_o the number of heterozygous loci in an individual. On the Y-axis is QT score in an individual ranging from zero to one. Numbers beside lines denote number of loci in the sample of 10 loci under the simulation.

Formally this situation is described by case (4) of intralocus allele effects above and simulated for instance by Bezrukov (1989) and Kartavtsev and Zhdanova (2012) for combined H_o action. Turelli and Ginzburg (1983) showed that under any of the various modes of balance selection, increase of fitness is also expected with an increase of heterozygosity. Smouse (1986) developed a multiplicative model of overdominance, in which the inbred depression model is a special case. The author called it the model of adaptive distance. Ideologically this model is connected with the cited development in Turelli and Ginzburg (1983), which showed that marginal over-dominance is the most frequent consequence of multiple-locus polymorphism.

We will consider the model of adaptive distance (AD) in brief, as it is presented in Bush et al. (1987). The model is designed to describe a relation between $\boldsymbol{Y = Log(W)}$, where W is fitness, and $\boldsymbol{X}$, that equals AD in the form of allele frequencies: $\boldsymbol{P_A = f(A_1)}$ and $\boldsymbol{Q_A = f(A_2)}$.

Here three genotypes of a diallelic locus give scores for AD via allele frequencies: $P_A = f(A_1)$ and $Q_A = f(A_2)$. Smouse (1986) showed that when S_A and T_A are selection differentials (1-s and 1-t) for homozygotes A_1A_1 and A_2A_2, consequently, equilibrium allele frequencies are the functions of S_A and T_A. Thus, if we designate AD as follows:

Genotype	A_1A_1	A_1A_2	A_2A_2
AD	P_A^{-1}	0	Q_A^{-1}
W	$-S_A$	1	$-T_A$

Then the relationship between $\boldsymbol{X}$ and $\boldsymbol{Y}$ will be linear under equilibrium (Fig. 12.2.3). The coefficient of regression X_A is the so-called **segregation genetic load** for this locus (Morton et al. 1956):

$$\boldsymbol{a = [S_A T_A/(S_A + T_A)].} \qquad \textbf{(12.1)}$$

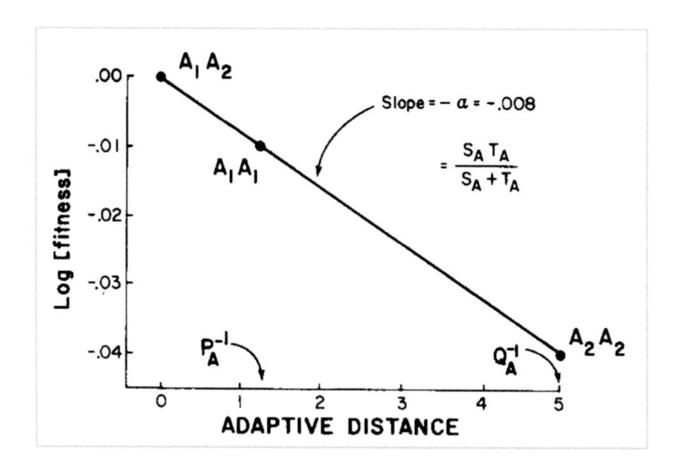

Fig. 12.2.3. Example plot of log (fitness) against adaptive distance (AD) for the three possible genotypes at locus A.
The AD value of the heterozygote is zero. The AD value for each homozygote is equal to the inverse of the equilibrium frequency of the corresponding allele. Here, $P_A = 0.80$ and $Q_A = 0.20$ (from Bush et al. 1987 with permission from Wiley & Sons publisher).

In this case, we can define the regression model for j-th individual as follows:

$$\boldsymbol{Y_j = 0 - aXA_j + e_j} \qquad \textbf{(12.2)}$$

Where $\boldsymbol{e_j}$ is the member that defines both the error of estimate in $\boldsymbol{Y_j}$ and deviation from the model. For the multiplicative model of overdominance, the AD values for different loci are additive. So, (11.2) it can be extended to:

$$Y_j = 0 - aXA_j - bXB_j - ... - kXK_j + e_j. \qquad \textbf{(12.3)}$$

A zero first member in the equation suggests using a K-locus heterozygote as a reference genotype.

Usual estimates of P_A and P_Q are given as

$$P = Z/N \text{ and } Q = (N - Z)/N, \qquad \textbf{(12.4a)}$$

Where Z is the number of alleles for A_1, and $(N - Z)$ is number for A_2. Expressions for P and Q, as well as their inverse values P_A^{-1} (1/P) and Q_A^{-1} (1/Q), are biased, and it is possible to avoid this by substitution:

$$P = (Z + 1)/(N + 1) \text{ and } Q = (N - Z + 1)/(N + 1). \qquad \textbf{(12.4b)}$$

Inverse values for these latter estimates have expectations (Smouse and Chakraborty 1986):

$$E[1/P] = 1/P[1 - Q]^{N+1} \text{ and}$$
$$E[1/Q] = 1/Q[1 - P]^{N+1}. \qquad \textbf{(12.5)}$$

These last expressions give a quantitative basis for empirically testing the overdominance hypothesis. In particular, in a study of pine tree *Pinus rigida* (Bush et al. 1987) this model better fit the empirical data. Some other relevant data will be considered in the next section.

12.3 EMPIRICAL RESULTS AND INTERPRETATION

In accord with our main scope, we will consider largely data for fish and shellfish species. Mollusks were analyzed mostly among these animals, although a considerable literature is available for other organisms, such as humans (Dubrova et al. 1990; 1994), fish (Mitton and Lewis 1989; Dubrova et al. 1994), and coniferous plants (Politov et al. 1992). From seventeen investigations of mollusks, only in five was the correlation between heterozygosity and growth rate not observed (Table 12.3.1).

Correlations as a rule are stronger among juveniles and weaker or even not seen in late ontogenetic phases. In the mollusks, this tendency is connected with the heterozygosity deficit index (D = –Fis), which in lower in adults (Levinton and Koehn 1976; Kartavtsev 1979; Koehn and Gaffney 1984; Zouros and Foltz 1987; Gaffney et al. 1990). In pink shrimp correlation of H_o with the two indexes was found only for males, the smallest group in this transition hermaphrodite: $r_p = -0.24$, $r_p = -0.26$ ($n = 55$, $P < 0.05$ and $P < 0{,}025$; Kartavtsev 1996a). There is association of the two indices W/L and H/L with H_o in two samples of the pacific mussel *M. trossulus*: $F = 3{,}126$ (d.f. = 2, 67, $P = 0.05$) and $F = 4{,}980$ (d.f. = 2, 36, $P = 0{,}012$), but not in another 15. Heterogeneity in this respect among 15 samples was statistically significant: $F = 1{,}535$ (d.f. = 33, 1234, $P = 0{,}027$; Kartavtsev 1995). Similar results were obtained by Gaffney (1990) for *M. edulis*. In salmon many QT & H_o associations or other types of heterozygosity gradation have been reported (Leary et al. 1985; 1987; Danzman et

Table 12.3.1. Heterozygosity and growth rate in the natural population of mollusks (Modified from Zouros 1987).

Species	N	k	*D*	r_p	Source	Note
Crassostrea virginica	400	5	–	+	1	Juvenile
Same	1500	7	–	+	2	Same
-”-	200	4	–	+	3	-”-
Mytilus edulis	650	5	–	+	4	-”-
Same	500	5	0	0/–	5	Mixture of different ages
-”-	300	7	–	+	6	Mixture of different ages
-”-	2739	5	–	0/+	7	Same, 6 samples
-”-	106	?	?	+	8	Same, + – increased density
Macoma baltica	150	6	–	+	9	?
Placopecten magellanicus	60	6	0/–	0	10	Single age
Pecten maximus	110	5	0	0	11	Mixture of different ages?
Taias hemastoma	290	6	–	+	12	?
Mulinia lateralis	110	6	0	+	13	Mixture of different ages?
Same						
-”-	80	18	?	+	14	Juvenile
Mytilus edulis	3726	17	?	+	15	Juvenile
Same	> 3000	5	–	+	16	Mixture of different ages
Same	551	5	–?	0	17	Same
Mytilus trossulus	1234	15	–	+/0*	18	Same

Note. N is the number of animals studied, k is number of loci analysed, D is an index of excess (+) or deficiency (–) of heterozygotes (D = –Fis), r_p is the coefficient of correlation between heterozygosity level and growth rate (only statistically significant values are shown): “+” it is positive, “–” negative, “0” absence of correlation. Sources: (1) Singh and Zouros 1978, (2) Zouros et al. 1980, (3) Singh 1982, (4) Koehn and Gaffney 1984, (5) Diehl and Koehn 1985, (6) Zouros et al., in press (from Zouros 1987), (7) Gaffney 1990, (8) Gentini and Beaumont 1988, (9) Green et al. 1983, (10) Zouros and Foltz 1984, (11) Beaumont et al. 1985, (12) Garton 1984, (13) Garton et al. 1984, (14) Diehl et al. 1986, (15) Koehn et al. 1988, (16) Gardner and Skibinski 1988, (17) Gosling 1989, (18) Kartavtsev 1995 (*denotes variance analysis).

al. 1989; Mitton and Lewis 1989; Kartavtsev 1990). For example, the heterozygosity rank of parents influences the viability of larvae in pink salmon *Oncorhynchus gorbuscha* (Fig. 12.3.1).

Two further QT & H_o associations are shown in Figs. 12.3.2–12.3.3. Obviously, in pink salmon fry there is negative relationship between an individual’s deviation from the group centre in QTs score (P) and H_o (Fig. 12.3.2). Morphologically average individuals are more heterozygous than small individuals (Fig. 12.3.3). Multiple-locus heterozygosity correlates with QT in many fish species (Mitton and Lewis 1989). In salmon, which are better investigated than other fishes, QT & H_o associations are detected usually at early stages of the life cycle or in the group that ranked in accord with environment quality rank (Altukhov and Varnavskaya 1983; Altukhov 1989; Leary et al. 1985; 1987; Danzman et al. 1989; Kartavtsev 1990; 1992b; 1998; Kartavtsev et al. 1990; Kartavtsev et al. 1991), but not in the adults from the wild (Beacham and Withler 1985).

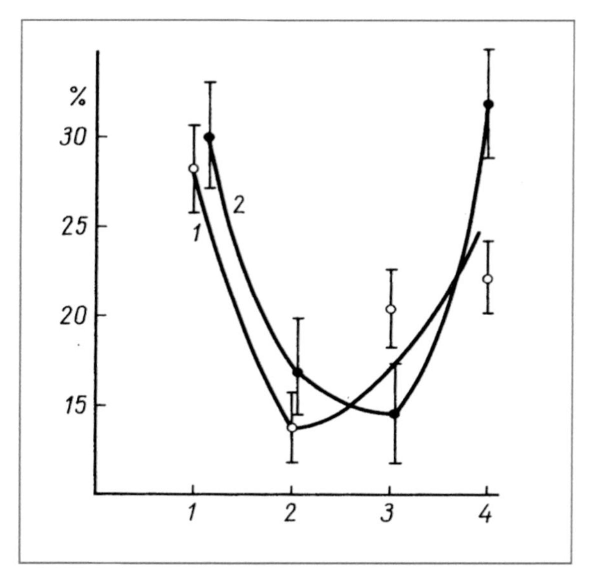

Fig. 12.3.1. Association of heterozygosity rank of parents and the death rate of larvae in pink salmon (from Kartavtsev et al. 1990).
On the X-axis is heterozygosity rank (number heterozygous loci in P1 + P2), on the Y-axis is death rate. Ranks are defined as the sum of heterozygous loci in the two parents: 1–0 (♀) + 0 (♂); 2–2 (♀) + 0 (♂); 3–0 (♀) + 4 (♂); 4–2 (♀) + 4 (♂). Significance, curve 1: t_d (1–2) = 8.7, P < 0.001; t_d (2–4) = 5.9, P < 0.001, n = 5184; curve 2: t_d (1–2) = 8.2, P < 0.001; t_d (3–4) = 8.1, P < 0.001, n = 2732.

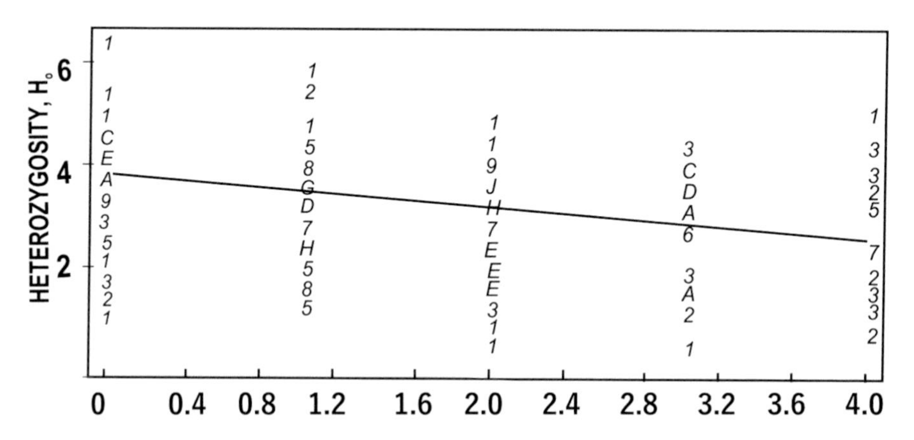

Fig. 12.3.2. Correlation of H_o variability with deviation from group torus at 10 morphology traits in pink salmon fry, taken in di-allele crosses.
On the X-axis is H_o, on the Y-axis is posterior probability (P) of an individual's exclusion from the certain H_o-group during discriminant analysis. Numbers and letters indicate frequency of occurrence (from Kartavtsev 1990).

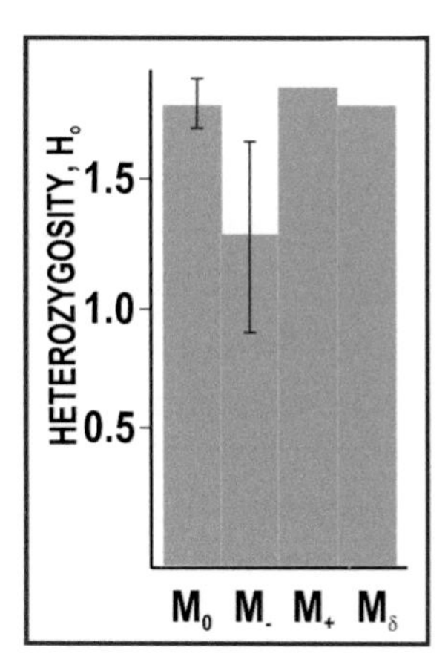

Fig. 12.3.3. H_o comparison in four morphologically distinct groups of pink salmon fry in experimental unit 2: M_0, M_-, M_+ are individuals with average, small and large morphology scores, M_σ is disproportional individuals. Lines are confidence limits (95%) (from Kartavtsev 1990).

Relationships of QT with heterozygosity are not always linear (Fig. 12.3.4; see also Fig. 12.3.1 above). Such non-linear relationships are presented in many papers (Zouros and Foltz 1987; Mitton and Lewis 1989; Dubrova et al. 1990; 1995).

Which of the two hypotheses, which we considered in the Section 12.2, fit the empirical results better? At first glance, the hypothesis of inbreeding depression is more attractive because of generality and well-known deleterious effects under

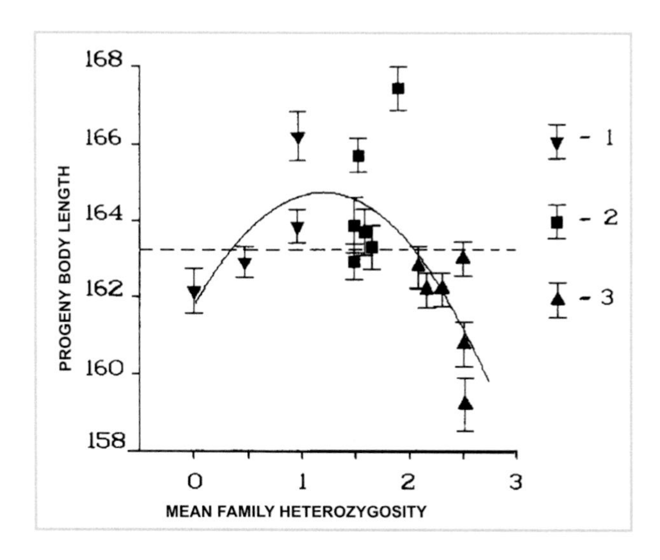

Fig. 12.3.4. The relationship between mean family heterozygosity and progeny body length of pink salmon: (1) families with low heterozygosity; (2) families with intermediate heterozygosity; (3) families with high heterozygosity. The broken line indicates the mean body length value for all the progeny examined. Linear fit (not shown) $y = 164.9 - 0.87x$; F (1; 2563) = 66.79, $P < 0.001$; $R^2 = 0.0253$; quadratic fit $y = 161.7 + 5.51x - 2.27x^2$; F (2; 2562) = 189.23, $P < 0.0001$, $R^2 = 0.1290$ (from Dubrova et al. 1995 with consent from Wiley & Sons publisher).

different inbreeding levels. However, empirical results don't seem to agree well with it. Disagreement is seen in the following 7 topics.

1. Correlations obtained belong in many cases to small samples of loci from the genome.
2. Power of correlations changes through the life cycle.
3. Ranking in accord with environmental stress strengthens the correlations.
4. Sign of correlation between QT and heterozygosity may differ among populations and species, i.e., it is not only positive, but is sometimes negative.
5. Variance of FT is correlated positively but not negatively with heterozygosity in some cases, as this hypothesis predicts.
6. True overdominance occurs in some cases.
7. Associations of FT and heterozygosity are nonlinear in many cases.

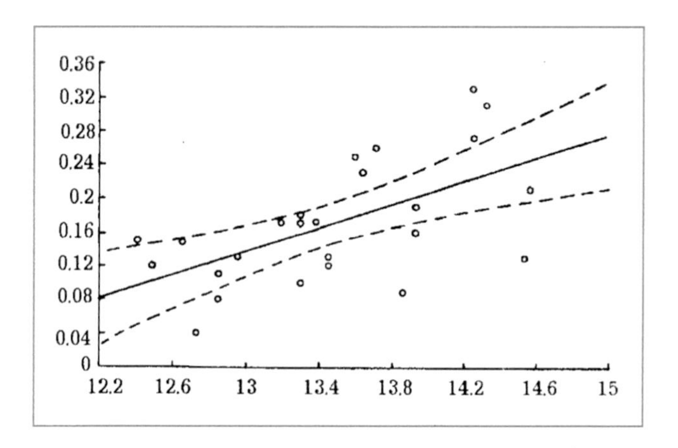

Fig. 12.3.5. Associated variation in expected heterozygosity (Hs) and body weight in pink salmon fry among groups that subjected to rearing in conditions of different density.
On the X-axis, square root of weight (m), on the Y-axis, Hs for two allozyme loci. Association is statistically significant: $r_p = 0.59$ ($n = 25$, $p < 0.025$). The regression line $y = -0.7537 + 0.06855x$ and its confidence limit (95%) are shown (from Kartavtsev 1998).

Does that mean that the balance hypothesis fits the empirical results better? I doubt this, although most of the seven topics above may be explained by this second hypothesis. What are the reasons for this scepticism? Before answering this question let us consider together the theoretical predictions of the relationship between FT and heterozygosity under different modes of gene action (Kartavtsev 2005a; 2009b; Kartavtsev and Zhdanova 2013) (Fig. 12.3.6).

For four types of intra-locus action (additivity, dominance, overdominance and neutrality) the expression of genotypic values (***Y***) across the dose of gene (***X***) are considered, based on a linear model. Analytically and with the utility of artificial numerical examples it is shown that under additivity, dominance and overdominance there is a positive relationship between ***X*** and ***Y***, while under neutrality there is no relationship between these variables. Averaging of single-locus genotypic values

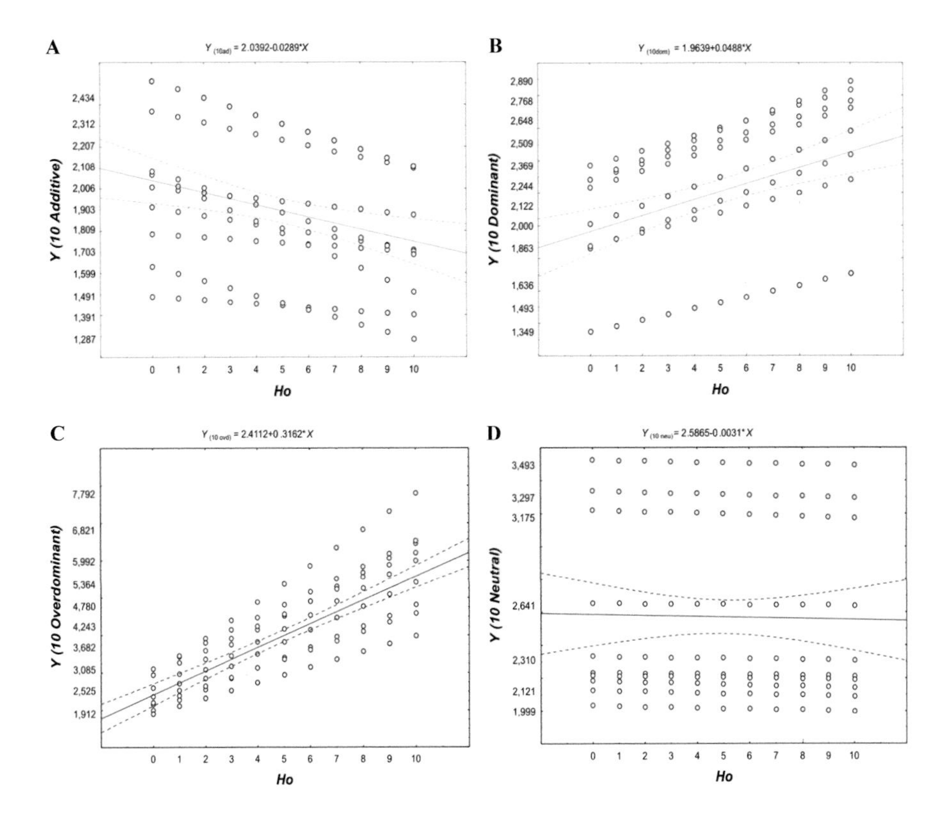

Fig. 12.3.6. Association of individual heterozygosity level (H_o) and mean genotypic value at ten-locus genotype scores (Y).
1–10 are H_o-classes, which represented by the number of heterozygous loci in an individual, H_o = 0, 1, ... 10. Curves depict linear regression fit in simulated examples from the source table. (A) Additivity, (B) Dominance, (C) Overdominance, (D) Neutrality (from Kartavtsev and Zhdanova 2012).

into multi-locus genotypes (as in Zouros' example above) does not give uniform results with different types of intra-locus action (Fig. 12.3.6). Under dominance and overdominance, the genotypic values and individual heterozygosity are positively correlated, but under additivity these two variables are negatively correlated. Under neutrality there is no relation between two variables with averaging, the same as in the single locus case.

Combining these results gives a clear view of the relationship between H_o and QT that is similar to many empirical data (Fig. 12.3.7).

Extended Data Analysis

The above analyses suggest a new view on the available empirical data and generalizations made about the relationship between heterozygosity and QT. First, it has become clear that the relationship varies depending on the type of intralocus interactions. When data on multiple-locus genotypes are pooled, averaging affects the

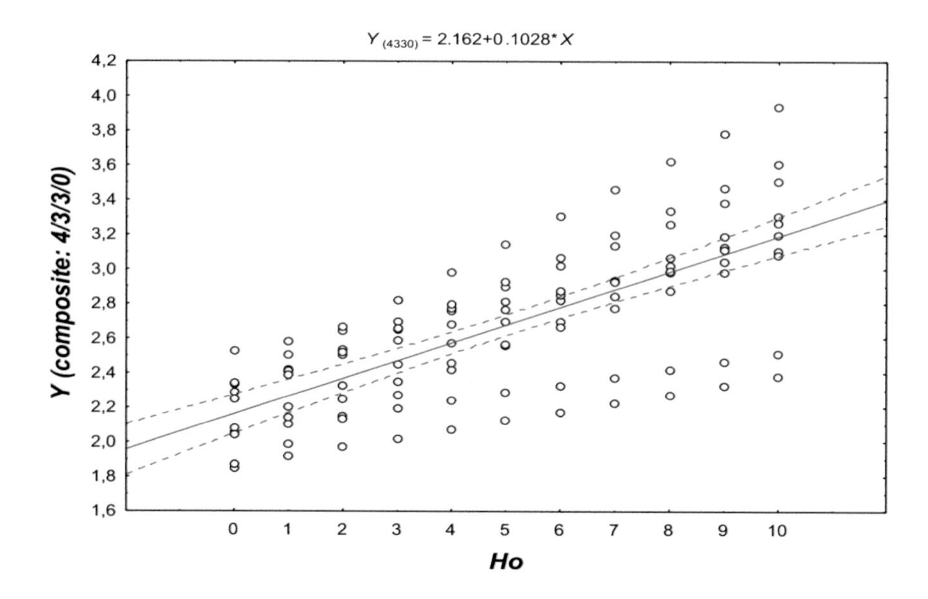

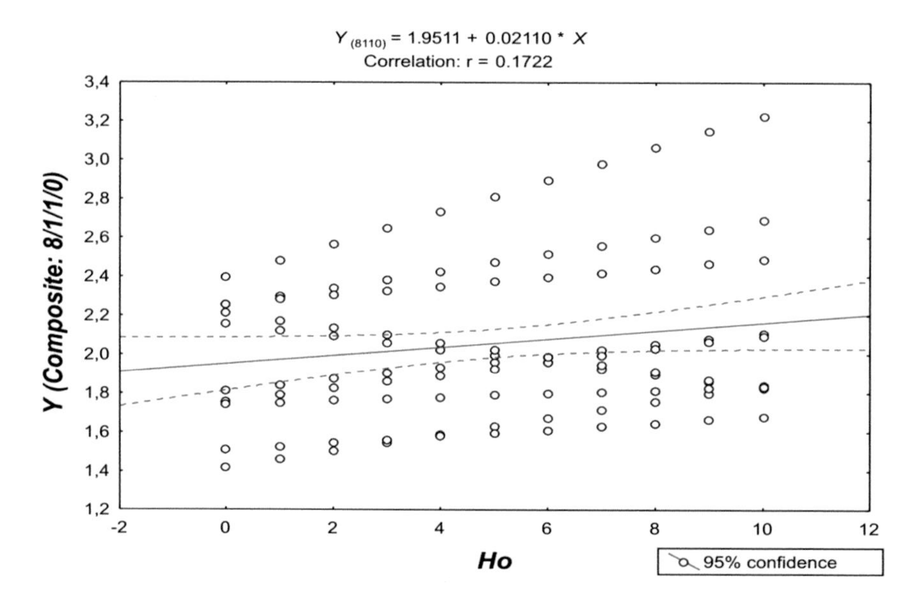

Fig. 12.3.7. H_o (X) vs. QT (Y) relation under combined intralocus effects and multilocus averaging. Combine effect shown for 10 loci composite case; with 4/3/3/0 for upper and 8/1/1/0 for bottom plot numbers of additive, dominant, overdominant and neutral loci taken for the numerical simulation runs. Correlation shown in bottom graph is not significant (from Kartavtsev and Zhdanova 2012).

relationship of H_o (X) vs. QT (Y) differently. In the case of additivity (1), averaging yields a negative association between H_o and the QT; in the cases of dominance (2) and overdominance (3), this relationship is positive; and in the case of neutrality (4), the relationship is absent (Fig. 12.3.6). The correlation coefficients for the numerical simulations of 10 runs of 10 loci are: 1. Additivity, $r = -0.29$ ($P < 0.01$), 2. Dominance, $r = 0.38$ ($P < 0.01$), 3. Overdominance, $r = 0.77$ ($P < 0.01$), 4. Neutrality, $r = -0.02$ (NS). Simulation data for two composite cases of intralocus effects yield associations of H_o and QT (Fig. 12.3.7) that are close to many experimental observations, with a weak correlation of individual heterozygosity and fitness traits, as obtained for allozyme loci (Zouros 1987; Kartavtsev 1990; 1992; Kartavtsev et al. 1992).

The numerical examples considered here agree with the calculations performed by Zouros (1987) for two examples of what the author presumed was dominance. However, a closer consideration of the ratios between genotypic values A_1A_1, A_2A_2, A_1A_2 and B_1B_1, B_2B_2, B_1B_2 (see Table 12.3.1) shows that Zouros analysed one example of dominance and one example of additivity. In these cases, there were, respectively, positive and negative associations between heterozygosity and QT value, as in the examples considered elsewhere (Kartavtsev 2005a; 2009b; Kartavtsev and Zhdanova 2013). Chakraborty and Ryman (1983) derived a similar analytical relationship between heterozygosities and QTs controlled by multiple-locus genotypes, on the basis of a mathematical combinatorial analysis and an assumption on the additivity of gene effects (in this case major interlocus effects are presumed to be additive). These authors demonstrated that the mean genotypic values in multiple-locus heterozygotes are higher than in composite homozygous classes. Correspondingly, this generates a positive correlation between individual heterozygosity and QT values. The deductions of Chakraborty and Ryman (1983), as was stressed above, strongly agree with both the numerical simulations and empirical observations. However, the interpretation of the results based on additive interlocus effects as proposed (Chakraborty and Ryman 1983) seems obscure. As demonstrated here and earlier (Kartavtsev 2005a; 2009b), this type of intralocus allelic interaction is of crucial importance for the H_o and QT association, and this was not addressed either by Chakraborty and Ryman (1983) or later (David 1997; Szulkin et al. 2010). The results discussed in the present chapter indicate the possibility of abundance if not prevalence in a whole genome the dominance in the allelic gene interactions, which jointly with overdominance gives positive H_o and QT correlations in most observations. Zouros (1987) also noted the importance of dominance in creating positive correlations due to multi-locus averaging. However, the difference between additivity and dominance is not absolutely distinct. This is why the term incomplete dominance is often used. If specific numerical data are used, negative or positive correlations will be generated for additivity and dominance, depending on whether the "middle" value in the inequality $(A_1A_1 + A_2A_2)/2 <> A_1A_2$ has been passed.

Chakraborty and Ryman (1983) noted that, conversely to the mean, the variance of QT in homozygous classes is higher than in heterozygotes, which causes a negative correlation between heterozygosity rank and QT variance in multi-locus data. This corresponds well with the results obtained in the current paper (Fig. 12.3.8) and in another simulation (Bezrukov 1989). If we consider these data in a wider context, taking into account different types of intralocus interactions and their combination in multi-locus genotype, it is clear that the interrelationship between H_o and QT variance may often be not only negative but also non linear (Fig. 12.3.8). The neutrality case was omitted in the latter figures for brevity because it does not add any substantial information.

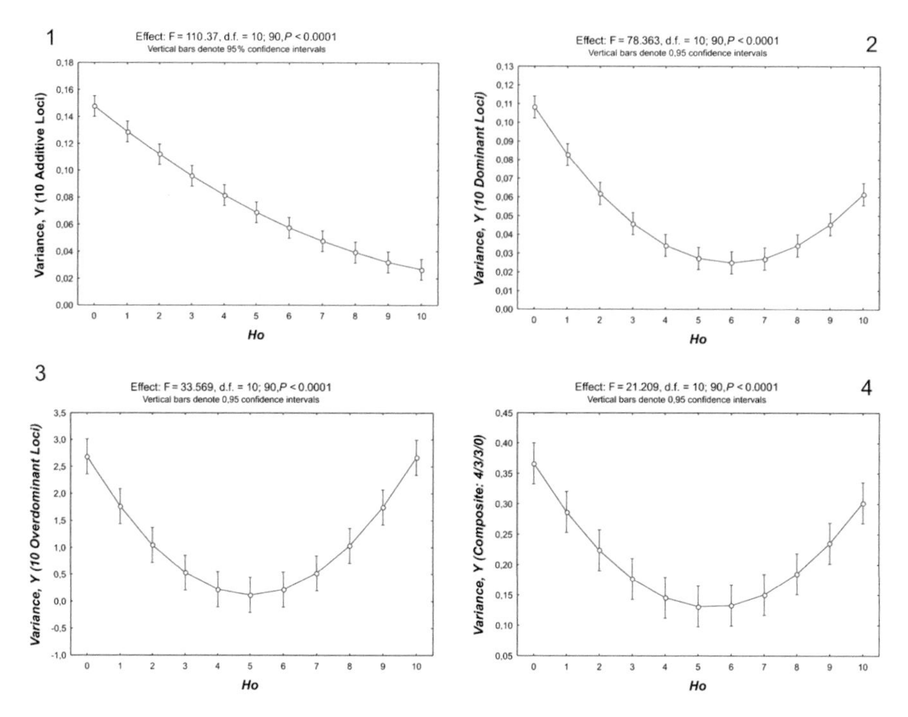

Fig. 12.3.8. H_o (X) and variance of QT (Y) relationship under 3 types of intralocus effects (1–3) and in one composed case with multilocus averaging.
(1) Additivity, (2) Dominance, (3) Overdominance, (4) Combined effect shown for 10 loci composite case; with 4/3/3/0 number of additive, dominant, overdominant and neutral loci in numerical simulation runs. Significance of associations is shown on top of the graphs (from Kartavtsev and Zhdanova 2012).

Three types of intralocus interactions (neutrality, additivity and overdominance) and the effect of heterozygosity on QTs were analysed in simulations by Bezrukov (1989). The predictions of the pattern of the relationship between heterozygosity and QTs in the cases of neutrality and overdominance fit the data presented above (see Figs. 12.3.6–12.3.8). However, in the case of additivity, an absence of correlation between heterozygosity and the QT values was demonstrated (Bezrukov 1989, Fig. 1). What caused this discrepancy? The mutual extinguishing of the positive and negative associations in intralocus effects and multiple-locus averaging is the likely cause. However, the details of this discrepancy remain obscure. In some situations, combinations of intra- and interlocus effects may definitely yield a zero association between H_o and QT. Indeed, combining effects decrease the correlation in two of our composite cases (Figs. 12.3.7–12.3.8), as well as in some cases in Bezrukov (1989; e.g., Fig. 2d, e). The analysis performed in two studies (Chakraborty and Ryman 1983; Bezrukov 1989) uses a polynomial expansion for the estimation of QT variation, depending on the heterozygosity rank in composite groups: from zero (different classes of homozygotes for a specific number of loci) to the maximum (an *n*-locus heterozygote). Therefore, in both studies, intralocus interactions are only declared, but their effects on the relationship between heterozygosity and QT values are not analysed. Hence the discrepancies in H_o and QT relationship described above may have occurred.

Thus, it is obvious that single-locus effects on the relationships between heterozygosity and QTs must not be ignored. The data obtained in this study also indicate that the superposition of intralocus effects and multiple-locus averaging may result in nonlinear relationships (Fig. 12.3.8). Nonlinear relationships between the values of heterozygosity and QT or their variances are also evident when other interactions between loci occur, e.g., in the case of epistasis (Dubrova et al. 1990; 1994) and if group effects on heterozygosity are considered (Dubrova et al. 1995; see also Figs. 12.3.1, 12.3.4, 12.3.5). A nonlinear relationship between the variances of heterozygosity and the QT may also result from combining intralocus effects in a multiple-locus genotype (Fig. 12.3.8 and Bezrukov 1989). Apparently, intra- and intergenic interactions are usually combined in an individual genotype of natural population. Therefore, the type of relationship between heterozygosity and QT values in each particular case is determined by a combination of effects, varying in different phases of the life cycle and in different generations. It is possible that on occasion, the positive associations between H_o and QT are basically inbreeding depression effects (in the cases of dominance prevailing in a set of loci), but in other cases the balance concept (with selective effects in action) may be more relevant. Researchers will rarely know precisely what combination of effects they are dealing with in natural populations, at least in higher organisms reproducing via crossing. On the other hand, it is obvious that a positive correlation between the values of heterozygosity and QT cannot be invariably generated due to multiple-locus averaging or any other single source.

General Context of the H_o and QT Relationship

As mentioned in the introduction to Section 12.2, recently a new vision of the impact of H_o on QT was suggested (David 1997). Generally, heterozygosity vs. fitness correlation (HFC) is the common name for the estimation of impact of heterozygosity at a marker gene and individual measures of life-history, morphological, physiological and other traits that are meaningful for fitness. Heterozygosity is hereafter attributed to only an individual and to several marker loci; this measure is called multiple locus heterozygosity (MLH, H_o in the text above or their derivatives), and is used in many papers, including recent overviews that are based on microsatellite data (Chapman et al. 2009; Szulkin et al. 2010).

Chapman et al. (2009; after Hansen and Westerberg 2002), when introducing the MLH vs. QT relationships, propose three hypotheses to explain HFC. The first is functional overdominance, which occurs due to the expression of marker loci themselves (such as allozyme or MHC loci) that may have direct effects on fitness. The second, called the 'local effect' hypothesis (Chapman et al. 2009; after David et al. 1995; Lynch and Walsh 1998; pp. 288–290; Hansson and Westerberg 2002), is based on the associative overdominance as the explanation (Ohta 1971), i.e., in this case the apparent increase of the QT estimate, jointly with increasing MLH is due to nonrandom association of these loci with loci affecting fitness; that is, the marker and fitness loci are in linkage disequilibrium (LD, David 1997). The third hypothesis, the 'general effect' hypothesis, is used to explain HFC detected by neutral markers (Chapman et al. 2009; after David et al. 1995; Lynch and Walsh 1998; Hansson and Westerberg 2002). In this case, the apparent increase of the QT estimate with increasing MLH is due to the nonrandom association of diploid genotypes in zygotes, and reflects the fitness cost of homozygosity at loci throughout the genome; that is, the marker and fitness loci are in identity disequilibrium (ID, David 1997; Szulkin et al. 2010). The third hypothesis differs from the second because of the absence of physical linkage requirements for the MLH–QT effects. All three hypotheses assume a direct connection of diversity of marker genes and diversity of loci that affect QT variation. It is evident that the first of these hypotheses is a case of the general balance model, while other two are similar to the inbreeding depression model as was defined in the Section 12.2.

Meta-Analysis of HFC

Concluding the discussion, it is relevant to present a summary of a recent comprehensive statistical meta-analysis performed by Chapman et al. (2009). They summarized the literature on QT vs. MLH (or similar measures), and attempted a multi-dimensional approach. The authors used all comparable statistical estimates, like r, t, F, etc., and called it the estimate of effect size. For comparison they also included previous meta-analyses based on single-dimension statistics (Coltman and Slate 2003). Their main conclusions are as follows (Chapman et al. 2009):

1. There are small but statistically significant and positive effect sizes for life-history, morphological and physiological traits;
2. Despite prediction of the theory, higher values of the effect sizes for life-history traits create no extra QT variation and thus no significant differences among trait types was observed;
3. For new proposed measures of genetic variation no advances obtain in comparison with MLH;
4. Presumed larger variation in inbreeding coefficient does not affect mean effect sizes;
5. The review showed that HFC investigations in general do not reveal patterns predicted by population genetic theory and are of small effect of total variance, less than 1%.

The data considered here and their interpretation make it clear why not only positive but also negative associations between heterozygosities for marker genes and values of adaptive traits have been found in many studies (Mitton and Grant 1983; Nevo et al. 1984; Zouros 1987; Zouros and Foltz 1987; Kartavtsev 1990; Kartavtsev et al. 1990; Altukhov 1989; Waples 1991). For negative correlations, the occasional prevalence of additive intralocus interactions may be the cause and the effects of autobreeding depression might also be a plausible explanation (Chapman et al. 2009; Szulkin et al. 2010). The strength of the relationship between heterozygosity for marker genes and the value of the QT or its variance should also be taken into account. The association between QT and H_o cannot be strong for marker genes, e.g., allozyme or microsatellite loci, as was deduced from general principles (Lewontin 1974; David 1997) and confirmed by empirical data (Kartavtsev 1992; 1996; Chapman et al. 2009; Szulkin et al. 2010). Most correlation coefficients fall within the range 0.2 to 0.4, which corresponding to factorial effects of 4–16% (Kartavtsev 1990; 1992; 1996) or less down to 1% (Chapman et al. 2009). Even in specially selected examples where intralocus effects generate a strong correlation between H_o and QT or QT variance (Kartavtsev and Zhdanova 2013), multiple-locus averaging and the combination of single-locus effects decrease the total correlation (Figs. 12.3.7–12.3.8). The weak correlation between sample heterozygosity for a set of loci and genomic heterozygosity also weakens the association between H_o and QTs (Chakraborty 1981; Mitton and Lewis 1989).

This study provides the basis for criticizing the oversimplified approaches to the analysis of the association between heterozygosity and QTs along with many others (Chakraborty 1981; Mitton and Lewis 1989; Waples 1991; Kartavtsev 1992; Dubrova et al. 1995; Chapman et al. 2009; Szulkin et al. 2010); it also enhances our prospects of studying the effects of heterozygosity for marker genes on the variability of adaptively important traits and heterosis as a whole in cases when planning crosses are unavailable and direct QTL analysis is absent. Absence of any significant HFC in the simulation presented in the current paper for neutral alleles (both for intralocus effects and for multiple-locus averaging) is supported by theoretical predictions (Ohta 1971), empirical testing (Hole 1989; Whitlock 1993), and another simulation (Balloux et al. 2004) for large panmictic populations. In other words, in our set of simulated "populations" HFC could not be attributed to ID or LD; they are caused by other factors as given above. Thus, generality of the conclusion that inbreeding depression is the main source of HFC (Szulkin et al. 2004) seems premature. Probably, more appropriate is the conclusion that neither of the theories is true. Such a suggestion follows from recent meta-analysis (Chapman et al. 2009; see their conclusion 5 above), from the current chapter and from earlier analysis of allozyme data and underlying models (Kartavtsev 2005a; 2009b, pp. 163–169).

Three other important matters should also be noted. Firstly, each individual marker locus (allozyme, DNA site or immunochemical) is expected to have only a small if any effect on fitness of an individual at the phenotypic level (Lewontin 1974). In our case (see Fig. 12.3.4), factorial influence at allozyme loci was shown to be only: $R^2 = 0.0253$ up to $R^2 = 0.1290$ or near 2–13%. Secondly, when the factorial influence is large, 0.4–0.6 or 40–60% (Zouros and Mallet 1989), it does not reflect H_0 but rather a group's heterozygosity. Thirdly, when complex analyses are performed, interlocus effects are found, and there is no simple H_0 & QT (or environment) interaction (Dubrova et al. 1995). The latter is remarkable because although heterozygosity and heterosis at QT may be correlated, the former is not necessarily the cause of the latter, as we showed in the beginning of this chapter (see also Strunnikov 1986).

Despite all these controversies it seems that we should accept the balance hypothesis as a working tool in searching further for the truth. One reason for this opinion is the generality of the selective models suggested (Turelli and Ginzburg 1983; Smouse 1986; Bush et al. 1987). When a simpler model (used as null-hypothesis) is rejected as inadequate, it is possible to accept a more complex one.

One more point, which is necessary to consider in closing the above discussion, is the conceptual necessity for a model that permits auto-regulation or GH at the population level. Available models of frequency dependent selection do not explain many things, including segregational load and QT variability. As a draft for such a model may be considered balanced selection, with changing vectors, in favour of heterozygotes for growth rate and against heterozygosity increase for viability. Also, at some stages of the life cycle selection is relaxed down to nil. Three types of facts lead to the suggestion of such a model. (1) In many cases (see Table 12.3.1; Altukhov 1989; Altukhov et al. 1991), multiple locus heterozygotes have advantages in respect to impact on QT (increased size or growth rate), like that obtained in pink salmon fry (Fig. 12.3.5). (2) At the same time viability is greatest in the average heterozygosity classes (see Figs. 12.3.2, 12.3.4), which are less variable at QT (see Figs. 12.3.2, 12.3.3, 12.3.8). (3) Let us take into account as well that the influence of heterozygosity upon QT may be lowered, both as a result of the relaxation of selection during ontogenesis, and because different modes of intra-locus and inter-locus gene action are combined (see above). These facts, in combination with earlier ideas (Livshits and Kobyliansky 1985; Altukhov 1989; Altukhov et al. 1991), permit us to suggest a *mechanism of balanced natural selection as regulator (1) of an optimal heterozygosity level in the species populations, protecting them from too large a segregational load, and (2) as a regulator of phenotypic homeostasis at QT*. If such a model can be developed, it will be the best development of the GH idea at the population level. Certainly, some of the premises developed by Lerner in 1950s, should be changed in accordance with modern genetic knowledge.

12.4 TRAINING COURSE, #12

12.1 Solve two examples in class to define the Ho & QT association with STATISTICA (1994) software.

12.2 Principles of ANOVA for determination of QT variation. Finding variances in the complete factor scheme of the analysis.

12.3 Principles of discriminant and factor analyses for investigation of QT variation.

12.4 The following terms are relevant to the subject and should be learned:

- **Genetic homeostasis** can be defined as the property of the population to equilibrate its genetic composition and to resist sudden changes.
- **Heterosis** hybrid vigour or increased "power" of F_1 offspring as compare to both parent forms.
- **Inbreeding depression hypothesis** suggests that the loci themselves are selectively neutral and they are only markers of genome heterozygosity.
- **Balance hypothesis or hypothesis of overdominance** predicts that the loci themselves are closely linked markers that influence H_o & QT association.

13

DNA POLYMORPHISM WITHIN AND AMONG POPULATIONS

MAIN GOALS

13.1 DNA Sequence Polymorphism. Nucleotide Diversity
13.2 DNA Polymorphism Estimated from Restriction Site Data
13.3 DNA Length Polymorphism. Theory and Observations
13.4 Divergence of Populations at the DNA Level
13.5 TRAINING COURSE, #13

SUMMARY

1. A nucleon and haplotype (necleomorph) correspond to a gene (locus) and allele in a common genetic sense.
2. Haplotypic diversity (h) is a measure of variability, which is equivalent to the expected heterozygosity (gene diversity) per locus.
3. Nucleotide diversity (π) is the mean number of nucleotide differences per site between two sequences.
4. The mean number of nucleotide substitutions for a randomly sampled pair of haplotypes from two populations or taxa is the difference between the mean number of nucleotide substitutions per site between i and j haplotypes (d_{ij}) and the mean weighted number of nucleotide substitutions within populations (d_x and d_y). This measure is equivalent to a genetic distance.
5. Phylogeny established for species and genes are not identical. Gene lineages usually have more ancient divergence then divergence of populations or species themselves.

6. Coding sequences of structural genes evolve slower than non-coding DNA sequences, which mark silent polymorphism. Alleles detected by enzyme markers may have higher levels of nucleotide substitutions than the other DNA markers' alleles. For example, for the *Adh* gene S and F enzyme alleles in *D. melanogaster* are caused by a longer history of divergence for the former.

13.1 DNA SEQUENCE POLYMORPHISM. NUCLEOTIDE DIVERSITY

In earlier chapters, we learned that protein markers have indicated large amounts of genetic variation in natural populations. However, the precise estimate of genetic variation would be better represented at the level of DNA markers.

For genetics the primary interest is **DNA Sequence Polymorphism** due to the ability to detect a single nucleotide substitution. How and what should we measure? Let us accept that nucleotide sequences are known for a set of loci or alleles sampled from a population. Then there are several ways to estimate the extent of DNA polymorphism.

Number of Polymorphic Sequences

One way to start is to measure the proportion of different or polymorphic sequences (k) in the whole sample. Such a measure has the same statistical properties as that of the number of alleles in a sample, and is dependent on sample size. That is why, for comparison of the levels of polymorphism, say between two populations or loci by using this measure, it is necessary to have the samples of the same or comparable size. There is another problem concerning analysis of DNA sequence. For example, if a long DNA fragment is studied, all sequences examined may differ from each other. In such cases, the proportion of polymorphic sequences is not at all a suitable measure of polymorphism; all samples are polymorphic, i.e., there is 100% polymorphism.

Number of Polymorphic Nucleotide Sites

A more appropriate measure of DNA polymorphism is the proportion (p) of polymorphic nucleotide sites per nucleotide sequence (Nei 1987): $p_n = s_n/m_T$, where s_n and m_T are the number of polymorphic sites per sequence and the total number of nucleotides examined, respectively. This measure is better than proportion of different sequences because it is always applicable, even if all sequences examined are different. However, this measure also depends on sample size. In the case of the infinite-site model of neutral mutations (Kimura 1971), the expected value of p_n over independent populations (with the stochastic process) is:

$$E(p_n) = L[1 + 1/2 + 1/3 + \cdots + (n-1)^{-1}], \quad (13.1)$$

where n is the number of sequences examined, and $L = 4N\mu$ (Watterson 1975), and N and μ are the effective population size and mutation rate per nucleotide site per generation, respectively. If n is large, $n > 20$, equation (13.1) is approximately given by (Nei 1987):

$$E(p_n) = L[0.577 + \log_e(n-1)]. \quad (13.2)$$

If there is no recombination, the variance of p_n over the stochastic process is given by (Nei 1987):

$$V(p_n) = E(p_n)/m_T + \Sigma_{i=1}^{n-1} L^2\ 1/i^2. \quad \textbf{(13.3)}$$

Therefore, it is possible to estimate $L = 4N\mu$ by (Nei 1987):

$$L\hat{} = p_n /A, \quad \textbf{(13.4)}$$

where $A = 1 + 1/2 + \cdots + (n-1)^{-1}$. This L^ can be used as a measure of polymorphism. However, we must remember that equation (13.4) is valid only if neutral mutations are considered, and if the population is in equilibrium with respect to the effects of mutation and genetic drift.

Nucleotide Diversity

A better measure of DNA polymorphism than the former two is the average number of nucleotide differences per site between two sequences, or **nucleotide diversity**. This measure, as well as its estimates and variance, are defined below in the version given by Nei (1987):

$$\pi = \Sigma_{ij}\, x_i x_j\, \pi_{ij}, \quad \textbf{(13.5)}$$

where x_i, x_j is the population frequency of the ith and jth types of DNA sequences, and π_{ij} is the proportion of different nucleotides between the ith and jth types of DNA sequences. In a panmictic population, π is the usual heterozygosity at a nucleotide level. An estimate of nucleotide diversity can be found, either by equation

$$\pi\hat{} = [n/(n-1)]\ \Sigma_{ij}\, x_i\hat{}\, x_j\hat{}\ \pi_{ij} \quad \textbf{(13.6)}$$

or by

$$\pi\hat{} = \Sigma_{i<j}\, \pi_{ij}/n_c. \quad \textbf{(13.7)}$$

Here n, $x_i\hat{}$, $x_j\hat{}$ and n_c are the number of DNA sequences examined, the frequency of the ith and jth types of DNA sequence in the sample, and the total number of sequence comparisons $[n(n-1)/2]$, respectively. In equation (13.7), i and j refer to the ith and jth sequences rather than to the ith and jth types of sequences. If $\pi\hat{}_{ij}$'s are constant, the variance of $\pi\hat{}$ obtained by (13.6) is given by:

$$V(\pi\hat{}) = 4/n(n-1)\ [(6-4n)\ (\Sigma_{i<j}\, x_i x_j\, \pi_{ij})^2 + (n-2)\, \Sigma\, x_i x_j x_k\, \pi_{ij}\, \pi_{ik} + \Sigma_{i\ j}^{<}\, x_i x_j\, \pi_{ij}^2] \quad \textbf{(13.8)}$$

(Nei and Tajima 1981). This variance is generated from allele frequencies, and does not include the stochastic component.

If all mutations are neutral, the expectation of $\pi\hat{}$ over the stochastic process is given by:

$$E(\pi\hat{}) = L/[1 + (4/3)L] \approx L \text{ for } L \ll 1 \quad \textbf{(13.9)}$$

(Kimura 1968). Also, the variance of $\pi\hat{}$ during the random sampling or stochastic process is

$$V(\pi\hat{}) = [(n+1)/3n(n-1)\, m_T]\, L + [2(n^2 + n + 3)/9n(n-1)]\, L^2 \quad \textbf{(13.10)}$$

(Modified from Tajima 1983 by Nei 1987). As n increases, this equation approaches:

$$V(\pi^\wedge) = (1/3\ m_T)\ L + 2/9L^2. \qquad \textbf{(13.11)}$$

Let us consider an example (Nei 1987). Kreitman (1983) sequenced the alcohol dehydrogenase (*Adh*) gene region for eleven alleles of *Drosophila melanogaster*. The gene region studied contained 2,579 nucleotides (m_T), excluding deletions and insertions. Nine of the 11 sequences were different from each other, indicating a high degree of sequence polymorphism. There were 43 polymorphic nucleotide sites, so that $p_n = 43/2379 = 0.0181$. Under the assumption of neutrality, therefore, we have $L^\wedge = 0.0062$ from equation (13.4), since A = 2.929 for n = 11. From Kreitman's data, it is possible to compute the proportion of different nucleotides for all pairs of alleles. The results obtained are presented in Table 13.1.1. The nucleotide diversity for this gene section is therefore $\pi^\wedge = 0.0065 \pm 0.0017$; both equations (13.6) and (13.7) give same result. The standard error of the estimate was obtained by equation (13.8). It is possible to see that the estimate of nucleotide diversity is close to another estimation, taken from the number of polymorphic sites under the assumption of neutrality.

Table 13.1.1. Percent nucleotide differences between 11 alleles of the Adh locus in *Drosophila melanogaster* (Modified from Kreitman 1983).

Allele	**(1)**	**(2)**	**(3)**	**(4)**	**(5)**	**(6)**	**(7)**	**(8)**	**(9)**	**(10)**
(1) Wa-S										
(2) F1-1S	0.13									
(3) Af-S	0.59	0.55								
(4) Fr-S	0.67	0.63	0.25							
(5) F1-2S	0.80	0.84	0.55	0.46						
(6) Ja-S	0.80	0.67	0.38	0.46	0.59					
(7) F1-F	0.84	0.71	0.50	0.59	0.63	0.21				
(8) Fr-F	1.13	1.10	0.88	0.97	0.59	0.59	0.38			
(9) Wa-F	1.13	1.10	0.88	0.97	0.59	0.59	0.38	0.00		
(10) Af-F	1.13	1.10	0.88	0.97	0.59	0.59	0.38	0.00	0.00	
(11) Ja-F	1.22	1.18	0.97	1.05	0.84	0.67	0.46	0.42	0.42	0.42

Note. Frequency for x_i = 1/11 for all haplotypes except one, which has a sample frequency of 3/11 (see Table 13.1.1). The total number of nucleotide sites compared is 2,379.

At the *Adh* locus of *D. melanogaster*, there are also two alleles (electromorphs) that are detectable by enzyme electrophoresis, *S* and *F*. The alleles studied by Kreitman comprised six *S*- and five *F*-types. If we estimate nucleotide diversity for these two electromorphs separately, it becomes: $\pi^\wedge = 0.0056$ for *S*- and $\pi^\wedge = 0.0029$ for *F*-type. Thus, the nucleotide diversity for *S*- is almost twice as high as that for the *F*-type. These differences in nucleotide diversities within and between *S*- and *F*-alleles are depicted in Table 13.1.1 (inter allele values highlighted with cell filling). This suggests that the *S*- and *F*-alleles diverged a long time ago, before the occurrence of the alleles attributed solely to DNA sequence variation.

Ratio of Nucleotide Differences or *P*-Distance

A very simple and convenient measure of DNA polymorphism is the ratio of nucleotide differences between a pair of randomly chosen nucleotide sequences, or ***p*-distance**. This ratio may be estimated as (Nei and Kumar 2000, p. 33):

$$p\text{\^{}} = n_d/n, \qquad \textbf{(13.12)}$$

where n_d is the number of different nucleotides between DNA sequences X and Y, and n is the total number of analyzed nucleotides. Variance of p^, defined from (13.12) could be estimated from the expression:

$$V(p\text{\^{}}) = p\,(p - 1)/n. \qquad \textbf{(13.13)}$$

This is the usual binomial variance (see equations 9.4–9.5). In real calculations, population frequencies (probabilities) are substituted by sample estimates, ***p*^**.

Estimation of the Number of Nucleotide Substitutions

The value of *p*-distance gives a satisfactory estimate of the nucleotide substitution number when sequences DNA are close enough. However, if p is large, it underestimates the real number of nucleotide substitution because it does not take into consideration parallel and backward mutations. Nevertheless, when *p*-distance values are small $p < 0.2$ (Nei and Kumar 2000, p. 41, Fig. 3.1), p-distance estimate gives quite similar scores to other more advanced measures of nucleotide divergence.

For understanding the essence of nucleotide substitution we have to use specific models with mathematical foundation. Based on foundations of different measures, their mathematical expressions (models) and variances in four sources, we are able to outline below eight basic nucleotide substitution models (Nei 1987; Li 1997; Nei and Kumar 2000; Felsenstein 2004). Among the most frequently used are the Jukes-Cantor and Kimura two-parameter models, which correspondingly suppose equality of substitution rates for all four nucleotides and different ratios for transition (α) vs. transvertion (β) changes. At least eight different models are mentioned in the above sources: (1) Jukes-Cantor (*JC*), (2) Kimura two-parameter (*K2P*), (3) Equal-Input (*EI*), (4) Tamura (*Tr*), (5) Hasegawa-Kishino-Yano (*HKY*), (6) Tamura-Nei (*TrN*), (7) General Time Reversible (*GTR*) and (8) Unrestricted (*UR*).

In the *K2P* model equilibrium, frequencies of all four nucleotides are equal to 0.25. However, the algorithms suggested may be applied independently of initial frequencies (Nei and Kumar 2000, p. 38), and in this respect it is similar to the *JC* model and correspondingly is applicable to a wide variety of data as well in comparison with the other six models. We have to emphasize that in the *K2P* model transition vs. transvertion ratio is $R = \alpha/2\beta$, but some authors and many program packages use a different proportion: $k = \alpha/\beta$. We have to take this into consideration to avoid confusions in comparisons. Moreover, analysis of models and real data indicate the need for the thoughtful choice of the most suitable model for your particular case-study. As a rule, the best results are obtained if you spend some time finding an optimal model for your case, avoiding routine, default options, leading, say, to use of *K2P*. However, we have also to note that in more complicated models, having more parameters leads to greater SE of estimates (Nei and Kumar 2000). In a review, we have summarized data for over 18,000 different animal taxa with *p*-distances and its derivatives. The estimates most frequently used *K2P* (29%; Kartavtsev and Lee 2006), but quite frequently, uncorrected *p*^ or *p*-distance is used. There are examples of using the *HKY* (Williams et al. 2001; Kontula et al. 2003; Martinez-Navarro et al. 2005), the TrN model (Kontula et al. 2003; Bertsch et al. 2005) and the GTR model (Kontula et al. 2003; Kartavtsev et al. 2007a, etc.). Numerical simulation for the infinite nucleotide model showed that if substitution level is not high ($< 20\%$), all measures give similar values (Nei and Kumar 2000, Fig. 3.1). However, with the increase of substitution number and homoplasy, the *p*-distance more quickly bias the value of expected distance. An important correction of distance measures, connected with variation of substitution rate along the sequence, is the gamma (G) correction (Li and Zarkhih 1995; Nei and Kumar 2000; Creer et al. 2003; Felsenstein 2004). Popular now for determination of an optimal model is the MODELTEST 3.06 program (Posada and Grandal 1998) and later versions. The MEGA-5 program package, the tools of which we will consider separately (Chapter 15), also provides estimation of nucleotide substitution models (Tamura et al. 2011). Very useful information on the model properties and their applicability for a range of data are available in many papers (Nei and Kumar 2000; Hall 2001; Sanderson and Shaffer 2002; Felsenstein 2004). Various options (modes) for appropriate phylogenetic tree building are provided by several program packages, such as PAUP* (Swofford 2000), MEGA-4, MEGA-5 (Kumar et al. 1993; 2011), MrBayes (Ronquist and Hulsenbeck 2003) and others. Convenient interface and variable statistic utilities, including DNA sequence analysis and frequency variability estimations, are given in the ARLEQUIN package (Schneider et al. 2000). Very convenient and well done for the "common" user is the handbook by Hall (2001; 2011).

13.2 DNA POLYMORPHISM ESTIMATED FROM RESTRICTION SITE DATA

Techniques for DNA Polymorphism Analysis

Beyond laborious DNA-sequencing, there are several other techniques to detect DNA variability. The extent of DNA polymorphism is usually studied by using the **restriction enzymes** or endonucleases and the technique of the **polymerase chain reaction** (PCR). DNA polymorphism is usually studied for a given fragment of DNA sequence. This fragment may include any region of DNA, such as a structural gene or coding region, intron, flanking region or intergenic region. Any short DNA fragments, including only one type of DNA region, as well as long DNA fragments with many kinds of sequences can be investigated.

Nei and Tajima (1981) called the entire DNA segment under investigation a **nucleon**, and a particular DNA sequence (or restriction site sequence) for the segment studied, a **nucleomorph**. Nucleon and nucleomorph correspond to gene (locus) and allele in the common genetic sense, respectively. In the literature, however, a nucleomorph is more frequently called a **haplotype**.

Let us consider three common techniques of DNA detecting polymorphism: RFLP, RAPD and AFLP.

RFLP, Restriction Fragment Length Polymorphism

RFLP is a technique based on the reaction of specific endonucleases to cut (restrict) DNA in a certain place, and is used to detect molecular genetic markers. There are several specific endonucleases (restrictases) recognizing four, six, eight and ten base-pairs. Using specific enzyme, polymorphisms in the location of particular restriction sites can be detected via differences in the length of DNA fragments. Concentration of fragments is low in a probe, which is why the DNA of these fragments is usually first subjected to PCR for the amplification of their number. After that, electrophoresis is performed to detect the mobility of DNA fragments and their length differences. RFLPs are used as **molecular markers of DNA variability**. The more restrictases we use, the more DNA sites (loci) may be scored. The technique is able to provide potentially thousands of markers by usage of hundreds of enzymes available now. Tetra-, hexa- and octa- (4, 6, 8) base-pare specific restrictases are widely used for DNA, including mtDNA.

By varying the restriction enzyme combinations, one may identify more than two differently sized fragments. Whenever the recognized fragments differ in length, we say the lengths are polymorphic. Probes of repeated nDNA sequences or mtDNA usually reveal RFLPs (Fig. 13.2.1). Variability of mtDNA belongs to a haploid genome, so single bands are seen in all individuals, after electrophoresis of the fragments and staining (Fig. 13.2.1). Nuclear DNA probes of individuals may be different and can behave as the codominant products under electrophoresis and be adequately stained in the gel, e.g., have two bands in heterozygote state (Fig. 13.2.2). The RFLP markers

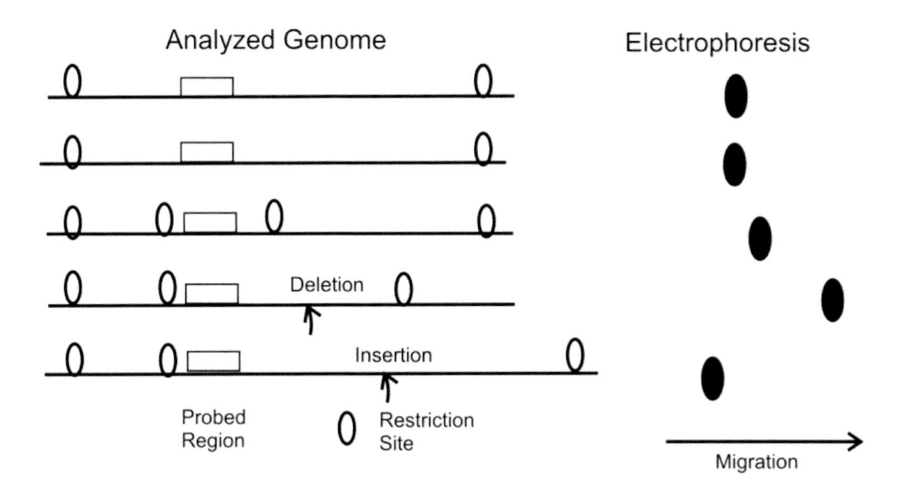

Fig. 13.2.1. The schemes of experimental design which allow us to obtain RFLPs.

Fig. 13.2.2. Illustration of single site variation and codominant type of inheritance in nuclear DNA. Two bands (AB) are seen in heterozygotes (A_1A_2) and different single bands (A and B) are seen in homozygotes (A_1A_1 and A_2A_2).

appear to be easily interpreted in genotypic terms (Fig. 13.2.2). In the example, there are two length types of the restriction fragments and three morphs/genotypes in the sampled population (Fig. 13.2.2; see A, B and AB phenotypes). The RFLPs are widely used as markers for intraspecies and interspecies genetic variability.

RAPDs, Randomly Amplified Polymorphic DNAs

RAPDs are molecular polymorphisms identified at random, and used as molecular markers to detect variability for genetic mapping. By this technique, after the isolation of the DNA, the unidentified fragments of total DNA are subjected to restrictases action. The complexity of eukaryotic nuclear DNA is sufficiently high that by chance pairs of sites complementary, say, to single octa- or decanucleotides may exist in the correct orientation and close enough to one another for PCR and amplification (Fig. 13.2.3).

With some randomly chosen decanucleotides no sequences are amplified. With others, the same length products are generated from DNAs of different individuals. With still others, patterns of bands (such as those illustrated) are not the same for every individual in a population (Fig. 13.2.4). The variable bands are commonly called

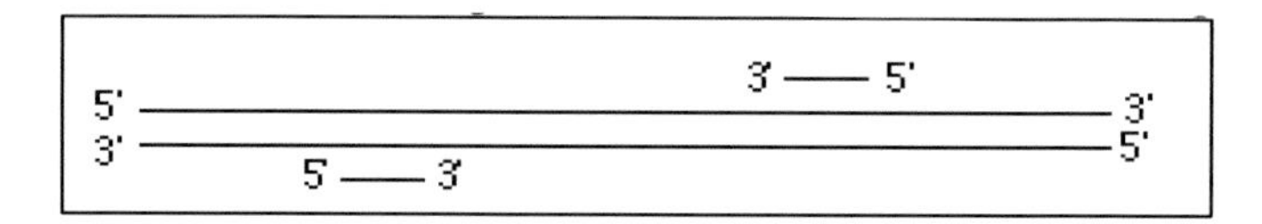

Fig. 13.2.3. DNA and fragments that can randomly be restricted in an individual.

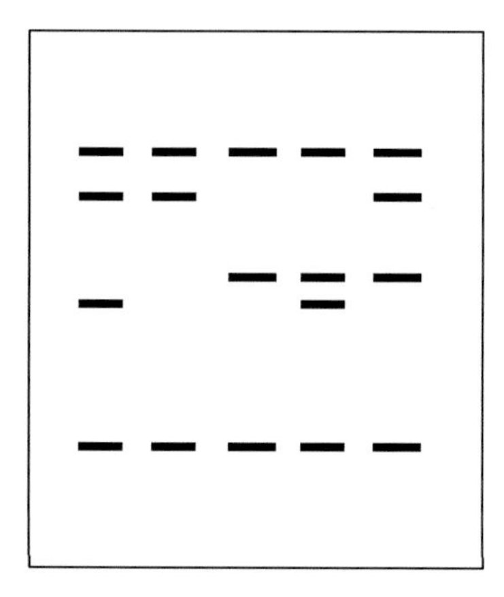

Fig. 13.2.4. Scheme of electrophoresis of RAPD fragments.

randomly amplified polymorphic DNA bands. Three of the bands in the diagram are RAPD bands (Fig. 13.2.4). RAPDs are used as genetic markers both for intra- and interspecies analysis. RAPDs are dominant, however, in the sense that the presence of a RAPD band does not allow distinction between hetero- and homozygous states, so their application for population genetic analysis is restricted. Breeders are able to identify RAPD bands closely linked to the marker they wish to transfer. Scoring individuals for the linked RAPD marker is helpful both to speed the breeding process and map other genes. For example, RAPD markers linked to genes of interest can serve as starting points for the technique "chromosome walks" to isolate those genes.

AFLP, Amplified Fragment Length Polymorphism

AFLP is a variant of RAPD. In the first step of AFLP analysis, genomic DNA is digested with both a restriction enzyme that cuts frequently (*Mse*I, 4 bp recognition sequence) and one that cuts less frequently (*Eco*RI, 6 bp recognition sequence) (Fig. 13.2.5).

The resulting fragments are ligated to end-specific adaptor molecules. A preselective **PCR amplification** is done using primers complementary to each of the two adaptor sequences, except for the presence of one additional base at the 3' end. Which base to choose is determined by the researcher. Amplification of only 1/16th of *Eco*RI-*Mse*I fragments occurs. In a second, "selective", PCR, using the products of the first as template, primers containing two further additional bases, chosen by the researcher, are used. The EcoRI-adaptor specific primer used bears a label (fluorescent or radioactive). Gel electrophoretic analysis reveals a pattern (fingerprint) of fragments representing about 1/4000 of the *Eco*RI-*Mse*I fragments.

AFLPs can be codominant markers, like RFLPs. Codominance results when the polymorphism is due to sequences within the amplified region. Yet, because of the number of bands seen at one time, additional evidence is needed to establish that a set of bands results from different alleles at the same locus. If, however, the polymorphism is due to presence/absence of a priming site, the relationship is of dominance. The non-priming allele will not be detected as a band. Compared to RAPDs, fewer primers should be needed to screen all possible sites.

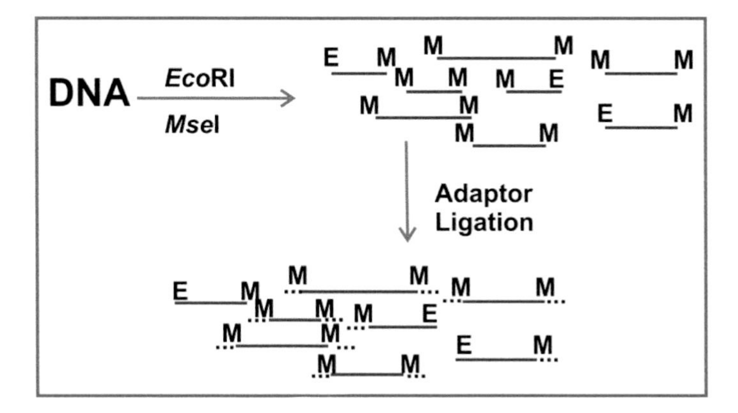

Fig. 13.2.5. AFLP, Amplified Fragment Length Polymorphism example.

Number of Haplotypes and Haplotypic Diversity

Let us focus on the amount of DNA polymorphism for a certain region of nuclear DNA or the entire mtDNA in a population. A certain number of individuals are sampled from the population, and their DNAs are examined by using restriction enzymes. These DNAs are classified into different haplotypes according to their restriction site patterns. When different haplotypes are identified, their relative frequencies in the sample can be computed the same as the allele frequencies at an enzyme locus. We may designate the population frequency of the *i*th haplotype by x_i and the sample frequency by x_i^. DNA polymorphism can be studied at two different levels, i.e., at the nucleon and nucleotide levels. If the nucleon level is considered, DNA polymorphism due to both the nucleotide substitution and deletion/insertion may be studied without distinction. When the nucleotide level is considered, then only nucleotide substitutions are taken into account. As seen earlier for DNA sequence analysis, the simplest measure of DNA polymorphism at the nucleon level is the number of haplotypes (k) observed in the sample. Also, as in the case of the number of different DNA sequences, this number depends on sample size. However, more suitable as a measure of nucleon polymorphism is nucleon or **haplotypic diversity** (h) (Nei and Tajima 1981). This measure is equivalent to the expected heterozygosity (or gene diversity) used in biochemical population genetics. It is defined by the same equation as 5.1a and 5.1 (Chapter 5, Training Course #5), and estimates may be obtained in several ways (Nei 1975; 1987). The statistical properties and the expectation of h under the infinite-allele model are also the same as those for expected heterozygosity.

Mean Number of Restriction-Site Differences

The amount of DNA polymorphism can also be measured by the mean number of differences in restriction-sites between two randomly chosen haplotypes (Nei and Tajima 1981). The number is defined as

$$v = \Sigma_{ij} x_i x_j v_{ij}, \qquad (13.12)$$

where v_{ij} is the number of restriction-site differences between the *i*th and *j*th haplotypes, and summation is taken over all combinations of haplotypes. The estimate of v and its sampling variance are given by equations (13.6) and (13.8), respectively, if we replace π_{ij} by v_{ij}. For neutral haplotypes due to nucleotide substitution, the expectation of v under the infinite-site model of mutations is

$$E(v) = M, \qquad (13.13)$$

where $M = 4Nv$ and v is the mutation rate per nucleon at the restriction site level. On the other hand, the stochastic variance of v^ is approximately given by (13.10) if we replace L by M and eliminate m_T. Elimination of m_T is necessary because we are now studying variation per nucleon rather than variation per nucleotide site (Tajima 1983).

Number of Polymorphic Restriction Sites

A quantity closely related to $\boldsymbol{v}$ is the number of polymorphic restriction sites (s_r), i.e., the number of restriction sites that are polymorphic in a sample of n nucleons. This measure is highly dependent on sample size, so it is not very useful for comparisons.

Nucleotide Diversity

A major problem with the two above measures is their dependence on the size of the DNA fragment studied; these measures increase as fragment-size increases. In practice, fragment-size varies greatly with the gene or gene region studied, so that they cannot be used as general measures of polymorphism.

To avoid this problem we should measure polymorphism at the nucleotide level. When most restriction-site polymorphisms are due to nucleotide substitution, as such a measure can represent the nucleotide diversity by expression (13.5). When the restriction enzyme technique is used, $\boldsymbol{\pi}_{ij}$ in (13.5) may be estimated by p^ = 1 – S^1/r or by another expression (5.50; Nei 1987), where S^ = m_{XY}/m^, m^ = $(m_X + m_Y)/2$ and m_X, m_Y, are numbers of restriction sites for DNA sequences X and Y, m_{XY} is number of restriction sites shared by two sequences. Thus, if sample haplotype frequencies are obtained, $\boldsymbol{\pi}$ can be estimated by (13.6) or (13.7).

As mentioned earlier, the expected value of $\boldsymbol{\pi}$ under the infinite-site model of neutral mutations is $L = 4N\mu$, whereas the expected value of v is $M = 4Nv$. Therefore, $\boldsymbol{\pi}$ can be estimated from v if we know the relationship between μ and v. It was shown (Nei and Tajima 1981) that

$$\boldsymbol{v = 2\Sigma_{i=1}^{s} m_i r_i \mu,} \tag{13.14}$$

where m_i and r_i are the number of restriction sites and the number of nucleotides in the recognition sequence of the ith endonuclease, respectively, and s is the number of enzymes used. Therefore, $\boldsymbol{\pi}$ can be estimated by (Nei 1987)

$$\boldsymbol{\pi = v/R,} \tag{13.15}$$

where $\boldsymbol{R = 2\Sigma_{i=1}^{s} m_i r_i}$. If we use (13.4), $\boldsymbol{\pi}$ can also be estimated from $\boldsymbol{s_r}$ (Nei 1987), i.e.,

$$\boldsymbol{\pi\text{^} = s_r/(AR),} \tag{13.16}$$

where $A = 1 + 1/2 + \cdots + (n-1)^{-1}$. Another formula for the estimation of π also was derived (Engels 1981). It is given by

$$\boldsymbol{\pi = (nc - \Sigma c_i^2)/[rc\,(n - i)].} \tag{13.17}$$

Where $\boldsymbol{c_i}$ is the number of members of the sample that have a cleavage at site i ($i = 1,\ldots, s_r$) and $c = \Sigma c_i$. This formula does not require the assumption of neutrality as emphasized by the author (Engels 1981), but it depends on the assumption of linkage equilibrium among restriction sites. By contrast, $\boldsymbol{\pi}$^ given by (13.6) or (13.15) does not require the assumption of either neutrality or linkage equilibrium. However, when the number of DNA sequences examined is large, say $n > 100$, then equation (13.17) seems to be easier to compute than (13.6) or (13.15). A formula for computing the variance of π^ obtained by (13.17) was also presented by Engels (1981), but it depends on the assumption of linkage equilibrium, so it is expected to give an underestimate.

Let us consider the example (Nei 1987). In three species of *Drosophila* (*D. melanogaster, D. simulans* and *D. virilis*), using restriction endonucleases *Hae*III (GGCC), *Hpa*II (CCGG), *Eco*RI (AATTC), and *Hind*III (AAGCTT), the genetic variability of mtDNA was studied (Shah and Langley 1979). The researchers identified seven haplotypes, and the frequencies of these haplotypes are given for each species in Table 13.2.1. In *D. simulans,* only five nucleons (DNA sequences) were sampled, and no polymorphism was found. In *D. melanogaster,* there are four haplotypes, and the estimate of nucleon diversity (h) is 0.71 ± 0.04 (Table 13.2.2). From this value, we can estimate $M = 4Nv$ by $M = h/(1 - h)$ under the assumption of no selection. It becomes 2.46. On the other hand, the estimate of M obtained from the number of haplotypes (k) [by using expression (8.16; Nei 1987)] is 1.95, which is considerably smaller than the estimate obtained from nucleon diversity. This difference apparently occurs because nucleon diversity (heterozygosity) tends to give an overestimate of M when it is based on a single locus (Zouros 1979).

To compute the mean number of restriction site differences, it is convenient to make a table of the numbers of restriction-site differences (v_{ij}) for all pairs of haplotypes (Table 13.2.3). These numbers are obtainable from Shah and Langley's (1979) Figure 1. From the values of v_{ij} for *D. melanogaster* in Table 13.2.3, we obtain v^ = 1.22 by using (10.5) with π_{ij} replaced by v_{ij}. Under the assumption of neutral haplotypes, this is an estimate of $4Nv$. It is still smaller than the estimate obtained from the number of haplotypes. The mtDNA polymorphism in *D. melanogaster* is caused by the polymorphism at three restriction sites. That is, $s_r = 3$ in this species.

Table 13.2.1. Haplotype frequencies in samples of mtDNAs from three species of *Drosophila; n* is number of mtDNA sequences sampled (Shah and Langley 1979).

Haplotype	m	m_a	m_b	m_c	s	v	v_d	**(*n*)**
D. melanogaster	0.1	0.3	0.5	0.1				10
D. simulans					1.0			5
D. virilis						0.6	0.4	10

Table 13.2.2. Haplotypic diversity (h^), number of haplotypes (k), average number of restriction-site differences (v^), number of segregating sites (s_r) and nucleotide diversity (π) in *Drosophila melanogaster* and *D. virilis* (Nei 1987).

	h^	k	v^	s_r	π
D. melanogaster					
Estimate	0.71 ± 0.12	4	1.22 ± 0.27	3	0.008 ± 0.002
$M = 4Nv$	2.46	1.95	1.22	1.06	
$L = 4N\mu$	0.017	0.013	0.008	0.007	0.008
D. virilis					
Estimate	0.53 ± 0.09	2	0.53 ± 0.09	1	0.004 ± 0.001
$M = 4Nv$	1.144	0.43	0.53	0.35	
$L = 4N\mu$	0.0088	0.003	0.004	0.002	0.004

Table 13.2.3. Restriction-site differences (v_{ij}) and $S\hat{}_{ij}$ values for pairs of haplotypes. The figures above the diagonal are v_{ij}'s and those below the diagonal are $S\hat{}_{ij}$'s (Nei 1987).

Haplotype	m	m_a	m_b	m_c	s	v	v_d
m		1	1	1	11	13	14
m_a	0.93 1.00		2	2	10	12	13
m_b	1.00 0.94	0.93 0.94		2	12	14	15
m_c	1.00 0.94	0.93 0.94	1.00 0.89		12	14	15
m_c	0.43 0.84	0.46 0.84	0.43 0.80	0.43 0.80		14	15
v	0.18 0.71	0.20 0.71	0.18 0.67	0.18 0.67	0.22 0.59		1
v_d	0.18 0.67	0.20 0.67	0.18 0.63	0.18 0.63	0.22 0.56	1.00 0.92	

Note. The upper $S\hat{}_{ij}$ value for each pair of haplotypes is for *Hae*III and *Hpa*II ($r = 4$), whereas the lower $S\hat{}_{ij}$ value is for *Eco*RI and *Hin*dIII (r = 6).

If we equate this number to the expectation in (13.2) with L replaced by M, we have $3 = 2.83M$ since $n = 10$. Therefore, we have another estimate of $M = 1.06$, which is close to the estimates from $v\hat{}$.

Let us now relate $4Nv$ to $4N\mu$ under the assumption that all evolutionary changes in these two species occurred by neutral nucleotide substitution. The average numbers of restriction sites for *Hae*II, *Hpa*II, *Hin*dIII and *Eco*RI for the four haplotypes of *D. melanogaster* are 3.7, 4, 4.6 and 4, respectively. Therefore, we have the relationship $v = 2\Sigma\, m_i r_i\, \mu = 2 * [(3.7 + 4) * 4 + (4.6 + 4) * 6]\, \mu = 164.8\mu$. Similarly, we obtain $v = 180\, \mu$ for *D. simulans* and $v = 100.8\, \mu$ for *D. virilis.* The average of these estimates is 149 μ. Note, however, that the G + C content of mtDNA in *Drosophila* is about 0.22, even if the A-T rich region is excluded (Kaplan and Langley 1979), and thus *Hae*III (GGCC) and *Hpa*II (CCGG) do not produce many restriction sites. At any rate, if we use the relationship $v = 149\mu$, an estimate of $4N\mu$ can be obtained from the $4Nv$ value. It ranges from 0.007 to 0.017, but the latter value, which was obtained from h, is probably an overestimate for the reason mentioned earlier. For estimating nucleotide diversity, $\boldsymbol{\pi}$, we must first compute the proportion of shared restriction sites for each pair of haplotypes by using the formula $S\hat{}_{ij} = 2m_{ij} = (m_i + m_j)$, where m_i, and m_j are the numbers of restriction sites for the ith and jth haplotypes, respectively, and m_{ij} is the number of shared restriction sites (see above). The estimate of π_{ij} is then given by $\pi_{ij} = (-log_e S\hat{}_{ij})/r$. When two or more enzymes with the same r value are used, $S\hat{}_{ij}$ should be computed by pooling m_i, m_j and m_{ij} over all enzymes. If r is not the same, they should be computed separately. In Table 13.2.3, $S\hat{}_{ij}$'s are given separately for *Hae*III and *Hpa*II ($r = 4$) and for *Eco*RI and *Hin*dIII ($r = 6$). From these values, $\boldsymbol{\pi}\hat{}$ can be estimated in the same way as $v\hat{}$. In *D. melanogaster*, $\boldsymbol{\pi}\hat{}$ becomes 0.0080 for the enzymes with $r = 4$ and 0.0076 for the enzymes with $r = 6$, the average is 0.008. This value is another estimate of $4N\mu$ and is close to the values obtained from $\boldsymbol{v}\hat{}$ and s_r. In

D. virilis, the same computations were done, and the results obtained are presented in Table 13.2.2. All estimates of genetic variability in this species are smaller than those *in D. melanogaster.*

13.3 DNA LENGTH POLYMORPHISM. THEORY AND OBSERVATIONS

DNA data indicate that a substantial part of its polymorphism is due to deletion and insertion (Langley et al. 1982; Bell et al. 1982; Chapman et al. 1986; Lewin 1987; Zhimulev 2002; Klug and Cummings 2002). DNA RFLP is also generated by unequal crossover (Coen et al. 1982; Zhimulev 2002; Klug and Cummings 2002). Such polymorphism is usually detected by the presence of gaps between two DNA or restriction-site sequences investigated. One method for measuring this polymorphism is to compute haplotype diversity. Obviously, this measure depends on the length of DNA studied. A more suitable measure is the average number of gap nucleotides (nucleotides in the gaps) per nucleotide site between two randomly chosen haplotypes. Let m_{gij} be the number of gap nucleotides between the *i*th and *j*th haplotypes, and m_{Tij} be the total number of nucleotides compared, including gap nucleotides. The number of gap nucleotides per nucleotide site (g_{ij}) between the two haplotypes is then given by $g_{ij} = m_{gij}/m_{Tij}$ and the average of g_{ij} weighted with m_{Tij}, for all combinations of haplotypes is (Nei 1987)

$$G = \bar{m}_g / \bar{m}_T, \qquad (13.18)$$

where $\bar{m}_g$ and $\bar{m}_T$ are the means of m_{gij} and m_{Tij}, respectively.

Note that $\bar{m}_g$ in (13.18) is equivalent to (13.6), so that the variance of $\bar{m}_g$ can be obtained by replacing π_{ij}, by m_{gij}, in (13.8). The approximate variance of g may be obtained by assuming that $\bar{m}_T$ is constant, since $\bar{m}_T$ is usually much larger than $\bar{m}_g$.

Let us consider another example (Nei 1987). In *D. melanogaster*, RFLP identified six DNA-length haplotypes in the alcohol dehydrogenase gene region (Langley et al. 1982; Table 13.3.1). These haplotypes are apparently caused by deletions from the most common haplotype (N1), which is about 12 kb long. All deletions were observed in the noncoding regions of DNA. The haplotype diversity for this polymorphism becomes 0.682 ± 0.111, which is a little lower than the haplotype diversity (0.853) obtained for non-deletion restriction site polymorphism for the same gene region (Langley et

Table 13.3.1. Deletion/insertion polymorphisms in the alcohol dehydrogenase gene region of *Drosophila melanogaster* (Nei 1987; data from Langley et al. 1982).

Haplotype	Observed number	Deletion/insertion				
		Δa 20 bp	Δb 550 bp	Δc 900 bp	Δd 180 bp	Δf 30 bp
N1 (Standard)	10	–	–	–	–	–
N2	2	+	–	–	–	–
N3	1	–	+	–	–	–
N4	1	–	–	+	–	–
N5	2	–	–	–	+	–
N6	2	–	–	–	–	+

al. 1982). To estimate *g*, we must first compute g_{ij}'s for all pairs of haplotypes. For example, g_{12} is given by $mg_{12}/m_{T12} = 20/12000 = 0.002$. In the present case, all deletions/insertions are nonoverlapping, so that $m_T = 12{,}000$ for all g_{ij}'s. Once g_{ij}'s are computed, *g* can be obtained by (13.18). It becomes $g = 0.0175 \pm 0.009$. This is about three times as high as the nucleotide diversity (0.006) for the same DNA region (see Table 13.3.1).

Some Comments on DNA Polymorphism

Nucleotide Diversity. In previous sections, we met several ways of measuring DNA polymorphism. If we exclude the effect of deletions and insertions, the most suitable measure of DNA variability, as noted earlier, is nucleotide diversity. This measure is not sensitive to fragment length and sample size, so it is a comparable measure of DNA polymorphism for different genes and populations.

Nucleotide diversity has been studied for many genes (Table 13.3.2). Most estimates were obtained using restriction site polymorphisms. The nucleotide diversity varies from 0.002 to 0.019 in eukaryotes, and is rather similar for mitochondrial and nuclear genes: mean $\pi = 0.007$, i.e., 0.7% for both groups (Nei 1987; Table 10.6; means are given in my recalculation: Kartavtsev 2009a). If we remember that a typical gene consists of thousand nucleotide pairs, this result suggests that the nucleotide sequences of two genes randomly chosen from a population are rarely identical. MtDNA, as already introduced, is maternally inherited in most vertebrate animals, and exists in the haploid form. Because of this, the *Ne* for mtDNA is expected to be about one-fourth that of nuclear genes.

However, the mutation rate for mtDNA is apparently considerably higher than that for nuclear genes (Chapter 6). These two compensating factors probably make $\boldsymbol{\pi}$ for mtDNA nearly equal to that of nuclear genes. In some cases, however, nucleotide diversity seems to be very low. For example, mtDNA in Indians from Venezuela is virtually monomorphic (Johnson et al. 1983). Apparently, this population went through a bottleneck relatively recently. Unlike eukaryotic genes, the hemagglutinin gene of the influenza A virus shows an extremely high nucleotide diversity. This high nucleotide diversity is obviously due to an unusually high mutation rate in this organism.

Silent Polymorphism. It is of interest to examine polymorphism that does not manifest phenotypically even at the simplest level of amino acid sequences in proteins. Ample literature exists on the subject, to which we refer the reader for further information (Nei 1987; Swofford et al. 1996; Li 1997; Graur and Li 1999; Zhimulev 2002; Barns 2003). Here, we confine ourselves to discussing several simple issues. Neutrality theory predicts that this so-called silent polymorphism occurs more frequently than polymorphism implemented at the level of amino acid sequences, because silent mutations would undergo less strong selection than non-silent ones. Conversely, if polymorphism is generally maintained by directional selection (in case of a negligibly small drift effect), then silent polymorphism would be less frequent than non-silent polymorphism. One of the ways to test these assumptions is examining polymorphism at the first, second and third codon positions in functional structural genes, as well as in pseudogenes, which currently are nonfunctional.

Table 13.3.2. Estimates of nucleotide diversity (π) or the proportion of nucleotide differences between a selected pair of DNA sequences (π_{ij}) (from Nei 1987 with adds).

DNA or gene region	Organism	Method	n	bp	π or π_{ij}
	Nucleotide diversity (π)				
mtDNA	*Man*	*R*	100	16,500	0.004
mtDNA	*Chimpanzee*	*R*	10	16,500	0.013
mtDNA	*Gorilla*	*R*	4	16,500	0.006
mtDNA	*Peromyscus*	*R*	19	16,500	0.004
mtDNA	*Fruitfly*	*R*	10	11,000	0.008
β-globin	*Man*	*R*	50	35,000	0.002
Growth hormone	*Man* [a]	*R*	52	50,000	0.002
Adh gene region	*Fruitfly*	*R*	18	12,000	0.006
Adh coding region	*Fruitfly* [b]	*S*	11	765	0.006
H4 gene region	*Sea urchin* [c]	*S*	5	1,300	0.019
Hemagglutinin	*Influenza Virus*	*S*	12	320	0.510
	Selected pair of DNA sequences				
Insulin	*Man*	*S*	2	1,431	0.003
Immuno. C_κ	*Rat*	*S*	2	1,172	0.018
Immuno. IgG2a	*Mouse*	*S*	2	1,114	0.100
mtDNA, Cyt-b	*Fish (2 species)* [d]	*S*	2	–	0.335*
mtDNA, CR	*Fish, Pterois miles* [e]	*S*	–	–	1.900
mtDNA, 16S rRNA	*Copepods (2 species)* [f]	*S*	–	350	0.435*
mtDNA, Cyt-b	*Copepods, Tigriopus* [g]	*S*	–	–	0.260

Note. * Average is given in my recalculation; *n,* sample size; *Bp,* base pairs; *R,* restriction enzyme technique; *S,* DNA sequencing. Source: [a] Chakravarti et al. (1984); [b] Kreitman (1983); [c] Yager et al. (1984); [d] Baker et al. 1995; [e] Kochzius and Blohm 2005; [f] Bucklin and Wiebe 1998; [g] Willett and Burton 2004; see Nei (1983) for others.

Li and co-authors (Li et al. 1981; 1984) were among the first to address these issues. In their studies of the myoglobin gene in comparison with four pseudogenes in human, mouse, rabbit and goat, these authors have shown that (1) the nucleotide substitution rate in functional genes is the highest at the third codon position, and (2) the nucleotide substitution rate in pseudogenes is twice as high as the corresponding parameter, even at the third codon position. The *R* index, $R = \sigma^2/\mu$ (Li 1997, p. 232), estimated for 20 loci, supports these conclusions for more extensive data for non-synonymous (*N*) and synonymous (*S*) substitutions, the means being $R_N = 8.26$ and

$R_S = 14.41$ (Li 1997, Table 8.8). The differences in the nucleotide substitution rates (r) of nuclear genes in humans (47 genes) and *Drosophila* (32 genes) are even more striking: $r_N = 0.74$ (0.67; in brackets is standard deviation); $r_S = 3.51$ (1.01) and $r_N = 1.91$ (1.42), $r_S = 15.6$ (5.5), respectively (Li 1997, Tables 7.1, 7.6). In *Tigriopus californicus,* the simple proportion *N/S* (d_n/d_s) for 3 pair-wise interpopulation comparisons were 0.017, 0.018 and 0.025 (Willet and Burton 2004).

Recently a novel method to address this issue was presented (Tennessen 2008), which demonstrated that divergence in bacteria-killing activity among animal antimicrobial peptides is positively correlated with the log of the d_N/d_S ratio. The primary cause of this pattern appears to be that positively selected substitutions change protein function more than neutral substitutions do. Tennessen (2008) thus believes that the d_N/d_S ratio is an accurate estimator of adaptive functional divergence. Earlier, using another gene set, the substitution rates for *N*- and *S*-codons in evolution were also shown to differ: 8.26 and 14.41, respectively (Gillespie 1989). On average, the nucleotide substitution rate in pseudogenes is 4.7×10^{-9} per nucleotide per year, and is thought to be close to the neutral process (Nei 1987; see also Chapter 6). Analysis of another multigene family, amylases, revealed clear differences in *p*-distances for synonymous (1) and nonsynonymous (2) nucleotide substitutions in the sequences in three *Drosophila* species: (1) $p = 0.398 \pm 0.043$ and (2) $p = 0.068 \pm 0.008$ (Brown et al. 1990). Analysis of extensive data showed that for a randomly selected coding sequence, the ratio of synonymous and nonsynonymous substitutions is approximately 25 : 75%, while this proportion is reversed (69 : 31%) for the third position (Li 1997, Table 1.4). Note that the *N/S* ratio is significantly higher in human and close anthropoid ape species than in other monkey groups, owing to greater *Ne* (Wu et al. 2000) and inevitably stronger natural selection. The increased proportion of nonsynonymous substitutions in hominids is attributed to the rapid adaptive evolution in this group.

A search for selective response in the electron transport system in *Tigriopus californicus* showed no impact on two mtDNA genes *Cyt-b, Cyt-c* and one nDNA gene RISP, with some impact on nuclear *Cyt-c* (Willet and Burton 2004). In this and other cases, evidence suggests that nuclear vs. non-nuclear interactions may play an important role in selective response (Gerber et al. 2001; Willet and Burton 2004). The above evidence suggests that (1) genes and their regions with and without functional significance accumulate mutations and diverge at different rates and (2) the presence of purifying selection on coding sequences of structural genes is a well-established fact. However, we have to keep in mind that measuring selection is quite complicated, and frequently we may have not enough information for a proper solution (Ohta and Gillespie 1996). At 4 out of 5 loci considered above, there was no sign of selective impact on protein coding genes (Willet and Burton 2004). In other cases of positive impacts (e.g., Plotkin et al. 2004), most evidence may be more apparent than real (Hahn et al. 2005). Compared data sets, taken from GenBank for instance, may be not randomized enough for neutrality tests, as may be the case for some evidences in Rand and Kann (1998). However, the general conclusion that mildly deleterious mutations prevail at mtDNA genes is reasonable (Rand and Kann 1998).

Other discussion on these questions was considered before in Chapter 6.

13.4 DIVERGENCE OF POPULATIONS AT THE DNA LEVEL

DNA Divergence Among Populations

To estimate the amount of DNA divergence between two closely related populations or species, it is necessary to consider the effect of polymorphism. This can be done in a way similar to the estimation of genetic distance.

Nucleotide Differences Among Populations. Let us suppose that there are m different haplotypes for a particular DNA region in populations X and Y, and let x_i^ and y_i^ be the sample frequencies of the ith haplotype for populations X and Y, respectively. The average number of nucleotide substitutions for a randomly chosen pair of haplotypes (d_x) in a population X can be estimated by (Nei 1987)

$$d_x\hat{} = [n_x/(n_x - 1)]\ \Sigma_{ij}\ x_i\hat{}x_j\hat{}\ d_{ij}, \tag{13.19}$$

where n_x is the number of sequences sampled, and d_{ij} is the number of nucleotide substitutions per site between the ith and jth haplotypes. d_{ij} may be estimated via the equation $d\hat{}_{ij} = (-log_e S\hat{}_{ij})/r$ or by other methods [e.g., Nei 1987, expression (5.3)]. When all DNA sequences are different, $x_i\hat{} = 1/n_x$ (other notations are given in Section 13.2). The average number of nucleotide substitutions (d_y) for Y can be estimated in the same way. On the other hand, the average number (d_{xy}) of nucleotide substitutions between DNA haplotypes (nucleomorphs) from X and Y can be estimated by (Nei 1987)

$$d_{xy}\hat{} = S_{ij}\ x_i\hat{}x_j\hat{}\ d_{ij}, \tag{13.20}$$

where d_{ij} is the nucleotide substitutions between the ith haplotype from X and the jth haplotype from Y. The number of net nucleotide substitutions between the two populations (d_A) is then estimated by (Nei 1987)

$$d_A = d_{xy}\hat{} - (d_x\hat{} + d_y\hat{})/2 \tag{13.21}$$

If the rate of nucleotide substitution is constant, and is l per site per year, and the time since divergence between the two populations is T, then the expected value (d_A) of d_A^ is

$$d_A = 2\lambda T. \tag{13.22}$$

The rationale of equations (13.21) and (13.22) is that in the presence of polymorphic sequences the average time since divergence between the alleles (nucleomorphs) from X and the alleles from Y is longer than the time of population splitting, as seen from Fig. 13.4.1. Therefore, to estimate the population splitting time (T), we must subtract the average nucleotide difference for polymorphic alleles at the time of population splitting from d_{xy}. This average nucleotide difference is estimated by $(d_x\hat{} + d_y\hat{})/2$ under the assumption that the expected value of d_x^ or d_y^ is the same for the entire evolutionary process (Nei 1987, Chapter 13).

Let us consider one more example (Nei 1987). The alcohol dehydrogenase gene for two sibling species of *D. melanogaster*, i.e., *D. simulans* and *D. mauritiana*, was sequenced (Bodmer and Ashburner 1984; Cohn et al. 1984). The sequences in the two groups are slightly different from each other, and this difference is apparently due to

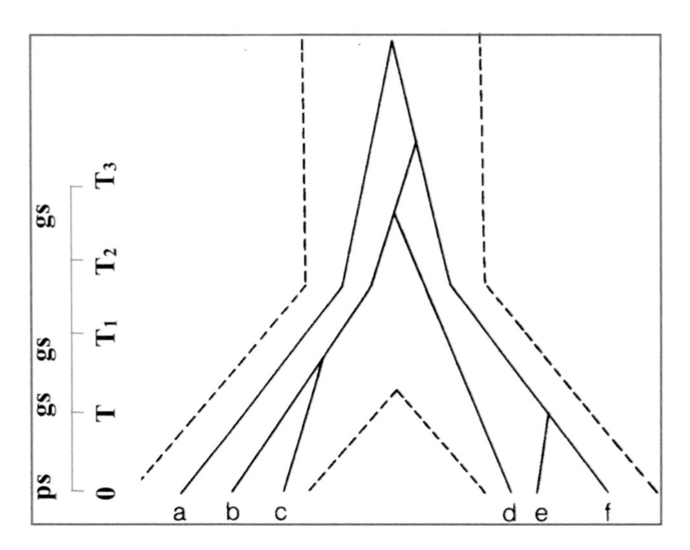

Fig. 13.4.1. Diagram showing that the time of gene splitting (gs) is usually earlier than the time of population splitting (ps) when polymorphism exists. a–f are different gene lineages, part of which (e.g., a, b and f) diverged long before population splitting. T denotes several time intervals of divergence (Modified from Nei 1987).

polymorphism within species that maintain different alleles with specific nucleotides. Stephens and Nei (1985) studied the interspecific and intraspecific differences of these sequences as well as those for *D. melanogaster* (Kreitman 1983), considering the shared region of 822 nucleotides. The results are presented in Table 13.4.1. The d_A values suggest that *simulans* and *mauritiana* are more closely related to each other than to *melanogaster*, and that speciation occurred first between *melanogaster* and the *simulans-mauritiana* group and then between *simulans* and *mauritiana.*

If we assume a constant rate of nucleotide substitution, we can estimate the time of divergence between these species. The average rate of nucleotide substitution for coding regions of genes seems to be about 2×10^{-9} per site per year ($0.71 \times 0.88 \times 10^{-9} + 0.29 \times 4.65 \times 10^{-9} \approx 2 \times 10^{-9}$) (see for comparison Chapter 6). If we use this value, the estimate of the time of divergence between *simulans* and *mauritiana* becomes $T = 0.0086/(2 \times 2 \times 10^{-9}) = 2.1 \times 10^6$ years from equation (13.22). This divergence time is only slightly larger than the divergence time (1.8 MY) of two polymorphic alleles ($d_x = 0.0073$) in *simulans*. On the other hand, the divergence times between

Table 13.4.1. Estimates of interpopulational (d_{xy}), intrapopulational (d_x or d_y) and net (d_A) nucleotide differences among three species of *Drosophila* for the alcohol dehydrogenase gene (822 nucleotides examined) (Stephens and Nei 1985).

Species	*D. melanogaster*	*D. simulans*	*D. mauritiana*
D. melanogaster	0.70 ± 0.41	1.73 ± 0.28	2.36 ± 0.24
D. simulans	2.45 ± 0.49	0.73 ± 0.30	0.86 ± 0.22
D. mauritiana	2.96 ± 0.54	1.47 ± 0.38	0.49 ± 0.24

Note. The figures on the diagonal refer to d_x (or d_y), and those below the diagonal d_{xy}. The figures above the diagonal represent the values of $d_A = d_{xy} - (d_x + d_y)/2$. All values should be divided by 100.

melanogaster and *simulans* and between *melanogaster* and *mauritiana* are estimated to be 4.4×10^6 and 5.8×10^6 years, respectively, the average being 5.1 MY. Therefore, the divergence between *melanogaster* and the *simulans-mauritiana* group seems to have occurred about three times earlier than the divergence between *simulans* and *mauritiana.*

Genealogical Relationships of Genes within and between Populations

The phylogenetic relationships for one gene or DNA fragment may be inferred from the DNA polymorphism in nucleotide sequences or restriction sites. If a phylogenetic tree is constructed on the basis of genes sampled from several populations connected by migration, theoretically we deal with mixed genealogies (Malecote 1973). MtDNA in human races, which today are actively intermixing, provides an example of such genealogical mixture. The mode of clustering of members of various races in a phylogenetic tree unequivocally demonstrates its "mixed" branches (Nei 1985; Fig. 13.4.2).

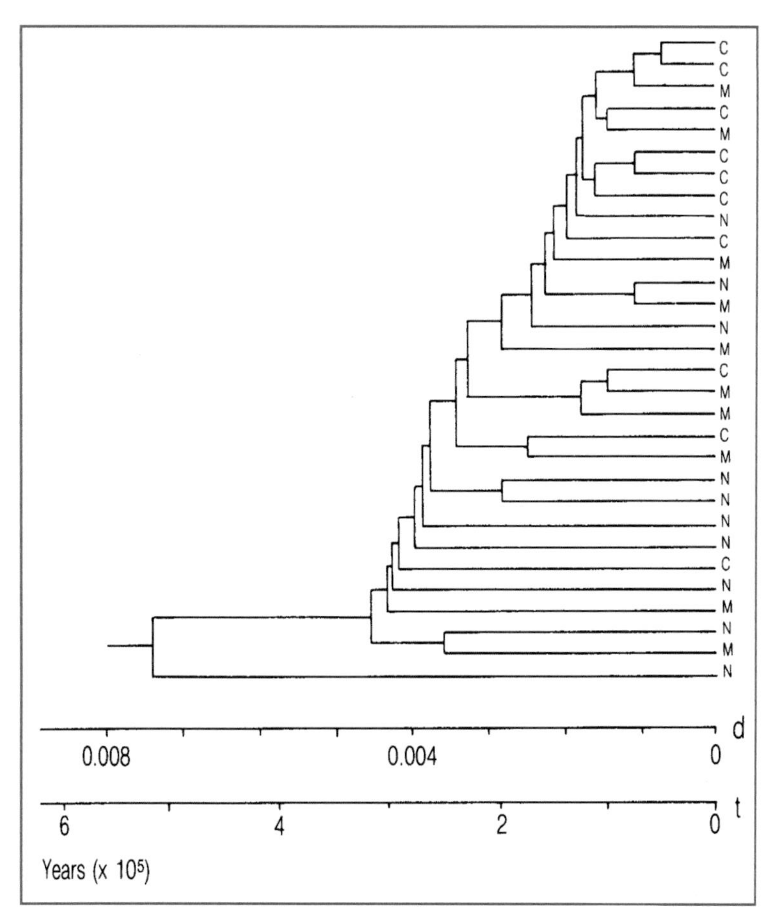

Fig. 13.4.2. Phylogenetic tree of mtDNAs from 10 individuals belonging Caucasoid (C), Negroid (N) and Mongoloid (M) races (from Nei 1987).

Similar results were obtained in other studies of humans (Cann 1982; Cann et al. 1982; Ingman and Gyllensten 2007); this ambiguous clustering was interpreted as showing migration among races (Cann 1982). The most ancient mtDNA divergence, dating back some 300,000 years, occurred in the members of Mongoloid and Negroid races. Apparently, the mtDNA divergence preceded the divergence of the races themselves, as follows from the estimates of their divergence based on other genes (Nei 1987). Note, however, that genealogical mtDNA mixing for various populations is expected, even in the absence of migration, if the ancestral population was polymorphic and the time since divergence was relatively short (Nei 1987). Lineage sorting of individuals by populations, which are isolated following it yields a phenomenon of older age of gene genealogies than population lineages. Later, evolution will convert this to difference between gene and species trees.

The evidences given in Sections 13.1 and 13.3 above on the intraspecies differentiation at DNA markers demonstrates a relatively low percentage of nucleotide divergence within species. Yet, as was noted in Chapter 7, the available data suggest that the divergence of populations within a species in some cases produces stable, geographically distinct spatial groups, phyletically marked by mitochondrial genes. This was found for the bottle-nosed dolphin, *Tursiops truncates* (Dowling and Brown 1993), Canadian goose, *Branta canadiensis* (Van Wagner and Baker 1990), in fish, *Fundulus heteroclitus* and *Stizostedion vitreum* (Gonzalez-Willasenor and Powers 1987; Billington and Strange 1995), and in a number of other organisms (Stepien and Faber 1998). Thus, migration and gene flow can be restricted, while intraspecific phyletic groups are real, stable population units of species, detected in the analysis of spatial genetic differentiation of particular generations or their mixtures (Avise and Walker 1999; Avise 2000). In Chapter 7 of this book, we considered the question whether and to what extent these data are associated with the genetics of speciation.

Introgression of Mitochondrial DNA

Ordinarily, different species are reproductively isolated, and there is no gene flow between them in accordance with the orthodox BSC. Recently, however, a number of authors have reported that mtDNA can cross species boundaries.

Investigation of mtDNA genotypes, in combination with nuclear DNA markers or isozyme loci, has sometimes demonstrated the ability of mtDNA to introgress from one species to another species, if the hybrids between these species and their progeny are fertile, and in this case, make an impact on the nuclear vs. cytoplasmic background. This introgressive hybridization requires successful backcrosses of the ancestral hybrid female with males of the parental species or other taxa. This introgression is independent of recombination and segregation events, occurring in the nuclear genome, if natural selection, maintaining nuclear—cytoplasmic compatibility, is absent (Takahata and Slatkin 1984; Nei 1987). However, evidence of this kind appears often, indicating the operation of subtle mechanisms that maintain the interaction of nuclear and, for example, mitochondrial genes (Gerber et al. 2001). A large number of cases of mtDNA introgression (see below and in Chapter 10) show that this selection, if it exists at all, is not sufficiently strong to prevent hybridization and introgression. Thus, the instances of possession of foreign mtDNA in natural species hybrids, identified by other methods, may be a proof of hybridization of closely related species (taxa). Such interspecific transfer of mtDNA has been found in species of invertebrates (*Drosophila*) and vertebrates (*Mus* and *Rana*) (Yonekawa et al. 1981; Powell 1983; Ferris et al. 1983; Spolsky and Uzzell 1984; Yonekawa et al. 2000; Suzuki et al. 2007; 2008). Literature on the topic was already considered with the aim of comparative analysis (Campton 1987; Avise and Wollenberg 1997; Yonekawa et al. 2000; Suzuki et al. 2007). Based on an analysis of the nuclear—cytoplasmic equilibrium (Clark 1984; Asmussen et al. 1987), an original method has been developed for testing the direction of hybridization and the intensity of introgression (Avise 2001). In this section, we only briefly touch upon the issue of mtDNA introgression, to elucidate its relationship with the species status in the BSC.

Firstly, let us consider some examples. In southern Denmark, there is a hybrid zone between two house mouse species, *Mus musculus* and *M. domesticus* (previously assigned to two distinct subspecies). Northern Denmark is inhabited mainly by *M. musculus*. Examination of mtDNA of these species has shown (Ferris et al. 1983) that, in contrast to the Eastern European form of *M. musculus*, in Denmark *M. musculus* to the north of the hybrid zone also possesses mtDNA of *M. domesticus*, which shows a 5% divergence of mtDNA nucleotide sequence from *M. musculus*. This part of *M. domesticus* is restricted to northern Denmark and some Swedish regions. Because of this, mtDNA introgression from *M. domesticus* to *M. musculus* seems to have appeared relatively recently. As reported, *M. musculus* species can be divided into at least three subspecies groups, *domesticus*, *castaneus* and *musculus*, which were genetically differentiated roughly one million years ago (Moriwaki et al. 2001). In Asia, the *castaneus* subspecies group inhabits mainly Southeast Asia, Taiwan and mainland China south of the Yangtze River, whereas the *musculus* subspecies group inhabits mainland China north of the Yangtze River, northeast China, Far East Russia, the Korean Peninsula and Japan. The former group has w1- and p-haplotypes of beta-hemoglobin (Hbb) alleles, the latter group has d-haplotype Hbb alleles. Recent molecular analysis of this Hbb DNA revealed that the nucleotide sequence of Hbbp-b1 gene is almost identical to that of Hbbd-b1, and that of Hbbp-b2 gene to that of Hbbw1-b2. This result strongly suggests that the p-haplotype is a recombinant between the b1-d and b2-w1 genes, probably between the two subspecies groups, *castaneus* and *musculus*. Comparisons of the nucleotide sequences of genomic DNA between Hbbp-b1 and Hbbd-b1 and also those between Hbbp-b2 and Hbbw1-b2 suggest that the recombination occurred more than 0.1 million years ago. Further analysis of the nucleotide sequence in the DNA stretch between b1 and b2 genes in p-haplotype samples obtained from several localities geographically apart demonstrated their common break point near b1 gene ORF. Those findings allow us to infer that the wild mouse populations carrying p-Hbb haplotype began to spread geographically quite recently in an evolutionary sense. Although at this moment it is still not possible to pinpoint the area where the d-w1 recombination occurred, the broader distribution of p-Hbb haplotype could suggest that it was somewhere in the central region in Asia (Moriwaki et al. 2001). Interestingly, the nuclear genes did not show evidence of introgression. It may well be that introgression at nuclear genes in these mammalian species is prevented by sterility or non-viability of hybrids, which is caused by the nuclear genes, whereas mtDNA, which does not affect fitness, can be inherited and transmitted independently.

Data presented here lets us conclude that despite gene flow determined at such markers as mtDNA genes, species hold their integrity; probably, that may occur because of the impact of other genes, having more serious influence on reproduction and adaptation. We have discussed this problem from another side, when considered speciation, hybridization and introgression earlier in the book (Chapters 7 and 10).

13.5 TRAINING COURSE, #13

13.1 Estimate p_n and p using GENEPOP (Raymond and Rousset 1995) or ARLEQUIN software and examples therein.

13.2 Visit the internet, and use two sequences from gene bank data base to apply GENEPOP or ARLEQUIN software.

13.3 Few terms for consideration:

Nucleotide diversity is the average number of nucleotide differences per site between two sequences.

The entire DNA segment under investigation is a **nucleon**, and a particular DNA sequence (or restriction site sequence) for the segment studied is a **haplotype** or a **nucleomorph**.

RFLP, Restriction Fragment Length Polymorphism, is a technique based on reaction of specific endonucleases to cut (restrict) DNA in a certain place, and is used as molecular genetic markers.

RAPD, Random Amplified Polymorphic DNA, is a molecular polymorphism identified at random and used as a molecular marker to detect variation and for **genetic mapping**.

AFLP, Amplified Fragment Length Polymorphism. AFLP is a variant of RAPD.

Silent polymorphism is DNA polymorphism that is not expressed at the amino acid level.

14

PHYLOGENETIC TREES

MAIN GOALS

14.1 Types of Phylogenetic Trees
14.2 Distance Matrix Methods for Tree Building
14.3 Parsimony and Probability Methods for Tree Inference
14.4 Population Genetic Theory and Tree Construction
14.5 TRAINING COURSE, #14

SUMMARY

1. There are two types of trees, which are called **species trees** and **gene trees**. They can coincide but are not necessarily identical.
2. A species tree is always an expected distance tree. A gene tree can be either an expected distance tree or a realized distance tree. In the expected distance tree, the lengths of branches, leading to two new species from a common ancestral species, are equal. Correspondingly, time since divergence between a pair of genes must be identical. When the rate of substitutions in genes is constant, then the expected evolutionary distance is also equal along each branch. This type of a tree is conveniently called an **expected distance tree**. However, the number of mutations or substitutions that actually occur (are realized) are calculated for a particular gene, they may differ between branches. This type of tree is called a **realized distance tree.** Realized trees can be obtained, for instance, when Fitch and Margoliash or parsimony methods are used.
3. Interrelationships of phylogenetic lineages of organisms or genes can be visualized in the tree-like mode with a root. This type of tree is called a **rooted tree**. Interrelationships can also be shown in a form of a **net** or **graph**, which is called an unrooted tree.
4. The three most popular methods of phylogenetic tree reconstructions are **Distance Matrix Methods**, **Maximum Parsimony Methods** and **Probability Tree Methods**: **Likelihood** and **Bayesian Inference**.

5. The simplest of the Distance Matrix Methods is the Unweighted Pair-Group Method of Analysis (UPGMA) or the method of mean distances. Frequently the Neighbor Joining, NJ method is very efficient.
6. A tree requiring the minimal number of substitutions is called a **Maximum Parsimony Tree (Maximum Economy Tree).**
7. Building a correct phylogenetic tree is a very difficult task, especially when the number of taxa is large. Nevertheless, the possibility to obtain gene genealogies reveals a new era in phylogenetic research that is called **Molecular Phylogenetics**.

14.1 TYPES OF PHYLOGENETIC TREES

Among the most important achievements of molecular evolutionary genetics is the discovery of the approximate constancy of the rate of amino acid or nucleotide substitution. This discovery gave us a new tool for phylogenetic tree construction. As stressed in Chapter 4, the steady rate of amino acid or nucleotide substitution holds only approximately, and there are some assumptions and complications. That is why reconstruction of a phylogeny and estimation of divergence time require a professional approach and knowledge of the whole complexity of the matter studied. However, compared with morphological or paleontological approaches, molecular data exhibit more reproducible patterns of evolutionary change. Molecular divergence may be measured in a single, comparative scale, as we summarized in Chapters 1 and 7. Consequently this can provide comparative values, and permit us to evaluate genealogy on the quantitative genetic basis of estimated co-ancestry, unlike with morphological and paleontological data, where there are many experts' inferences. Therefore, molecular phylogenetic approaches may give clearer representation of evolutionary relationship of organisms than, say, morphological characters. For a tree built on the basis of morphological traits, it is very difficult to give an evolutionary time scale. On the contrary, it is easy to convert molecular divergence into time estimates. This knowledge led to formation of the new field of genetics and evolutionary biology that is **Molecular Phylogenetics**. A very important property, which some critics of molecular phylogenetics do not comprehend (Pavlinov 2005), is that genes and genomes are transmitted through gender lines or via germ-plasm without environmental modifications and thus comprising a real basis for homology-orthology estimates of traits for the reconstruction of phylogeny. All this does not mean that other approaches should be rejected as outdated. The complexity of life, species types and modes of their evolution means that no single approach will always work, even the most modern approaches. This understanding was evident when trees based on allozymes and morphology together predicted relationships in some cases better (more explanatory in a complex background) than simple molecular genetic approaches (Hillis et al. 1996; Avise and Wollenberg 1997).

There is a long-standing controversy between supporters of phenetic and cladistic approaches (Bailey 1967; Sneath and Sokal 1973; Eldredge and Cracraft 1980). A significant part of this controversy is the ideology and methodology regarding the items' classification: individuals, taxa, etc., that for convenience are called OTUs (Operational Taxonomic Units). In this chapter, our main task is the construction of phylogenetic trees rather than taxonomy. That is why we will not go into the depths of this controversy. We will, however, meet various tree-making methods, some of which are favoured by different branches of numerical taxonomy. We will consider methods basically in terms their applicability to molecular data. However, many are also suitable for morphological characters.

Phylogenetic Trees: Species Trees and Gene Trees

There are two kinds of trees: *species trees* and *gene trees* (Tateno et al. 1982). We saw a kind of species tree in Chapter 7 (Fig. 7.1.1, Dobzhansky's scheme of divergence in time). As noted, the two types of trees do not necessarily coincide. One reason for that was explained in the previous chapter (Fig. 13.4.2). Moreover, even the branching pattern or tree topology constructed from genes may differ from that of the species tree. There are three possible different relationships between the species and gene trees for the case of three species (Fig. 14.1.1). In relationships (a) and (b), the topologies of the species and gene trees are the same, but in relationship (c) they are different. The probability of occurrence of the relationship of the mode (c) is reasonably high when the period between the first and second species splitting ($t_1 - t_0$) is short.

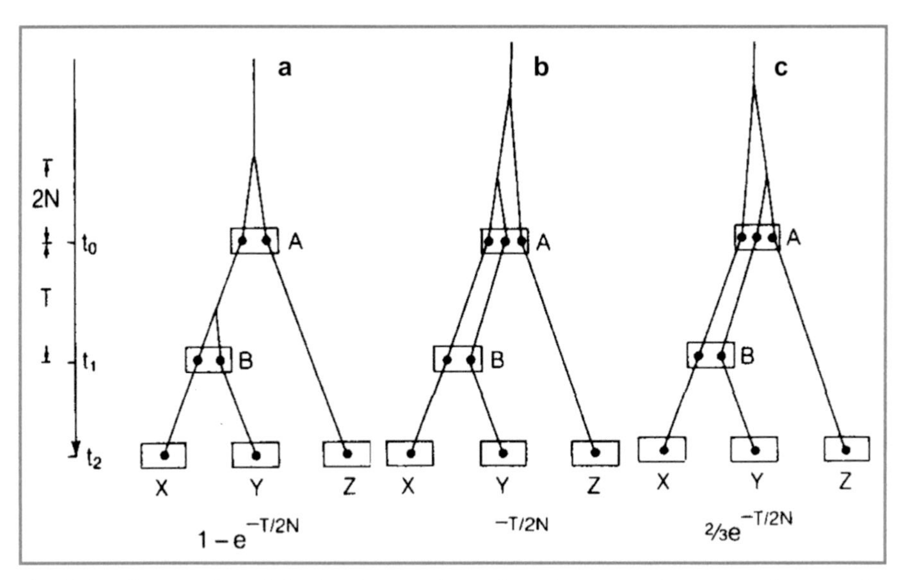

Fig. 14.1.1. Three possible relationships between the species and gene trees for the case of three species (X, Y and Z) in the presence of polymorphism, t_0 and tn are the times of the first and second species separation, respectively. The probability of occurrence of each tree is given underneath the tree. $T = t_1 - t_0$, and N is the effective population size (from Nei 1987 with permission).

Expected and Realized Distance Tree

A quantitatively estimated species tree is always an expected distance tree. In contrast, a gene tree may be either an expected distance tree or a realized distance tree. Let us define the difference between these trees. When investigating the species trees, there is a general interest to find out the time since divergence between each pair of species. In the species tree, the lengths of the two branches leading to two new species from the common ancestor must be equal. The same is true for the gene tree, the time since divergence between a pair of genes should be identical. When the rate of substitutions in a gene is constant, the expected evolutionary distance should also be the same along each branch. Thus, it is necessary to construct a tree, each branch of which is proportional to evolutionary time or expected evolutionary distance. This type of tree is called an **expected distance tree**. However, the number of mutations or nucleotide substitutions that really occur (are realized) in each of the two evolutionary lineages may not be the same because of stochastic error or varying rate of mutations/ substitutions in these lineages. Estimates of the actual number of mutations realized in each branch of a tree will be different. Consequently, the length of each branch may be different in this case. This kind of tree is called a **realized distance tree**.

Rooted and Unrooted Trees

Relationships in lineages of organisms or genes may be visualized in a treelike form with a root; this is the most common mode, and is called a **rooted tree** (Fig. 14.1.2a). However, it is also possible to construct a tree without a root, as an unrooted tree or graph (Fig. 14.1.2b).

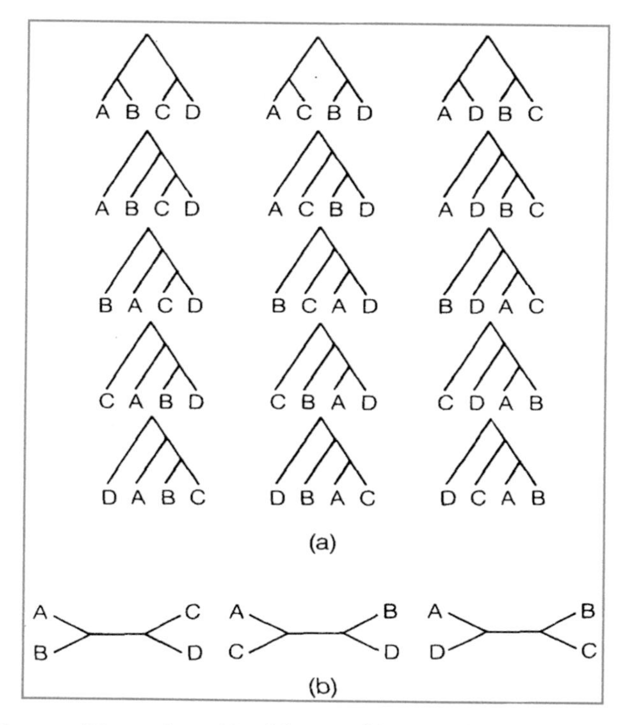

Fig. 14.1.2. Fifteen possible rooted trees (a) and three possible unrooted trees (b) for four species or OTUs (from Nei 1987 with permission).

There are many possible rooted and unrooted trees for a given number of species (n). Thus, for the case of $n = 4$, there are 15 possible rooted trees and 3 unrooted trees (Fig. 14.1.2). The number of possible bifurcating trees increases with increasing n by following equation (Nei 1987):

$$1 * 3 * 5 \ldots (2n - 3) = [(2n - 3)!]/[2^{n-2} (n - 2)!]. \quad \textbf{(14.1)}$$

Phenetic and Cladistic Trees

In numerical taxonomy, there are two different approaches to the construction of trees (Sneath and Sokal 1973; Bailey 1967). One is the phenetic approach, by which a tree is constructed considering the phenotypic similarities of the OTUs without considering the evolutionary history of the species. This approach is acceptable when the rules of evolutionary change of the characters used in the investigation are not quite certain. This approach is also reasonable when we are interested in the taxonomy, which are mostly based on phenotypic traits. Trees constructed by this approach may not represent the evolutionary relationships; that is why it is called a **phenogram**. In this case, tree construction is merely a way of grouping organisms. However, some of the statistical methods developed in phenetics can be used for constructing phylogenetic trees. Still, the criticisms of the phenetic approach have been discussed thoroughly (Farris 1977; Felsenstein 2004).

The second approach is called the **cladistic** approach. Under this a tree is constructed by considering various possible ways the traits may have evolved, following certain rules and choosing the best possible tree; thus, a phylogeny or history of OTU formation is inferred. A tree constructed by this approach is often called a **cladogram** (dendrogram or phylogram), in contrast with a phenogram. Some may conclude that this approach is better than the phenetic approach in constructing phylogenetic trees. In fact this is not necessarily true. The assumptions required for cladistic methods are not always met with molecular data (Nei 1987; Nei and Kumar 2002), as with other data (Bailey 1967). So, the superiority of this approach is not guaranteed.

Reconstruction of a Phylogenetic Tree

Generally, it is very difficult to reconstruct a true phylogenetic tree through which the descendant species or other OTUs originate and diverge. There are three most popular types of methods for reconstruction of a phylogenetic tree: 1. **Distance Matrix Methods** (DMM); 2. **Maximum Parsimony Methods** (MPM); and 3. Probability Methods, which include **Maximum Likelihood** (ML) and **Bayesian Inference** (BA). In DMM, evolutionary or genetic distance (similarity) is computed for all pairs of species or populations, producing a matrix of OTUs or OPUs (see explanation below in Section 14.2), and a phylogenetic tree is then constructed by considering the relationships among all pair-wise distance measures in this matrix. When the distance matrix is obtained, we should visualize the information in it. There are several different ways to do this, and one is constructing a tree. In MPM, the nucleotide or amino acid sequences of ancestral species are inferred from those of descendant species. A tree is produced by minimizing the number of branch changes for the whole phylogenetic tree. Such a tree is called a minimal tree or maximum parsimonious (parsimony) tree. For constructing a maximum parsimonious tree there are several algorithms. Also popular now are so-called probability or compatibility methods: maximum likelihood and Bayesian inference, which are used for tree building and give high resolution for many complicated cases (Hall 2001; 2011).

14.2 DISTANCE MATRIX METHODS FOR TREE BUILDING

Unweighted Pair-Group Method

The simplest of the distance matrix methods is the unweighted pair-group method (UPGM) of analysis or average distance method. This method was originally developed for constructing a phenogram (Sokal and Michener 1958; Bailey 1967). However, this technique can be used for constructing a phylogenetic tree as well, if the expected distance values are not very different in lineages investigated. Computer simulations have shown that when distance estimates are subject to large stochastic errors, UPGM is even superior to other distance matrix methods in obtaining the true tree (Tateno et al. 1982; Nei et al. 1983; Sourdis and Krimbas 1987). The UPGM is intended to estimate a species tree or expected gene tree. To achieve this purpose, it is better to use linear distance measures, such as the number of amino acid substitutions or Nei's standard genetic distance, D_n.

Let us consider the algorithm of UPGM. In UPGM, a certain measure of distance (or similarity) is computed for all pairs of OTUs or operational population units (OPUs), i.e., taxa or populations. Using a matrix of all pair-wise distances, a tree is built up by first joining the two OTUs with the smallest distance. A new distance matrix is then constructed, with the joined taxa now considered as one OTU. If the united taxa were indexed i and j, then for all $n \neq i,j$ the distance from n to the cluster (node) is $r_n(i,j) = (r_{in} + r_{jn})/2$. In other words, the distance from each OTU n to the cluster is the arithmetic average of the distances from OTU n to each of the OTU members in the cluster. All other distances are kept the same as in the initial matrix. The new distance matrix is again examined for the smallest score, and this forms the next joining point. This procedure is repeated until all OTU are clustered into a tree.

The UPGM algorithm is operating in an example below. Let the distance values obtained be presented in the following matrix.

OTU (OPU)	1	2	3
2	r_{12}		
3	r_{13}	r_{23}	
4	r_{14}	r_{24}	r_{34}

Here, r_{ij} denotes the distance between the ith and jth OTUs. Let the distance between OTUs 3 and 4 (r_{34}) be the smallest among all six distance values in the matrix. These two OTUs are then clustered in a single OTU (OTU 3-4) with a branching point located at distance $r_{34}/2$. It is assumed that the lengths of the branches leading from the branching point to OTU 3 and OTU 4 are the same. New distances between this combined OTU and the other OTUs are:

OTU (OPU)	1	2
2	r_{12}	
(34)	$r_{1(34)}$	$r_{2(34)}$

Here, $r_{1(34)}$ and $r_{2(34)}$ are given by $(r_{13} + r_{14})/2$ and $(r_{23} + r_{24})/2$, respectively. Finally, let r_{12} be the smallest distance. Put $r_{12}/2$ on the tree then and define the last cluster as: $r_{(12)(34)} = (r_{1(34)} + r_{2(34)})/2$, and put it also on the tree. We have the following tree in this case for the four OTUs:

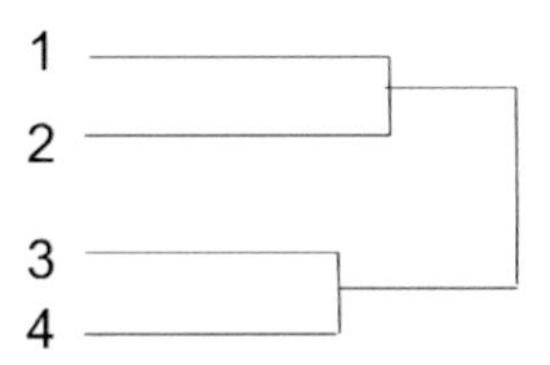

Theoretically under the UPGM technique for $n = 4$ OTUs, 15 types of trees are possible (see Fig. 14.1.2). Expression 14.1 helps to imagine how many types of tree are theoretically possible, say, for an animal genus or family of with $n = 25$ OTUs or twenty-five species (not very many), which equals 3^{49}! That is one reason why it is difficult to define the true tree topology, but there are others. We will see them below.

The UPGM is a simple method, but it has some very appropriate statistical properties. The main underlying assumption of UPGM when applied to molecular data is that the expected rate of gene substitution is constant. If the distance measure used is exactly linear with evolutionary time without error, UPGM gives the correct topology and exact branch lengths. If the distance measure is subject to stochastic errors, however, both the topology and branch lengths may be biased, even if the substitutions rate is constant. In real investigations, the correct topology is usually not known. Thus, it is difficult to evaluate the amounts of topological errors and branch length errors. It is known, however, that once the correct topology is obtained, UPGM gives the best least-squares estimates of branch lengths (Chakraborty 1977). One way to examine the accuracy of the topology of a phylogenetic tree from existing data is to evaluate the standard error (SE) of each cluster or node of the tree. In the case of a UPGM tree, this SE can be evaluated easily, if we assume a constant rate of substitutions. Obviously, the SE of a node leading to two extant OTUs (e.g., a in Fig. 14.2.1) is equal to half the SE of the distance between the two OTUs. We may also use the SEs of distance measures in the original matrix and extend them to branching points. However, not all measures have SEs. There is a way to estimate variances and SEs from distances and the numbers of OTUs (Nei et al. 1985; Nei 1987, pp. 295–296).

Let us consider a numerical example for a UPGM tree for hominids (Nei 1987). A segment (895 nucleotides) of mtDNA was sequenced (Brown et al. 1982) and compared in the human, chimpanzee, gorilla, orangutan and gibbon. From these data, we can estimate the number of nucleotide substitutions (*d*) and their standard errors by using the respective equations (Nei 1987, expressions 5.3 and 5.4; see also 13.5 and 13.7 in

Chapter 13), if one used nucleotide diversity. The results obtained are presented in Table 14.2.1. It is obvious that the *d* value between the human and chimpanzee is smallest, so that the human and chimpanzee must form the first cluster at a branching point (*a*) at $d_{HC} = 0.094/2 = 0.047$ (Fig. 14.2.1). The human and chimpanzee are now combined into a single OTU, (*HC*). The *d* values between this OTU vs. the gorilla, orangutan and gibbon become (0.111 + 0.115)/2 = 0.113, (0.180 + 0.194)/2 = 0.187 and (0.207 + 0.218)/2 = 0.212, respectively. The other distance values remain unchanged. The smallest *d* value in the new *d* matrix is that (0.113) between (*HC*) and the gorilla. Thus, the gorilla joins (*HC*) with a branching point of $d_{G(HC)} = 0.113/2 = 0.056$. If this type of calculation is repeated, we finally obtain the phylogenetic tree given in Fig. 14.2.1.

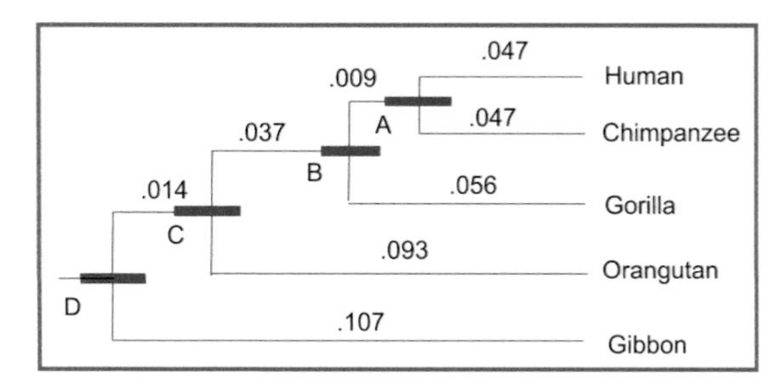

Fig. 14.2.1. Phylogenetic tree reconstructed by UPGM from the distance matrix in Table 14.2.1. The hatched bar represents one standard error on each side of the branching point (from Nei 1987 with permission).

Table 14.2.1. Proportion of different nucleotides (*p*) (above the diagonal) and estimates of the number of nucleotide substitutions (*d*) per site and their standard errors (below the diagonal) obtained from the nucleotide sequence data for five primate species (Brown et al. 1982).

	Human	*Chimpanzee*	*Gorilla*	*Orangutan*	*Gibbon*
Human		.088	.103	.160	.181
Chimpanzee	.094 ± .011		.106	.170	.189
Gorilla	.111 ± .012	.115 ± .012		.166	.189
Orangutan	.180 ± .016	.194 ± .016	.188 ± .016		.188
Gibbon	.207 ± .017	.218 ± .017	.218 ± .017	.216 ± .017	

Neighbour Joining Method (NJ)

NJ (Saitou and Nei 1987) is close to UPGM because it is also a DMM; thus, it uses distance-similarity matrix scores. NJ is building a tree by reducing in dimension the original matrix at each step; it continues to build the tree based on a series of matrices of decreased size. NJ differs from UPGM because it does not forms clusters or nodes on each step, but directly computes distances for internal nodes. Starting with the initial matrix, NJ at the beginning computes for each OUT its divergence in an unrooted tree from the rest of the OTUs as a sum of individual distances. Then NJ uses this "net's" divergence for computation of a corrected distance matrix. After that, NJ find the pairs of OTUs with the least corrected distances and computes distance from each of these OTUs to the nodes that joins them. Under this approach, the distance from each pair of OTUs pair to the node cannot be equal. On the next step a new matrix is created, in which a new node is substituted for the older one for the given pair of OTUs. NJ does not suppose that all OTUs are equally distant from the root.

NJ, similarly to MP method, is a method of minimal branch lengths or evolutionary changes, but unlike the MP it is not based on finding a tree with least overall distance. There are cases for which shorter trees then NJ are defined (Hillis et al. 1996). Some authors (Hillis et al. 1996; Swofford et al. 1996) believe that the best use of NJ is as a starting point for other methods, which use predicted models of nucleotide substitutions, such as ML or BA (see Section 14.3 and Chapter 15). However, experience has shown (Kartavtsev et al. 2007b; Sasaki et al. 2007) that NJ trees themselves frequently give satisfactory reconstruction of molecular phylogenetic interrelationships. Taking into account a very efficient computational algorithm, NJ usage may be very important when OTU number and sequence lengths are large. Probably, this namely quality and the property of minimal branch lengths gave quite a lot of popularity to the NJ approach in molecular phylogenetics.

Fitch and Margoliash Method (FMM)

One of the realized distance methods is that of Fitch and Margoliash (1967a). Suppose the number of substitutions distinguishing sequence i and j is d_{ij}. If the tree relating species sequences 1, 2 and 3 has branch lengths A, B and C (Fig. 14.2.2), then the branch lengths can be estimated from

$$\begin{aligned} A &= (d_{12} + d_{13} - d_{23})/2 \\ B &= (d_{12} + d_{23} - d_{13})/2 \\ C &= (d_{13} + d_{23} - d_{12})/2 \end{aligned} \qquad \textbf{(14.2)}$$

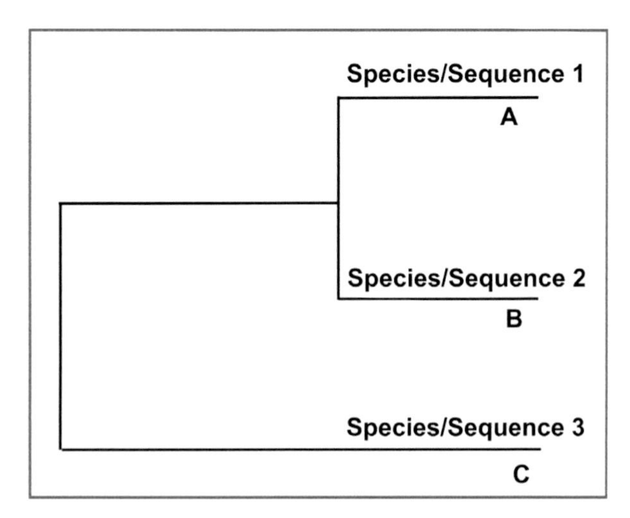

Fig. 14.2.2. A simple phylogenetic tree. A, B and C represent branch lengths from the most recent common ancestor.

These relations were found by solving the equations $d_{12} = A + B$, $d_{13} = A + C$ and $d_{23} = B + C$. If there are more than three sequences, the tree is built up by considering three units at a time, beginning with the two most closely related sequences and joining the remaining sequences (C is the composed group now). If sequences 1 and 2 are the most similar, then the distances from sequence 1 to the remaining group is the average of the distances from sequence 1 to each member of the group. In this way, only three distances are considered at a time, and expressions 14.2 allow us to estimate the branch lengths. This procedure is continued until all OTUs are united in a single tree. Unlike UPGM, the FMM does obtain a root of the tree as the average value. To find this root we must enter an "external" taxon or calculate the root as (A + B + 2C)/4 in the final phase of building the tree. The topology of the FMM tree is always the same as in UPGM, but the lengths of branches are different.

The procedure described above is only a part of FMM. Considering the probability that the topology of the tree obtained is not correct, it is necessary to analyse other possible topologies before making a final decision. Because the number of possible trees is very high with a large number of OTUs (see above), under the FMM application only the first few trees are considered (usually 2-3). A decision on the best tree is made using additional criteria, like: (1) score for cophenetic correlation coefficient, r ($r > 0.95$–0.99) or/and (2) so-called, percent standard deviations, s_{FM} (Nei 1987; expr. 11.9), etc. One of the trees is judged to be more correct if the r is higher or s_{FM} is smaller; s_{FM} is the deviation between observed and patristic tree branch length. One can find more details elsewhere (Nei 1987, pp. 299–301; see also Training Course #14).

Transformed Distance Method

This is a method derived mostly from FMM. Farris (1977) showed that when the rate of substitution varies among evolutionary lineages, the transformation of a distance may improve the tree topology for an average distance method. The algorithm for the transformation is

$$d'_{ij} = (d_{ij} - d_{ir} - d_{jr})/2, \qquad \textbf{(14.3)}$$

where r refers to an outgroup (reference) OTU. Due to stochastic errors, differences in backward and parallel mutations along different branches, errors in both the tree topology and branch length may be introduced. That is why it is recommended to use two or more different reference OTUs. When several trees are constructed, the true tree is then chosen by another its property (e.g., as shortest one) and accepted as the most consistent or consensus tree. How do we find a consensus tree? There are several ways. The simplest is to choose a tree that best fits all known facts (empirical approach). There are scientific tools to do this as well (Swofford et al. 1996; Hall 2001).

Farris' Distance-Wagner Method (FDWM)

This method was suggested for realized trees. However, in FDWM distance measures must be used that fit a triangle rule (hypotenuse is shorter than the sum of two legs). As in FMM, it is possible to build several trees by FDWM, but only one is chosen as correct. Let us consider the distance matrix in Table 14.2.1 for 5 OTUs. Let d_{ij} be the distance between the *i*th and *j*th OTUs. Let us further assume that distance d_{12} is smallest in the OTU matrix, so that OTUs 1 and 2 join first. The distance between this OTU (1,2) pair and each of the other OTUs is calculated as the averages between OTU (1) and OTU (3), and also OTU (2) and OTU (3). Suppose that now the distance between OTU (1,2) and OTU(3) is the smallest of all the distances in the matrix obtained by this way. Then OTU (3) joins OTU (1,2) as it is shown in Fig. 14.2.3a.

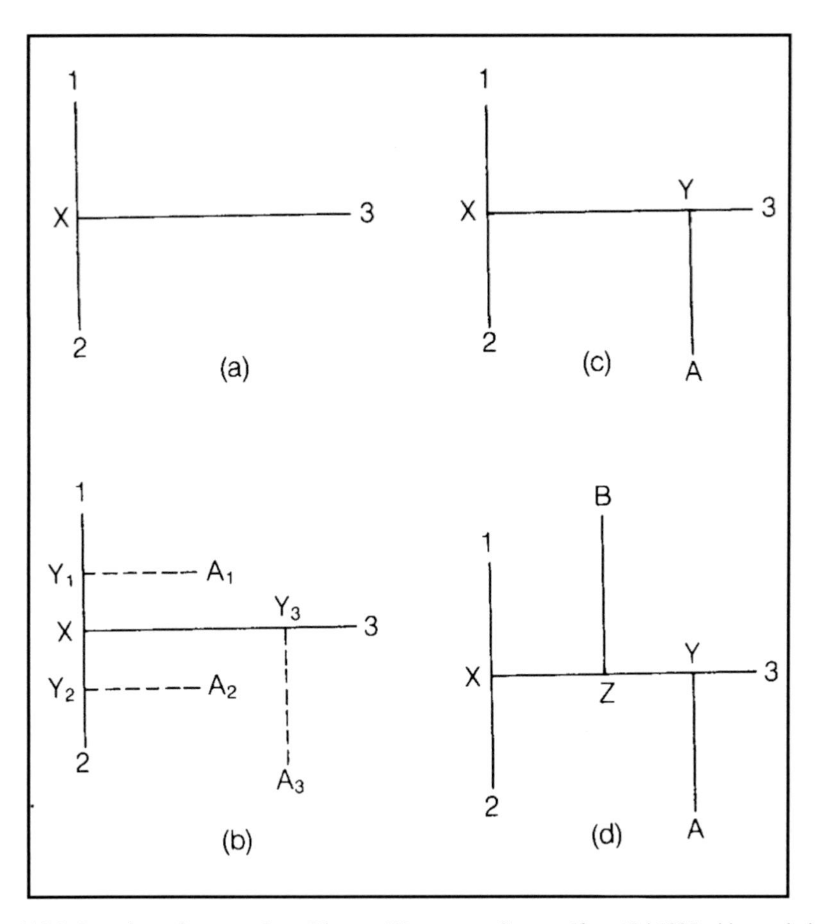

Fig. 14.2.3. Procedure of constructing a Distance-Wagner tree. See text (from Nei 1987 with permission).

This tree represents a network or unrooted tree because the approach itself does not define a root. The root is found later by entering an outgroup or by some other way. On Fig. 14.2.3, X is a branching point. The length of each branch is calculated using the following formulae:

$$L(3,X) = (d_{13} + d_{23} - d_{12})/2, \quad \textbf{(14.4a)}$$

$$L(1,X) = d_{13} - L(3,X), \quad \textbf{(14.4b)}$$

$$L(2,X) = d_{23} - L(3,X), \qquad (14.4c)$$

where $L(a,b)$ are the lengths between points a and b. In reality these lengths are calculated separately for each branch from the pair OTU (1,2) and the rest of the OTUs, and the OTU that shows the least distance to X is then chosen.

Now we come to the second step, under which one more OTU is added to the network. There are three possibilities for one OTU, say A, to be joined to the tree (Fig. 14.2.3b). It is necessary to calculate the lengths of branches $L(A_1, Y_1)$, $L(A_2, Y_2)$, $L(A_3Y_3)$. These calculations are performed for all the rest of the OTUs (4 and 5), and the OTU that shows the least distance is then chosen to join the network. As before, in reality $L(A_iY_i)$ is calculated by the following formulae.

$$L(A_1, Y_1) = \{L(A_1, 1) + L(A_1, X) - L(1, X)\}/2, \qquad (14.5a)$$

$$L(A_2, Y_2) = \{L(A_2, 2) + L(A_2, X) - L(2, X)\}/2, \qquad (14.5b)$$

$$L(A_3, Y_3) = \{L(A_3, 3) + L(A_3, X) - L(3, X)\}/2. \qquad (14.5c)$$

In these equations $L(A_1, 1)$ is obtained directly from the distance matrix; however values of $L(1, X)$ were already found from expressions (14.4). On the other hand, these values can be evaluated from the following formulae.

$$L(A_1, X) = L(A_1, 2) - L(2, X) = L_1 \text{ or } L(A_1, 3) - L(3, X) = L_2, \qquad (14.6a)$$

$$L(A_2, X) = L(A_2, 1) - L(1, X) = L_5 \text{ or } L_2, \qquad (14.6b)$$

$$L(A_3, X) = L_3 \text{ or } L_1. \qquad (14.6c)$$

Among L_1, L_2 or L_3, Farris' algorithm chooses the largest value and uses it for all $L(A_i, X)$ from equations 14.5. Let us assume that $L(A_3, Y_3)$ is smallest among $L(A_1, Y_2)$, $L(A_2, Y_2)$ and $L(A_3, Y_3)$. Then OTU A is joined with the branch 3-X, as shown on Fig. 14.2.3c. The last OTU (B, which is OTU 4 or 5) is then added to the network, as shown on Fig. 14.2.3d. The joining procedure is the same as before, except that five different routes of joining B must be considered. The procedure is repeated until all OTUs are clustered.

There are now modifications of Farris' method (Tateno et al. 1982; Faith 1985).

Comparative Properties of Distance Matrix Methods

When the substitution rate in genes is constant, the best results, as provided by numerical simulation (Tateno et al. 1982), in respect to the tree topology come from Farris' method and its modifications. This is especially true if d is not large, i.e., d = 0.09 or so. When d is larger, d = > 0.37, the best is UPGM (Nei 1987, p. 309). For estimation of the expected branch lengths, the best method is almost always UPGM. Blanken et al. (1982) showed that with heterogeneous substitution rates along the branches, the best results are from FMM, in a modification of Farris (1977). When there is a constant rate of substitutions they also note UPGM as the best technique for clustering (Blanken et al. 1982).

Nei et al. (1983) showed the three most important peculiarities of DMMs, which are partly illustrated in Fig. 14.2.4.

1. In all tree-constructing methods, the precision of tree topology and length of branches in a reconstructed tree (rooted tree) is very small, when the number of loci is less than 20, but gradually increases as the number increases.
2. When the expected number of substitutions per locus b = 0.1 or higher and 30 or more loci are used, then topology error, which is estimated by the index of distortion d_T (Robinson and Foulds 1981), is not large. However, the probability of finding a correct topology is less than 0.5, even with a sample size of 60 loci. Under small distances, that is, when d ≈ 0.004, the probability is much less.
3. UPGM and Farris' modified method exhibit better topology than Farris' method.

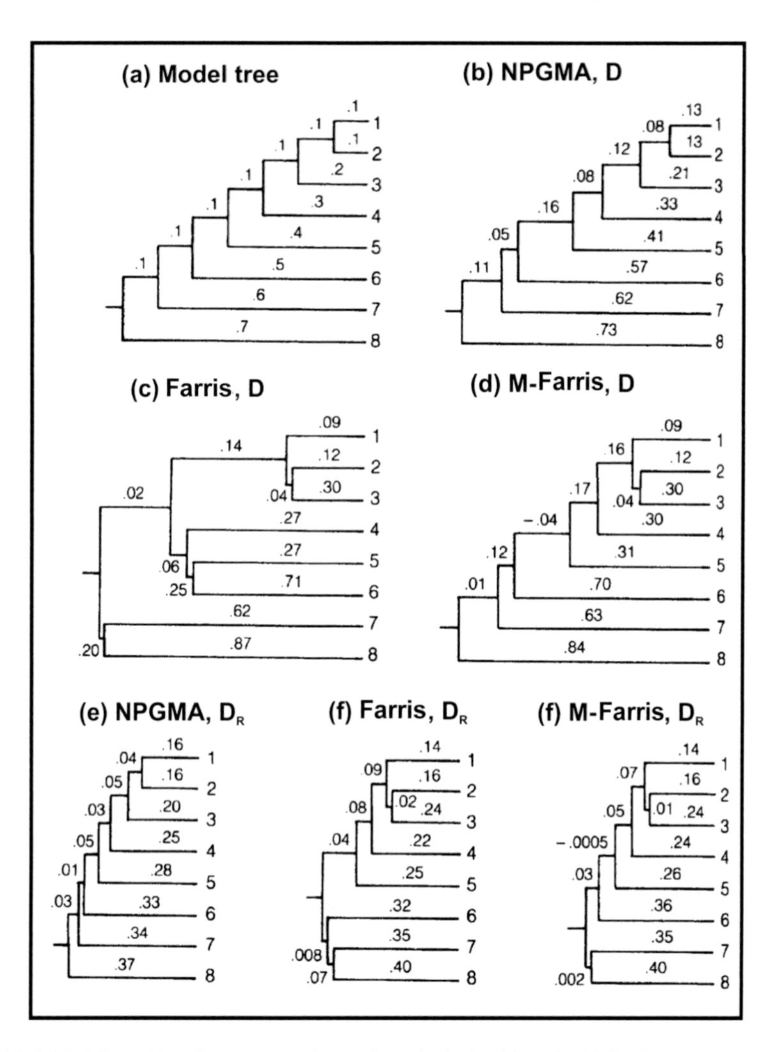

Fig. 14.2.4. Model tree (a) and reconstructed trees (b, c, d) obtained by using Nei's distance, D_n, and those (e, f, g) obtained by using Rogers' distance, D_r. These are based on the same set of gene frequency data from one replication of computer simulation with 50 loci (from Nei et al. 1983; adopted from Nei 1987 with permission).

14.3 PARSIMONY AND PROBABILITY METHODS FOR TREE INFERENCE

There are several variants of a parsimony tree (Felsenstein 1982; 2004; Swofford et al. 1996). The first suggestion of this approach appeared in a paper by Eck and Dayhoff (1966). The idea of the method is to introduce the amino acid or nucleotide sequence of the ancestor species, and then chose a tree that will require the minimal number of mutational changes. A tree defined by such a method is called the **Maximum Parsimony** (MP) tree. All MP trees are so-called minimal trees. The MP method is initially used for the topology building, and lengths of branches are not calculated, excluding some special cases.

The Parsimony Algorithm for Tree Building

Let us see the algorithm as described by Nei (1987). Firstly let us consider a topology for a certain group of OTUs and infer the ancestral sequences for such a tree. Then let us count the minimum number of substitutions that are required to explain the evolutionary changes in the tree. When this number is obtained, another topology is tested, and the minimum number of substitutions for this tree is obtained. The procedure is continued until all reasonable topologies are checked. In the end, the topology that requires the fewest substitutions is chosen as the correct or maximally parsimonious tree.

The minimum number of substitutions required for a given tree is defined for each nucleotide (or amino acid) site, and the total number of substitutions for all sites is then obtained. Let us define the first topology obtained by the modified Farris', or transformed distance method. Let us further analyse the six OTUs, which are assumed to be related by the rooted tree (Fig. 14.3.1a). Suppose as well that at a certain nucleotide position, the OTUs have the nucleotides listed at the tips of the tree. From these nucleotides, we can infer the nucleotides for the five ancestral sequences (nodes)

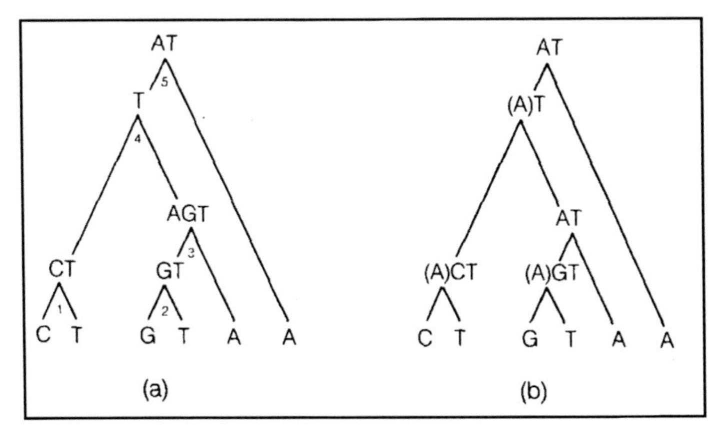

Fig. 14.3.1. Nucleotides in six extant species and the possible nucleotides in five ancestral species (Modified from Fitch 1971a; used with consent from publisher).

1, 2, 3, 4 and 5. The nucleotide at node 1 must be either C or T, if we consider the minimum number of substitutions. Similarly, the nucleotide at node 2 is inferred to be either G or T, whereas the nucleotide at node 3 must be A, G or T. Node 4 is expected to have T because its immediate descendant nodes are 1 and 3, both of which have T. Finally, we infer the nucleotide at node 5, which must be either A or T. It becomes clear that the minimum number of nucleotide substitutions for this set of 6 OTUs is obtained by assuming that all the ancestral nodes had nucleotide T. The minimal number is consequently four substitutions. However, this is not the only possible way of explaining the evolutionary change of the OTUs with four substitutions.

If we suppose that all middle nodes have A, the number of substitutions required is again four (Fig. 14.3.1b). There are nine more possibilities with the same minimum number of substitutions.

Four of them are as follows: (5-A, 4-T, 3-T, 2-T, 1-T), (5-A, 4-A, 3-A, 2-A, 1-T), (5-A, 4-A, 3-A, 2-A, 1-C), and (5-A, 4-A, 3-A, 2-G, 1-A). It is clear from above that the nucleotides at the ancestral node cannot always be determined uniquely, and the trees listed in Fig. 14.3.1 are equally parsimonious ones. However, it is possible to count the minimum number of substitutions required.

Alternative topologies for a given set of data can come from various methods. One is to use Fitch and Margoliash's (1967a) method mentioned earlier. Another one is Dayhoff and Park's (1969) method, in which a cut is made for each of the branches of the first tree, and each resultant part is grafted onto all branches of the other part. The minimum number of substitutions is examined for all topologies produced in this way, and the best one is chosen. This process is continued several times, and the best of all topologies examined is chosen as the maximum parsimony tree.

Informative Sites for Determining Topology

It may be suggested that any polymorphic (polytypic) site in which two or more different types of nucleotides (or amino acids) exist is sufficient to detect the tree topology. However, to build the maximum parsimony tree, the nucleotides that exist uniquely in an OTU (**singular nucleotides**; Fitch 1977) are not informative. Such nucleotides always may be assumed as having arisen by a single mutation in the immediate branch leading to the OTU in which it occurs, and so the nucleotide substitution is compatible with any topology. A nucleotide site is **informative** only when there are at least two different kinds of nucleotides, each of which is represented at least two times. For instance, the first nucleotide site in Fig. 14.3.2 is noninformative, whereas the fourth nucleotide site is informative.

Let us consider an example (Nei 1987). Figure 14.3.2 shows nucleotide sequences of mtDNAs from the human (H), chimpanzee (C), gorilla (G), orangutan (O) and gibbon (G) (Brown et al. 1982). In this figure, only sites at which at least two different kinds of nucleotides exist are given. There are 282 such polymorphic sites, but only 90 of them are informative for determining topology. They are marked with dots in Fig. 14.3.2.

Brown et al. (1982) considered six different topologies, which had previously been proposed by various authors for the evolution of this group of organisms; the best four are presented in Fig. 14.3.3.

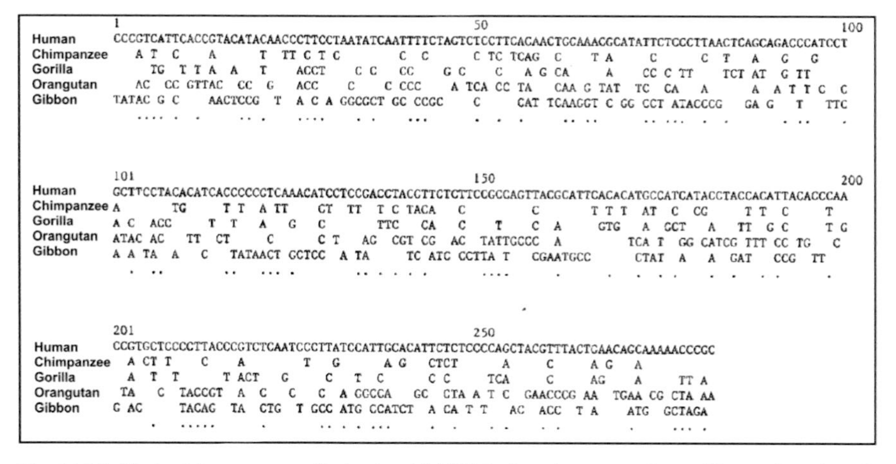

Fig. 14.3.2. Nucleotide sequences of mitochondrial DNAs from humans and apes. Here, only polytypic sites are presented. In chimpanzees, gorillas, orangutans and gibbons, the nucleotides that are identical with those of humans are not shown. Nucleotide sites with dots are informative sites (from Nei 1987; data fragment from Brown et al. 1982 with consent from Springer publisher).

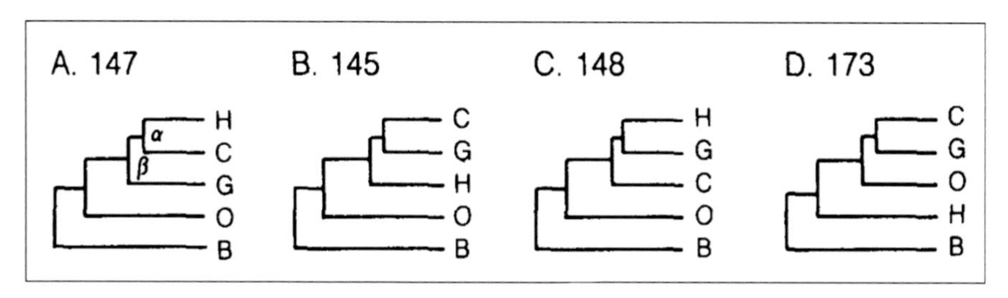

Fig. 14.3.3. Four possible phylogenetic trees (topologies) for the group of primates: human (H), chimpanzee (C), gorilla (G), orangutan (O) and gibbon (B).
The number given to each topology is the minimum number of nucleotide substitutions required for explaining the sequence differences (from Nei 1987 with permission).

In the construction of a maximum parsimony tree, it is convenient to classify the informative sites into different polytypic or polymorphic patterns. In the present case, there are 15 different polytypic patterns, as given in Table 14.3.1.

Polytypic pattern (HC)-(GOB) indicates that OTUs H and C share the same nucleotide, whereas G, O and B share a different nucleotide. Symbol G-(HC)-(OB) indicates that OTU G has a unique nucleotide, whereas H and C share a nucleotide and O and B share a different nucleotide. We note that some polytypic patterns are compatible with topology (A) in Fig. 14.3.3 under the assumption of a single nucleotide substitution, whereas others are not. For example, polytypic pattern "a" is compatible with topology (A) under the assumption of a single nucleotide substitution; in this case the single substitution may be assumed to have occurred between nodes a and β (Fig. 14.3.3A). However, polytypic pattern "b" requires at least two nucleotide substitutions to be accommodated with topology (A), whereas it requires only one to be accommodated with topology (C). So, pattern "b" is compatible with topology

(C) but not with (A). The list of compatible polytypic patterns for the four topologies is given in Table 14.3.2. For estimation of the minimum number of substitutions required for a topology, we must consider the minimum numbers for all compatible and incompatible polytypic patterns and get totals. The results obtained are given in Table 14.3.2. This calculation indicates that topology (B) requires 145 substitutions, which is the smallest number among the four topologies examined. In the present case, there are 11 more unrooted trees, but none of them has fewer substitutions than 145. Thus, topology (B) may be chosen as the most parsimonious tree.

Table 14.3.1. Polytypic patterns and their frequencies in informative sites for determining the parsimonious tree for the human (H), chimpanzee (C), gorilla (G), orangutan (O) and gibbon (B) (Nei 1987).

Polytypic pattern	Frequency	Polytypic pattern	Frequency
I. Informative sites			
a. (HC)-(GOB)	10	i. (GB)-(HCO)	8
b. (HG)-(COB)	5	j. (OB)-(HCG)	29
c. (HO)-(CGB)	2	k. G-(HC)-(OB)	3
d. (HB)-(CGO)	4	l. O-(HG)-(CB)	2
e. (CG)-(HOB)	10	m. O-(HB)-(CG)	1
f. (CO)-(HGB)	2	n. B-(HG)-(CO)	2
g. (CB)-(HGO)	4	o. B-(HO)-(CG)	1
h. (GO)-(HCB)	7	Total	90
II. Singular mutations			
p. Human	20	s. Orangutan	59
q. Chimpanzee	26	t. Gibbon	79
r. Gorilla	26	Total	210

Note. Singular mutations are listed separately.

Table 14.3.2. Compatible polytypic patterns with the four topologies given in Fig. 14.3.3 and the minimum number of nucleotide substitutions required for each topology (Nei 1987).

Topology	Compatible polytypic patterns	Number of compatible sites	Minimum number of substitutions
(A) HCGOB	a, j, k	42	147
(B) CGHOB	e, j, k, m, o	44	145
(C) HGCOB	b, j, k, l, n	41	148
(D) CGOHB	d, e, m, o	16	173

There is an opinion that for parsimony methods no assumption of approximate constancy of nucleotide or amino acid substitution is necessary. When the number of substitutions per site is small this is true. However, if this number is so large that parallel and backward mutations occur frequently enough, the parsimony methods may give large errors, if the substitution rate is not constant or varies among lineages. In such cases, the MP topology is not improved even if very large sequences of nucleotides or amino acids are examined (Felsenstein 2004). Moreover, if the compared number of nucleotides or amino acids is small and many backward and parallel mutations have occurred, MP methods may produce with high probability an erroneous tree, even under a constant rate of substitution (Peacock and Boulter 1975). Because MP trees are based on minimum estimates of realized numbers of substitutions, a statistical approach to test the variation in number of substitutions among different trees is difficult to develop. There are at least two sources of errors in the MP tree structure; they could come from the total number of substitutions and the heterogeneity of the substitution rate among different tree branches. To solve this problem Templeton (1983) proposed a nonparametric test for the case of restriction site data. However, it was shown (Nei and Tajima 1985; Li 1986) that parsimonious estimates of the number of restriction site changes for a polytypic site are subject to systematic error, depending on the tree structure and the number of changes. When four or more OTUs are involved, it is even unclear what kind of null hypothesis should be tested (Cavender 1981; Felsenstein 1985). These questions about the MP approach require thorough theoretical investigation, which has now been done (Deng and Fu 2000; Simmons and Miya 2004).

Parsimony methods usually do not provide estimates of branch lengths, as we noted before. However, it is possible to enter all ancestral nucleotides at each informative site and calculate the probabilities of having each nucleotide (Dayhoff and Park 1969; Fitch 1971a; Goodman et al. 1975). In this approach, the average number of substitutions for each branch in a tree can be estimated for all informative sites. After that we can add singular substitutions and obtain the total number of substitutions for each branch. When the parsimony tree is found, the branch lengths could be estimated by the method of Fitch and Margoliash (1967a). As for DMM, it is also difficult to find the correct topology for MP. Thus, for the primate group above, to achieve a high probability of correct topology (> 0.95) for the lineage human-chimpanzee-gorilla at mtDNA, a minimum 2,600 base pairs sequence is necessary. For nuclear DNA this number must be even larger because the rates of substitutions are usually lower for this kind of DNA (However, see discussion in Chapter 4).

Probability Methods for Tree Inference

Maximum Likelihood Method (ML)

The ML method for tree building was first applied by Cavalli-Sforza and Edwards (1967). A tree in this case was built from gene frequency data using the model of Brownian motion, i.e., following the Gaussian law. However, in the case of nucleotide sequence or amino acid data, the evolutionary change more likely follows the Poisson distribution or similar probability laws. Later this approach was used by other authors (Felsenstein 1973; 1981; 2004; Langley and Fitch 1974). Langley and Fitch (1974) also used the ML method for the estimation of expected branch lengths and consequent evolutionary times for a given topology when the observed number of substitutions for each branch is known. We will briefly consider both approaches, following Nei (1987).

Langley and Fitch's Method. Let us consider a phylogenetic tree as exemplified in Fig. 14.3.4. Here x denotes the observed number of substitutions for the ith branch, and t_i is the time interval. We are focused on estimating t_1, t_2 and t_3 by using the ML method. To apply this method, we must have a specific mathematical model of the change of genes or proteins. We assume that the rate of substitution (λ) is constant per year or per generation throughout the evolutionary process and that the number of substitutions for the gene or protein considered follows the Poisson distribution for any time interval. That is, the probability that x substitutions occur during evolutionary time t in one lineage is $exp(-\lambda t)\,(\lambda t)^x/x!$. Therefore, the likelihood of observing x_1, x_2... x_6 *for* the tree in Fig. 14.3.4 is

$$L = (e^{-v1}\, v^{x1}/x_1!) * (e^{-v1}\, v^{x2}/x_2!)\, [e^{-(v1+v2)}\, (v_1 + v_2)^{x3}/x_3!)]\, [e^{-(v1+v2)}\, (v_1 + v_2)^{x4}/x_4!)], \quad (14.7)$$

where $v_1 = \lambda t_1$. Since λ is usually unknown, we search for the ML estimates of v_i's rather than t_i's. They are given by the following simultaneous equations.

$$dLog_eL/dv_1 = -4 + (x_1 + x_2)/v_1 + (x_3 + x_4)/(v_1 + v_2) = 0, \quad (14.8a)$$

$$dLog_eL/dv_2 = -3 + (x_3 + x_4)/(v_1 + v_2) + x_5/(v_2 + v_3) = 0, \quad (14.8b)$$

$$dLog_eL/dv_3 = -2 + x_5/(v_2 + v_3) + x_6/v_3 = 0. \quad (14.8c)$$

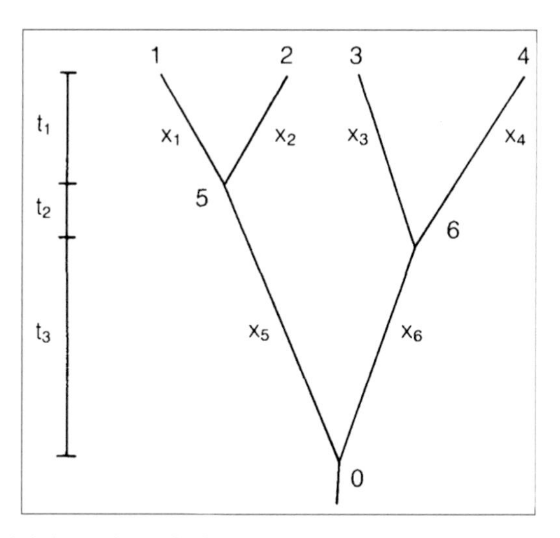

Fig. 14.3.4. Artificial phylogenetic tree for four OTUs (1, 2, 3 and 4).
t represents the evolutionary time, whereas x stands for the observed number of substitutions (from Nei 1987 with permission).

Solutions to these equations can be obtained numerically. The same approach may be used for any number of OTUs, because the number of parameters that must be estimated is always smaller than the number of observed scores.

We assumed that λ is constant in all lineages. However, that may be tested directly from v_i's by chi-square test (Langley and Fitch 1974). By these authors' testing in the lineages of 18 vertebrate species, high heterogeneity was found, which provided new evidence that the tested rate of amino acid substitution in a protein is not strictly a constant. Still, the number of changes appears to be a linear function of evolutionary time (when t is large).

Felsenstein's method. This method (Felsenstein 1981) is based on the ML observed nucleotide sequences in a group of OTUs. The ML is calculated for many different topologies, and the topology with the highest likelihood value is chosen as the true tree. This approach is even more complicated than the previous one. Felsenstein justifies his method by considering the extremal case of infinitely many nucleotides, but the number of nucleotides actually used is usually quite small. It is possible that the method is insensitive to these complicating factors, as is usually the case with many statistical methods. However, the truth of this should be proved. More details are presented elsewhere (Nei 1987; Swofford et al. 1996).

Bayesian (BA) Inference of Tree

BA tree is based on the notion of posterior probabilities, i.e., probabilities that are estimated using a certain model (prior expectations), based on information about the data. Bayesian analysis of phylogenies (Rannala and Yang 1996; Mau and Newton 1997; Mau et al. 1999) is similar to ML because the researcher postulates a model of evolution, and the program searches for the best trees that are consistent with both the model and with the data. There are also differences from ML. Thus, while ML seeks the tree that maximizes the probability of observing the data, given that tree, BA searches for the tree that maximizes the probability of both the tree, given the data, and the model for evolution. In principle, BA re-scales likelihoods to true probabilities in the way that the sum of the probabilities over all trees is 1.0 under the Bayesian approach; this technique allows using original probability theory to analyse the data. We might think of probability theory, and in particular of Bayes' rule, as providing us with a logic of uncertainty (e.g., such as in Gauss' theory). For example, given A we would "reason" about the likelihood of the truth of B (let's say B is binary for simplicity and is the heads rate of a single coin) via its conditional probability $P(B|A)$: that is, what is the probability of B given that A takes a particular value? An appropriate answer may be that B is true with a certain probability, say 0.5 for a coin's run sessions with A for heads and B for tails. The "machine learning" task is then to approximate $P(B|A)$ with some appropriately specified model, based on a given set of corresponding examples of A and B. Thus, if many coins are used and some of them are false, then expectation is biased for each from the theoretical expectation 1:1 and should be estimated empirically. The mathematical essence of the approach is given below following Tipping's (2004) representation.

Specifying a Bayesian Prior. Prior is a setting in the modelling procedure where Bayesian inference comes to the fore. We have to use some form of a parameterized model for the conditional probability:

$$\mathbf{P(B|A) = f(A; w),} \qquad \mathbf{(14.9)}$$

where w denotes a vector of all justifying parameters in the model. Then, given a set D of N examples of our variables, $D = \{A_n; B_n\}^N_{n=1}$. A conventional approach would involve the maximization of some measure of "precision" (or minimization of some measure of "inaccuracy") of our model for D with respect to the justifying parameters.

Of course, giving an identical solution for w as least-squares, maximum likelihood estimation will also result in overfitting. To control the complexity of the model, substituting the earlier formalization of weight penalty $E_W(\mathrm{w})$ (Tipping 2004), we may define a *prior distribution* that expresses our "degree of belief" over values that w might take:

$$p(\mathbf{w}|\alpha) = \Pi^M_{m=1} (\alpha /2\pi)^{½} \exp [(-\alpha/2)w^2_m]. \quad (14.20)$$

This choice of a zero-mean Gaussian prior expresses a preference for smoother models with smaller weights to be a priori more probable. Though the prior is independent for each weight, there is a shared inverse variance hyperparameter $\boldsymbol{\alpha}$, analogous to λ (See below and Tipping 2004), which establishes the strength of our "belief".

Posterior Inference. Given the likelihood and the prior, we could compute the *posterior distribution* over Bayes' rule:

$$p(w|t, \alpha, \sigma^2) = (likelihood \times prior)/(normalizing\ factor) = [p(w|t, \alpha, \sigma^2)\, p(w|\alpha)]/[p(t|\alpha, \sigma^2). \quad (14.21)$$

Because combining a Gaussian prior and a linear model within a Gaussian likelihood, the posterior is also conveniently Gaussian: $p(w|t, \alpha, \sigma^2) = N(\mu, \Sigma)$ with

$$\mu = (\Phi^T\Phi + \sigma^2\alpha I)^{-1}\Phi^T t, \quad (14.22)$$

$$\Sigma = \sigma^2 (\Phi^T\Phi + \sigma^2\alpha I)^{-1}. \quad (14.23)$$

So instead of "learning" a single value for w, we have a distribution for all possible values, and have updated our prior "belief" in the parameter values with connection to the information provided by the data t, and with more posterior probability (that assigned to values which are both probable under the prior and which "explain the data").

MAP Estimation: a Bayesian Short-cut. The maximum *a posteriori* (MAP) estimate for w is the single most probable value with the posterior distribution $\boldsymbol{p(w|t, \alpha, \sigma^2)}$. Because the denominator in Bayes' rule (14.21) is independent of w, this is equal to maximizing the numerator, or equivalently minimizing $E_{MAP}(w) = -\log p(t|w,\sigma^2) - \log p(w|\alpha)$. Retaining only those terms dependent on w gives:

$$E_{MAP}(w) = 1/(2\,\sigma^2)\{\Sigma^N_{n=1} [t_n - y(x_n; w)]^2 + [\alpha/2\, (\Sigma^M_{m=1} w^2_m)]. \quad (14.24)$$

The MAP estimate is therefore identical to the penalized least-squares estimate with $\lambda = \sigma^2\alpha$.

Unlike ML, which seeks the single most likely tree, BA searches for the best set of trees. As ML searches a landscape of possible trees, it moves from point to point seeking higher points (i.e., more probable trees). If there is more than one hill on the landscape, ML can get trapped on a hill even if there is a higher hill (i.e., a better set of trees) elsewhere. While other heuristic searches do not consider the same tree more than once, the Bayesian approach will often consider the same tree many times. Computationally this approach is realized with MrBayes program package (Ronquist and Hulsenbeck 2003).

MrBayes program (Ronquist and Hulsenbeck 2003) provides very efficient computations and well sustainable trees. It utilizes the Metropolis-Coupled Monte Carlo Markov Chain (MCMCMC) technique, which provides a set of independent algorithmic runs that sporadically exchange information. This technique allows leaps out a valley that would otherwise trap it on a suboptimal hill (not the best tree). The program output is a set of trees that are repeatedly visited and learned and which comprises the top of the hill. In essence, BA of phylogenetic trees works the same way as the machine learning probability approach. In this case, the model for learning is a tree with a certain branching pattern or topology and with predesigned branch length, a predesigned model of nucleotide substitutions (see Chapter 15), and specified distribution of substitution rates across the sites.

In the current Section 14.3, we omitted from consideration some methods (Swofford et al. 1996; Felsenstein 2004) or just mentioned others, like some ML methods. The field is too difficult to consider more fully in the context of this course. We turn to some methods again during the training course. Also there we will meet some software that are available for phylogenetic and population genetic research (see also Chapters 13 and 15). In their review Swofford et al. (1996) listed 42 software programs available for conducting phylogenetic and population genetic analysis both for Windows and Macintosh operation systems. Nowadays, the number of software and On-Line tools totals many hundreds. Our task is only to be familiar with 1–2. In future you may update this knowledge personally.

14.4 POPULATION GENETIC THEORY AND TREE CONSTRUCTION

Molecular phylogeny is based on the number of nucleotide substitutions in DNA or the number of gene substitutions in the genome. Nucleotide or gene substitution is a product of a complicated process of mutation, natural selection and genetic drift, so it is important to take into consideration these processes in tree construction. In common practice most molecular trees are built by using a single gene analysed in all species. These trees are often regarded as species trees. As mentioned earlier, however, a gene tree is sometimes quite different from a species tree. To evaluate the magnitude of difference between these trees it is necessary to know the forces that influence the gene substitution in populations. Such questions as neutrality, selective advantage, N_e size, mutation rate and migration rate all affect the precision of estimates, both for the tree topology and for time since isolation in different lineages.

Many authors have attempted to construct trees from data of protein electrophoresis, treating each allele as a frequency-less cladistic character. Many paid little if any attention to the dynamics of allele frequency change or gene substitution in populations and causes for it. However, the construction of a correct phylogenetic tree for a species is not possible without considering these causes. Recommendations to use the UPGM and Nei's standard genetic distances for allozyme data are based on the consideration of such causal sources (Nei et al. 1983; Nei 1987) and for DNA data is considered in Chapter 13 and in the current chapter. However, if there are strict bottleneck effects or inbreeding effects on genetic distances, as in the case of human populations, UPGM and any other technique is expected to produce a senseless tree (Nei and Roychoudhury 1982). We should meet these challenges with the understanding of the population genetic processes underlying phylogenetic change. This is quite a complicated field. Recent apparent and real controversies between phylogenetics and the biological species concept strongly support this notion (Avise and Wollenberg 1997; Harrison 1998; Chapter 7). Still, taking into account a lot of difficulties mentioned above and not mentioned in this brief chapter, surprisingly a good correspondence of protein-based and DNA-based trees has been obtained in many animal groups (Nei 1987; King 1993; Miya et al. 2003; Kartavtsev et al. 2007a,b; 2009, etc.).

14.5 TRAINING COURSE, #14

14.1 Estimate Percent Standard Deviation (s_{FM}) (distortion index, D_t or r) and time since divergence, t using GENEPOP or ARLEQUIN software and examples therein.

14.2 Visit the Internet and use three arbitrary sequences from the gene bank database to apply PHYLIP or ARLEQUIN software for constructing a phylogenetic tree.

14.3 Compare human-ape tree in Sibley and Ahlquist (1984), made on DNA hybridization data, with the two previous trees built with the usage of other gene markers (Fig. 14.2.1 and Fig. 14.3.3A). What should be a view of a consensus tree for the line HU-CH-GO? Draw it up. Consider in your decision the reasons that are explained below in point 14.5.

14.4 Do data on flu virus change in time if used records from paper by Fitch et al. (1991) (Fig. 14.5.1) agree with molecular clock theory?

14.5 **Percent Standard Deviation** (s_{FM}) for FMM and others.

This statistic is defined as follows:

$$s_{FM} = [1/\{n(n-1)\}(2\Sigma_{ij}\{(d_{ij} - e_{ij})/d_{ij}\}^2]^{1/2} * 100$$

for $i < j$, where n is the number of OTUs used, and d_{ij} and e_{ij} are the observed distance and patristic distances between OTUs i and j, respectively. The patristic distance between OTUs i and j is the sum of the lengths of all branches connecting the two OTUs. For example, the patristic distance between human and gorilla in the tree of Fig. 14.2.1 is a + c + d = 0.108. The tree with the smallest s_{FM} is as the best; $s_{FM} = 1.3$ for the human ape tree (Fig. 14.2.1, Table 14.2.1). For the other two variants, where GO-HU-CH changes, the $s_{FM} = 3.11$ (HU->CH-GO) and $s_{FM} = 1.88$ (CH->GO-HU). So, the best tree according to s_{FM} is as in Fig. 14.2.1. In Nei's (1987, p. 302) opinion s_{FM}, or s_o (Farris 1972) are meaningful when these statistics are quite large, and a large r does not grant the correct tree at all.

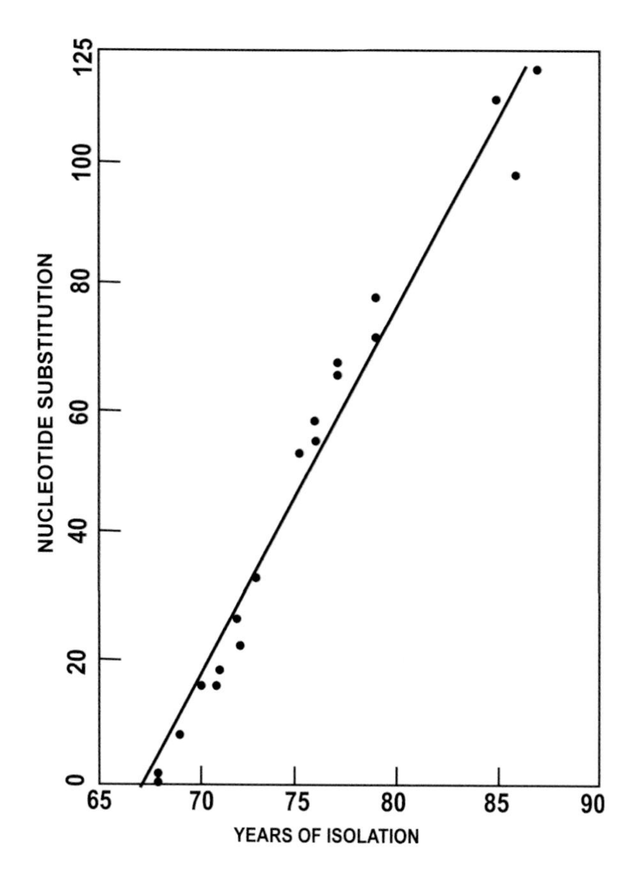

Fig. 14.5.1. Molecular clock in the influenza A hemagglutinin gene.
Number of nucleotide differences between the first isolate and each subsequent isolate as a function of years of isolation.

14.6 Few terms for learning:

Phenogram is a rooted tree which represents only phenotypic resemblance without reference to kinship.

Cladogram is a phylogenetic tree, which is constructed by considering genetic kinship and various possible ways of evolution, following certain rules and choosing the best possible tree.

A **nucleotide site** is **informative** only when there are at least two different kinds of nucleotides and one represented at least two times. The nucleotides that exist uniquely in an OTU (**singular nucleotides**) are not informative.

15

PRACTICAL TRAINING WITH NUCLEOTIDE SEQUENCES: EDITION, SUBMISSION, ALIGNMENT, TREE BUILDING AND ANALYSIS OF PHYLOGENY

MAIN GOALS

15.1 Editing Sequences and their Submission to a Gene Bank
15.2 Data Formats and Gene Banks Available
15.3 Alignment of Sequences
15.4 Searching for an Optimal Nucleotide Substitution Model
15.5 Tree Building Using MEGA-5 Program Package and Annotation on Programs PAUP, MrBayes and Some Others
15.6 Training Course, #15

SUMMARY

1. When the nucleotide sequence for a gene or its section is obtained using methods of molecular genetics, it is necessary to make the next step, which includes editing the sequence and submitting it to a gene bank. One of the most convenient program packages (PP) for editing sequences is **Chromas-pro** (http://www.flu.org.cn/en or http://www.technelysium.com.au/chromas.html). Popular as well for this purpose are Bio-Edit, MEGA-5 and many others.
2. Easy access, good opportunities for submission, retrieving and preliminary investigation of sequences are provided by GenBank (http://www.ncbi.nlm.nih.

gov). If sequences are not large, and are from few individuals, lines, etc., they can be submitted in an interactive mode through a Web-net and usage of the **Bankit** program utility.

3. Before beginning molecular genetic analysis, the alignment of sequences should be done. After that it is necessary to choose the optimal model of nucleotide substitutions for a set of analysed sequences. MODELTEST and MEGA-5 programs are convenient tools for this.
4. There are many PP that can be used for molecular phylogenetic tree building. Resources for tree building that are simple to use and easily accessed via the Web are provided by the PP **MEGA-4** and/or **MEGA-5** (http://www.megasoftware.net/). PP MEGA is examined in more detail, so that an inexperienced user can make a simple analysis and build a phylogenetic tree. An annotation of PAUP, MrBayes and PHYLIP software is also provided.

15.1 EDITING SEQUENCES AND THEIR SUBMISSION TO A GENE BANK

Original sequences obtained from a sequencing machine require editing and creation of a consensus sequence. Sequencing machines of such manufacturers as Applied Biosystems and Amersham MegaBase provide the output files containing image information in a mode of peaks (chromatograms) and their interpretation as letters for each of four main nucleotides C, T, A and G; they may be supplemented sometimes by the letter N, if interpretation is not clear, or by some other letters if it is ambiguous. Different software editors allow conversion of this information into another file that represents the digital letter sequence of corresponding nucleotides in the gene or its section. However, this information is not quite suitable for quantitative comparison of this and other sequences and for phylogenetic analysis. To be ready for such analysis sequences must be carefully edited. Different software processors allow conversion of edited, consensus sequences into another file that will represent the digital letter sequence of corresponding nucleotides in the gene or its section for further comparisons. Many requirements for the sequence processing are met by such program packages (PP) as MEGA-4 or MEGA-5 (http://www.megasoftware.net/), Bio-Edit (http://www.mbio.ncsu.edu/BioEdit/bioedit.html), DAMBE (http://dambe.bio.uottawa.ca/dambe.asp.), etc.; at least 30 are available nowadays from a web-list (http://evolution.genetics.washington.edu/phylip/software.html#methods). A very suitable PP tool for the primary editing is Chromas (Chromas-pro, http://www.flu.org.cn/en or http://www.technelysium.com.au/chromas.html). Our working Chromas version (Chromas-pro 2.31) lets us perform all necessary operations for sequence editing:

- Opens chromatogram files from Applied Biosystems and Amersham MegaBace DNA sequencers.
- Opens SCF format chromatogram files created by ALF, Li-Cor, Visible Genetics OpenGene, Beckman CEQ 2000XL and CEQ 8000 and other sequencers.
- View Genescan genotype files.

- Save in SCF or Applied Biosystems format.
- Prints chromatogram with options to zoom or fit to one page.
- Exports sequences in plaint text, formatted with base numbering, FASTA, EMBL, GenBank or GCG formats.
- Copy the sequence to the clipboard in plain text or FASTA format for pasting into other applications.
- Export sequences from batches of chromatogram files, with automatic removal of vector sequence.
- Reverse & complement the sequence and chromatogram.
- Search for sequences by exact matching or optimal alignment.
- Display translations in three frames along with the sequence.
- Copy an image of a chromatogram section for pasting into documents or presentations.
- Allows creation of a consensus sequence from two or more antiparallel reads.

Thus, the main set of editing options comprises: sequence comparison, primer removal in the beginning and in the end of the sequences, antiparallel sequences inversion, creation of consensus sequence from two or more chain reads and writing the whole obtained information as file in a mode that is convenient for subsequent computations.

When making a consensus sequence, it is very important to have high quality original chromatograms and check it using several repeat runs of the analysis; at least two is a general recommendation. A convenient approach is to sequence the DNA fragment for complementary chains using PCR in 5'-3' and backward 3'-5' directions, applying corresponding primers, say type F1 (Forward) and R1 (Reverse), etc. Sometimes, to achieve precise quality results we have to use four primers for PCR and two pairs of chains in the final building of the consensus sequence. Usually under such an approach, the consensus sequence is free of errors. The appearance of sequences as the chromatograms and their letter mode for Chromas PP (http://www.technelysium.com.au/chromas.html) is exemplified (Fig. 15.1.1.).

When a sequence is edited in Chromas or any other processor, we have to submit it to one of the recognized gene banks. Most recognized gene banks are connected via web-net. So, when submitting a sequence, for instance to GenBank (http://www.ncbi.nlm.nih.gov), one accepts that in a certain agreeable time his sequence will be open for any Internet user. When the sequence is placed into the gene bank there is a procedure of its quality control. When necessary, communication with the author(s) of the sequence is established to solve the questions in the list of records that were submitted and which accompany the main sequence information on the nucleotide content. At the GenBank site there are different utilities for submission of sequences as well as numerous tools for retrieving sequences, their testing and preliminary analysis. When an author is submitting one sequence of a single gene (section of it) or a set of such close sequences, the **Bankit** utility is very convenient, allowing sequence submission in an interactive mode. After submission, the author gets the number(s)

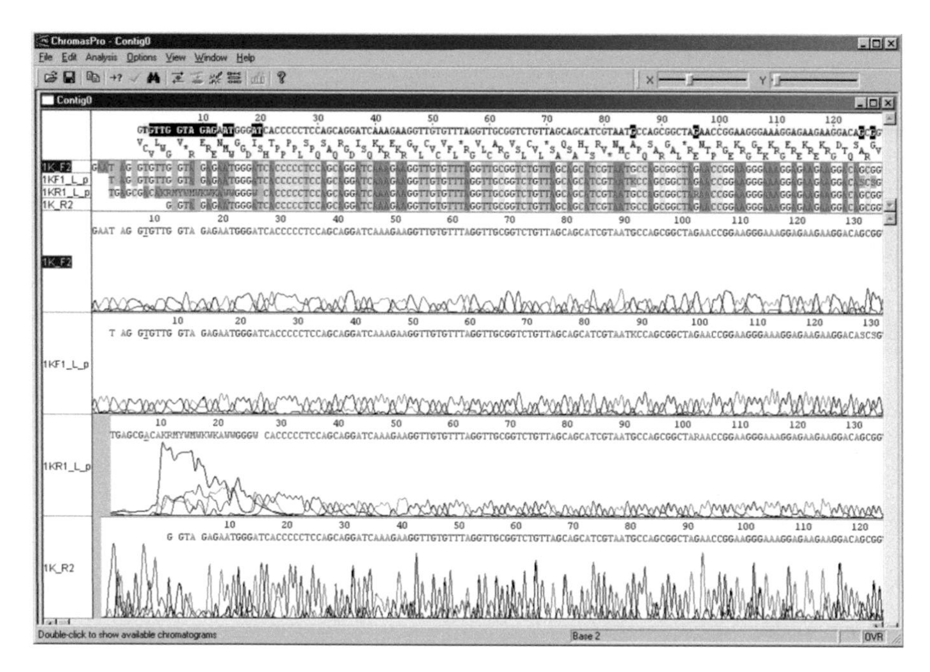

Fig. 15.1.1. Graphical and symbolic representation of a partial cytochrome oxidase 1 (Co-1) sequence in flounder, *Liopsetta pinifasciata*.
Sequencing was done on the machine ABI-3100 (Applied Biosystems, USA). Four repeats of single individual sequence (1K_F2, etc. left column) are represented with peaks and their letter interpretation. After inversion of antiparallel chains (1KR1_L_p and 1K_R2) and making complementation they have been automatically aligned. The consensus sequence that is being edited is shown on top. Lines on chromatogram and four different letters of nucleotides are given in different colours.

of the sequence, in the beginning a preliminary, registration number and then after checking the final access number(s) of GenBank. Part of a representative record of a submitted sequence to GenBank is shown in Fig. 15.1.2.

If the author intends to submit more complex sequences such as mtDNA, bacteria genome or other organisms' DNA, it is more convenient to use another GenBank utility, **Sequin**. This utility along with few others allows submission of large sequence files, which would be difficult to process with Bankit.

15.2 DATA FORMATS AND GENE BANKS AVAILABLE

Sequences submitted to GenBank will be available to all users in an agreed time span, usually after data publication in a journal within 6–12 months. Single sequences could be retrieved in several modes, GenBank, FASTA-format, etc. In the first case, the sequence mode looks like this (Fig. 15.2.1).

The whole record contains additional information, starting with the gene notation, access number, the encoded product, including the source organism and tissue (or other item) for DNA, taxonomic specification and authors' affiliation, etc. (Fig. 15.1.2). In

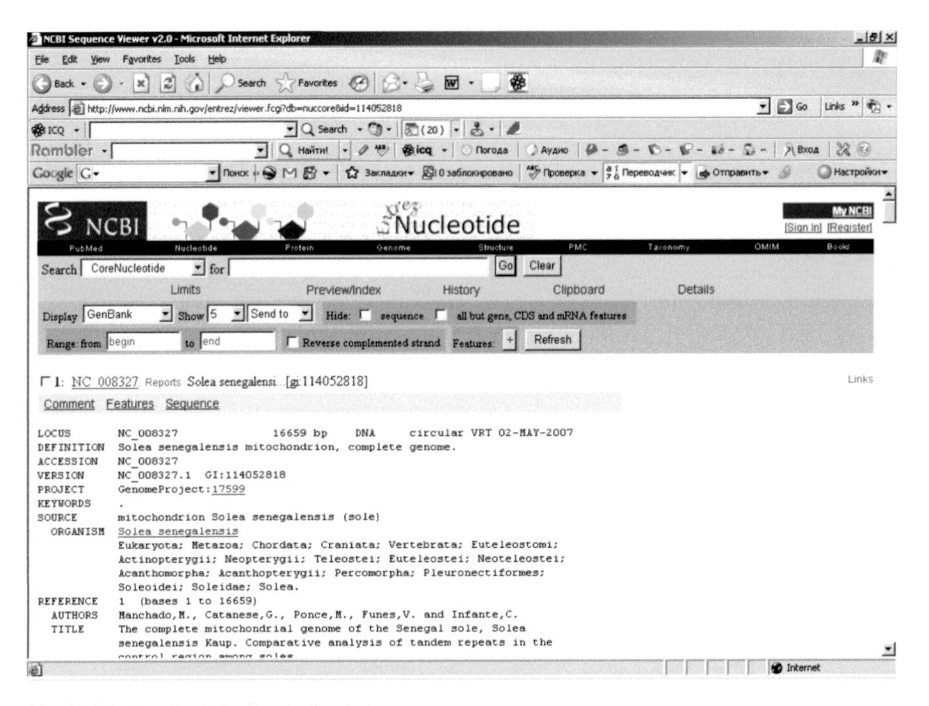

Fig. 15.1.2. Sample of the GenBank window.
Records for the complete mitogenome of one specimen of flatfish (Pleuronectiformes) are represented.

```
  1 gtgcctgagc cggaatagtc ggggacaggc ctaagtctgc tcattcgagc agagctaagc
 61 caacctgggt gctctcctgg gagacgacca aatttataac gtaatcgtca ccgcacacgc
121 ctttgtaata atcttcttta tagtaatacc aattatgatn cggagggttc ggaaactgac
181 ttattccatt aataattggg gcccccgnat atggccttcc ctcgaataaa taacatgagt
241 ttctgacttc tacccccatc ctttctcctc cttctagcct cttcaggncg tcgaagctgg
301 ggcagggaca ggatgaaccg tgtatccccc actagctgga aatctagcac acgccggagc
361 atcggtagac ctcaccattt tctctcttca ccttgccgga atttcatcaa ttctaggggc
421 aatcaacttt attactacta tcatcaacat gaaaccaaca gcagtcacta tgtaccaaat
481 cccactattt gtctgagccg tactaatcac cgcacgtcct tcttcttctt tcacactacc
541 acgtcactgg ccgctggcat tacaatgcta ctgactagac cgcaacacta aacacaaaca
601 cttctttgac cctgcyg
```

Fig. 15.2.1. Partial sequence of Co-1 gene obtained for a flatfish, *Pseudopleuronectes obscurus*.
In the left column, ordinal numbers for first nucleotides are given in a row. Nucleotides are grouped by 10 with a total number of 60 in a row.

FASTA format there is less information, but this file is more convenient for importing into other software and PP for further analysis, including molecular phylogenetic tree building.

Through the Internet, interactively or by e-mail, three gene banks are available; as mentioned they are connected. These are GenBank, NCBI (USA), EMBL in Europe (http://www.ebi.ac.uk/embl/) and DDBJ, DNA Data Bank of Japan (http://www.ddbj.nig.ac.jp/searches-e.html). There are also local DNA banks, e.g., Japan Bioresource Center Bank, RIKEN (http://www.brc.riken.jp/lab/dna/en/), «North» bank, NGB (http://www.ngb.se), etc. Beyond nucleotide sequences polypeptide sequences are also accepted for submission. However, the vast majority of registered sequences are DNA sequences because they are much easier to be obtained, and they can be translated if necessary into peptide sequences.

15.3 ALIGNMENT OF SEQUENCES

Sequences alignment is a very important procedure, which is necessary for their further quantitative analysis, allowing calculation of similarity/distance measures, estimates of homology/orthology and finally building phylogenetic trees. Without appropriate alignment, an inferred tree may be very far from reality. There are several alignment algorithms used by different sequence processors.

Alignment is aimed to detect homology/orthology of nucleotide (amino acid) sites in a set of sequences that represent specimens for comparison (e.g., populations' or taxa representatives). There is no exact solution for this task, so some procedure of optimization must be used. Optimization of homologous/orthologous site number in the sequences (of nucleotides or amino acids) is achieved by choosing appropriate values of penalties in an automated system like, for example, CLUSTAL-W PP. Software such as in CLUSTAL-W/X, Pileup in GCG or MUSCLE in MEGA-5, allow alignment to be performed quite quickly. The speed depends upon the length of the sequences, their number and the algorithm used by software. However, even in a modern PC, a set of 100 sequences of length 1000 nucleotides can take many hours using, say, CLUSTAL-W. So, more efficient algorithms like in MUSCLE (see below) are welcomed. We will consider both these options using MEGA.

In the beginning, a pair-wise alignment is usually performed. After that the whole set of sequences is aligned. Outcomes of the alignment procedure include: (i) identification of homology/orthology of sequences, (ii) identification of their potential identities and (iii) identification of potentially unknown sequences.

Global and Local Alignments and their Algorithms

Global Alignment (GLA) is the alignment procedure performed for the whole length of the sequence; this type of alignment is applied when similarity of the sequences compared is expected for some reason, e.g., all of them represent the same gene or its section. Global alignment maximizes the regions of similarity and minimizes the gaps, using scores of similarity matrices and gap parameters defined in the program.

Basically two algorithms are used for alignment: Niddleman and Wunsch (1970) (N-W) and Smith and Watterman (1981) (S-W). Their approaches are described briefly below, and for details it is best to consult the original sources (Niddleman and Wunsch 1970; Smith and Watterman 1981) and the book by Lukashov (2009). The N-W algorithm is used for GLA. The aim of the N-W alignment is a maximization of similarity, S for a pair of sequences by the formula:

$$S = max\ (x - \Sigma w_k z_k), \qquad \textbf{(15.1)}$$

where x is the number of similar nucleotides (amino acids), w_k is a penalty for a gap of k nucleotides in length and z_k is the number of gaps of k nucleotides in length.

Local alignment (LLA) is another procedure more focused on short sequences. For many practical needs the GLA is not necessary or even desired. When we have short, very similar sections of sequences, another algorithm is more efficient, i.e., S-W that minimizes the genetic distance between the pair of sequences:

$$D = min\ (y + \Sigma w'_k z_k), \qquad \textbf{(15.2)}$$

where y is the number of different nucleotides (amino acids), w'_k is a penalty for a gap of k nucleotides in length, analogous to that in (15.1) and z_k is same as in (15.1).

Below, in this section, we will consider the sequence alignment based on the CLUSTAL-W software, incorporated in MEGA (http://www.megasoftware.net/), and which is compatible with OS Windows, however there are other versions. To perform sequence alignment relative to each other one needs first to upload a set of selected sequences into the processor. Uploading can be done in three ways: (i) making a direct recording of nucleotides of the sequences in the specified window of the processor, (ii) importing the sequence from a previously made file and (iii) making a copy of the nucleotide sequence letters set and pasting it via the clipboard where necessary. In Fig. 15.3.1, the interface of CLUSTAL-W processor (Thompson et al. 1994) is shown; this version of CLUSTAL-W is integrated with MEGA-4 and MEGA-5 PP, and exemplified by partial sequences of the *Cyt-b* gene of flounders and two species of perch-like fish before and after alignment. MEGA-5 explorer's window is basically identical, but in the toolbar there is the Muscle icon and few other small changes.

In the current case, the sequences were uploaded via the clipboard (Fig. 15.3.1). Starting the run in MEGA-4, we choose from the main menu: **Alignment → Alignment explorer/Clustal → Create a new alignment**. In the last menu sub-option there are three possibilities: **Create a new alignment**, **Open a saved alignment session** and **Retrieve sequence from a file**. In MEGA-5 the Alignment explorer run starts in a slightly different way, and for the sequence alignment it contains more utilities in the Alignment toolbar of the explorer's menu. When sequences have been uploaded, a researcher usually will see many complications; he recognizes that sequences are different in size, the beginning and the end of different sequences do not match and moreover certain sequences have gaps caused by deletions/insertions in DNA, which are also located differently in the set of sequences. Alignment allows resolving of all these complications. Technically, to start execution of the Alignment explorer program we have to underline the whole sequence set and run **Align by ClustalW** utility in the Alignment option of the main menu of the explorer. As a result of these actions the dialog window appears (Fig. 15.3.2). These two actions are similar both for MEGA-4 and MEGA-5. Actually in Fig. 15.3.2, two windows are shown for economy of space; settings of these windows are used during two steps of the alignment.

Clicking on the option "execute" (OK), we run the alignment. Alignment is a delicate procedure and takes some patience. It is recommended to make several attempts with your set of sequences, varying gap penalties. The algorithm of alignment is such that with a higher penalty we get more gaps and relatively high homology/orthology of nucleotide sequences. However, if penalties are too high algorithm could not discriminate some nucleotides that are actually homologous but available only in some of the sequences sites; under certain penalties we even may meet a situation when all the nucleotides left will be the same. Conversely, penalties that are too small,

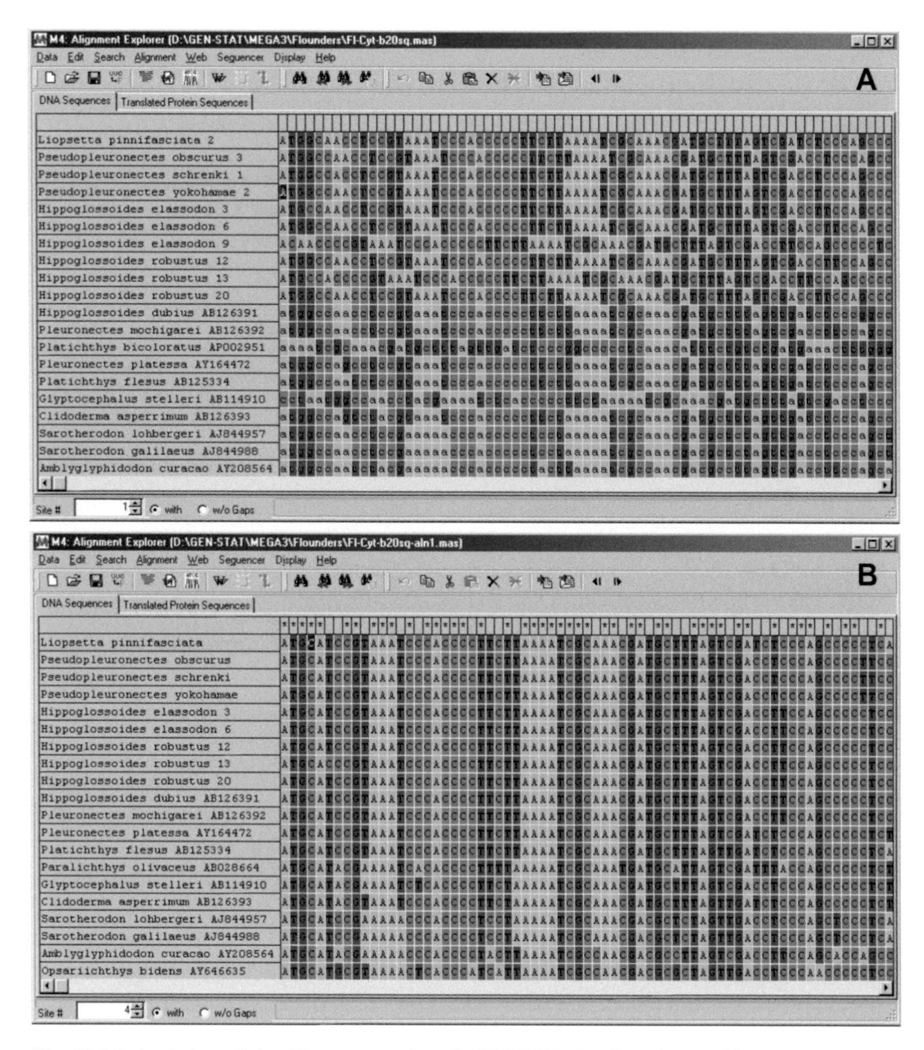

Fig. 15.3.1. A window of the Alignment explorer in CLUSTAL W of MEGA-4 with a sample of DNA nucleotide sequences at Cyt-b gene of some fish species before alignment (A) and after alignment (B). Different grey colours (different colour in original software) show sets of identical sites. An asterisk indicates sites that have 100% homology/orthology, i.e., identical in all sequences. After species names GenBank accession numbers are shown.

may give few gaps and a very large number of unique nucleotides; this will give in upper extreme ultimately non-similar sequences. In both these extreme cases we will not be able to get a satisfactory signal of the evolutionary history, and thus be unable to build a reliable tree. The author's experience with the sequences of different genes of mtDNA shows that penalties within the limit 15–30 units for the gap opening and 0.5–8.0 units for gap removal does well for the first step of the alignment (Fig. 15.3.2A).

In the opened window (Fig. 15.3.2A) **Gap Opening Penalties** were set to 15 units, and **Gap Extension Penalties** to 5 units, both for pair-wise and multiple alignment

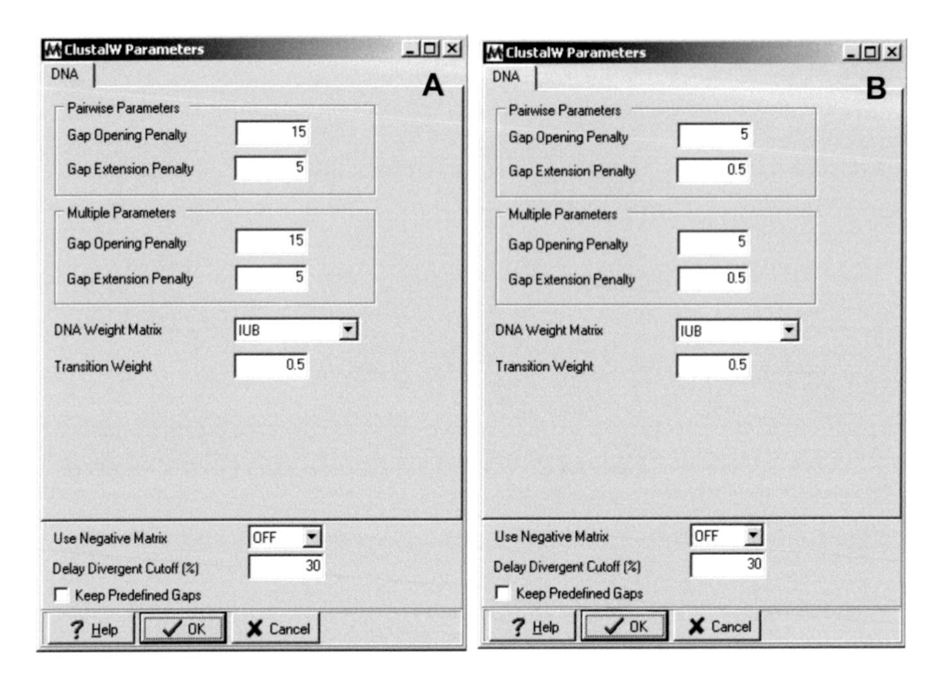

Fig. 15.3.2. Two dialog windows of the Alignment explorer in CLUSTAL-W of MEGA-4 exemplified different options for penalties under pair-wise alignment (**Pair-wise Parameters**) and multiple alignment (**Multiple Parameters**) at the first (A) and second (B) steps.

sessions. When the first alignment step of CLUSTAL-W has finished, the output window appears (Fig. 15.3.3), which contains gaps computed in accordance with the above settings. At this first step, the biggest (main) gaps are obtained; the sequence set now appears in the following mode (Fig. 15.3.3).

The file with sequences, like those represented in Fig. 15.3.3, is inspected for gaps visually, and the biggest gaps are manually removed. One may remove all gaps by the program processor, but this may lead to the loss of some potentially informative sites, which may be engaged in the alignment at the second step; so, this not desirable. After removal of big gaps, the dialog box of CLUSTAL-W is recalled and the program is run for the second step, now with decreased penalties (Fig. 15.3.2B). When the program stops now all the gaps are removed. Then the file is written in the necessary format for further computations and quantitative analysis. This analysis may include comparison of the nucleotide content, frequency of nucleotides according to their codon position, frequency of codon usage, genetic distance in a whole set of sequences or pair-wise distances among specified group of sequences (allowing comparisons within groups and among the set of groups), building of molecular phylogenetic (gene) trees and many other calculations are possible.

The MUSCLE (Multiple Sequence Comparison by Log-Expectation; Edgar 2004) utility in MEGA-5 PP is useful for making alignment when a large set of sequences is analysed. This utility is called from the MEGA-5 (MEGA-6) main toolbar menu, by clicking: **Align → Edit/Build Alignment → Create a new alignment** or **Open a saved alignment session** or **Retrieve sequence from a file.** When the Alignment

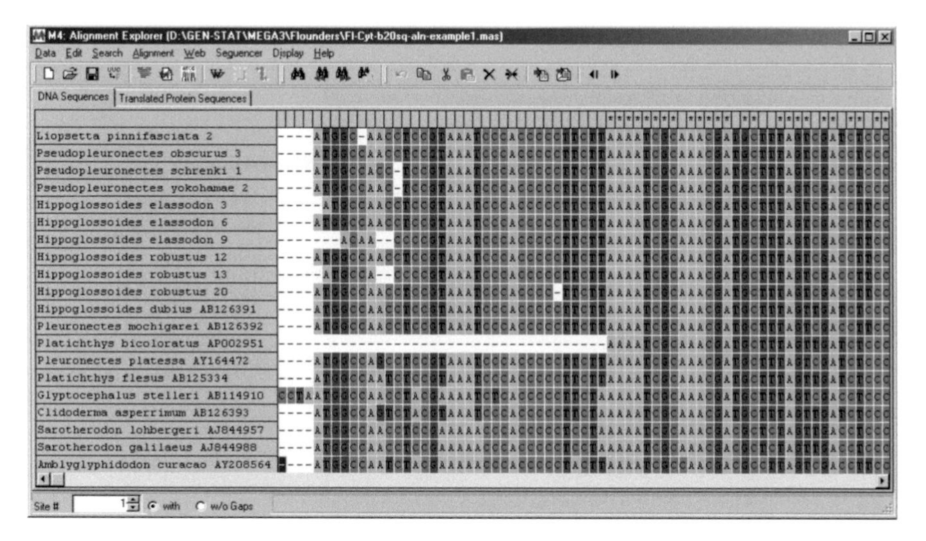

Fig. 15.3.3. A window of the Alignment explorer in CLUSTAL W of MEGA-4 with a sample of nucleotide sequences at Cyt-b gene after running the option "Alignment" and before gaps removal.
Gaps are seen as blank sectors (white colour). They will be removed manually and after that sequences will take the mode as in Fig. 15.3.1B.

Explorer has run, choose: **Alignment → Align by Muscle** or just directly click on the appropriate icon at the toolbar menu. You have to learn what penalties to set in this case; for the beginning, a good idea is to use the default settings. Hall (2011) also recommends default parameter settings in this case. In this approach for sites sorting the clustering algorithms are used, the UPGM by default or NJ, the request for which should also be set in a certain window. More details on running MUSCLE can be found in the original description (Edgar 2004) or in Hall's (2011) manual. Below a MUSCLE run is exemplified with cyprinid fish *Co-1* gene sequences (Fig. 15.3.4). For comparison, alignment of this file takes CLUSTAL-W nearly six hours while in MUSCLE it takes less than one minute.

15.4 SEARCHING FOR AN OPTIMAL NUCLEOTIDE SUBSTITUTION MODEL

As we agreed earlier, before the tree building a researcher should define which nucleotide substitution model best fits the data. Useful information on model properties and their applicability depending of a particular data set can be found in Chapter 13 and in other sources (Nei and Kumar 2000; Hall 2001; Sanderson and Shaffer 2002; Felsenstein 2004). One of the most developed software applications for this aim is MODELTEST 3.06 PP (Posada and Grandal 1998) and later versions, 3.6–3.7. MEGA-5 also contains a utility that allows testing your data set for the most optimal of several frequently used substitution models.

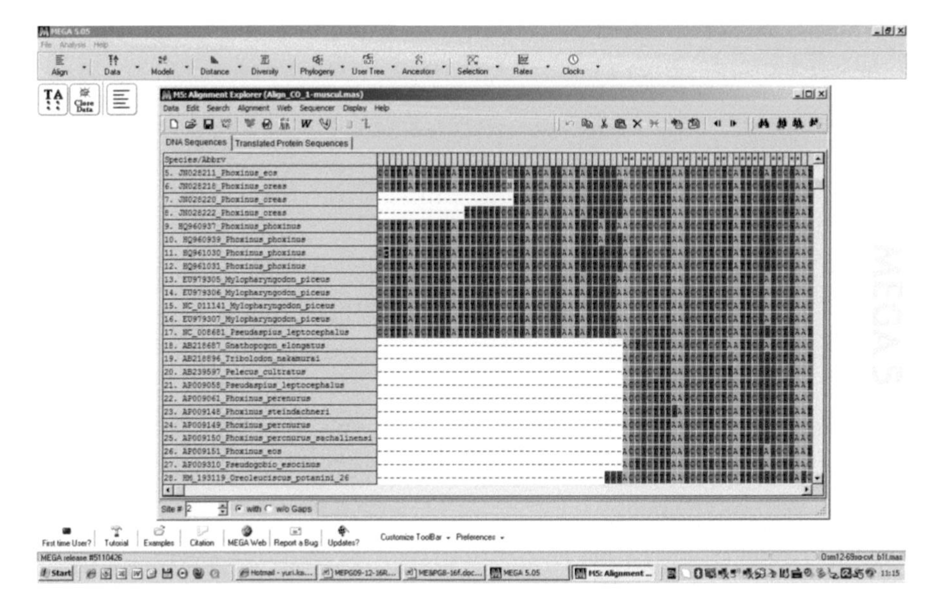

Fig. 15.3.4. A window of the Alignment explorer in MUSCUL of MEGA-5 with a sample of nucleotide sequences at Co-1 gene of cyprinid fish (Cypriniformes) after running the option "Alignment" and before gaps removal.
Notations and abbreviations are same as in Fig. 15.3.3.

To use MODELTEST one should first make acquaintance with PAUP PP because this program utilizes some PAUP modules. In principle the work with the PP is simple, and includes 5 steps:

1. You have to create a working file in **Nexus** (.nex) format that includes nucleotide sequences with necessary identifiers, corresponding to PAUP requirements;
2. Enter the Internet site of MODELTEST PP and download all recommended modules, and copy the "modelblockPAUPb10.txt" file as well, that is distributed jointly with the program into a previously created nex-file (this is attributed to PAUP 4b10 for Windows);
3. Run installed PAUP 4b10 PP (It is better to rename the original file with data) and execute the working file that was created earlier;
4. When the program stops, a new file named "model.scores" will appear in the folder of the PP installation;
5. Run from the OS DOS command line the program itself, say MODELTEST 3.7, using preferably the folder holding the executable file, modeltest3.7.win.exe. Corresponding identifiers in the command line will be as follows: modeltest3.7.win <model.scores> test.out (the last output file may have any arbitrary name).

When executed, the output file will have all necessary information and parameters of the best-fit model of substitution that will be given along with the other tested 67 models. Estimation of the model's fit is based on the likelihood ratio test and on the Acaike information criterion.

MEGA-5 model fit estimation is even simpler. When the working file has been uploaded and the program has run, one can open the option "**Models**" in the main menu or simply in the toolbar, and then choose in the menu bar the option "**Find Best DNA/Protein Models (ML)**". The corresponding window is shown in the Fig. 15.4.1. Several other options are also available for computation from this window menu. When this utility is executed there is another window to run the ML program for estimation of the best-fit model. The output file contains necessary information in a very convenient table form (Table 15.4.1). Important in the table are BIC, AIC and lnL scores; the lower the score the better the fit of the model to the data. In the example, the best fit is the HKY+G+I model (Table 15.4.1, first line; see notations in the footnote).

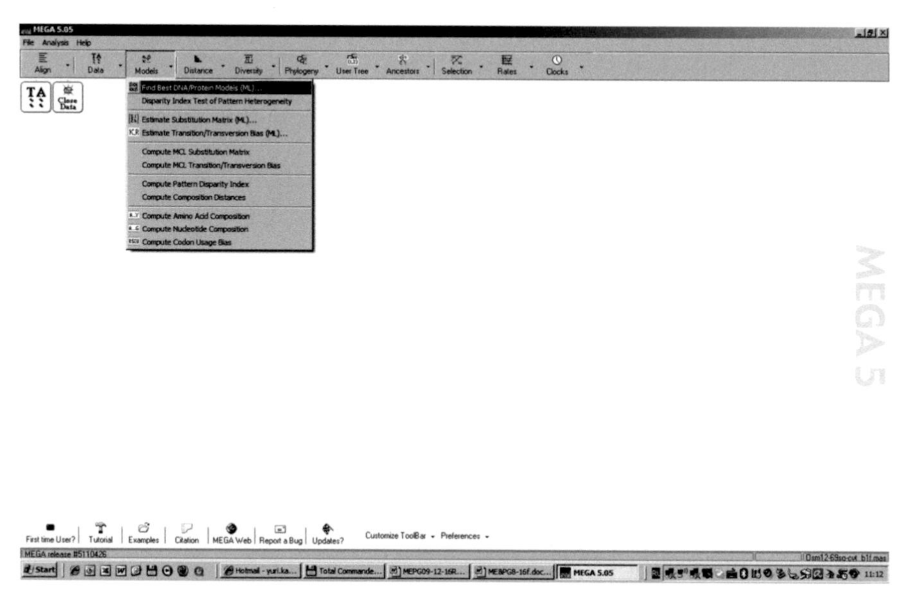

Fig. 15.4.1. A window of MEGA-5 showing main menu and toolbar with opened option "**Models**" and underlined in black color sub-option "**Find Best DNA/Protein Models**".

The analysis involved 69 nucleotide sequences. All codon positions were included: 1st+2nd+3rd+Noncoding. There were a total of 1037 positions in the sequence set analysed. Non-uniformity of evolutionary rates among sites was modelled by using a discrete Gamma distribution (+*G*) and by assuming that a certain fraction of sites are evolutionarily invariable (+*I*).

Table 15.4.1. Sample of output MEGA-5 table with maximum likelihood model fit estimation of nucleotide substitution in the set of 69 sequences of cyprinid fish for the *Co-1* gene section (Author's unpublished data).

Model	#Parameter	BIC	AICc	lnL	Invariant site ratio, I	Gamma parameter, G	R	Nucleotide Frequencies			
								A	T	C	G
HKY+G+I	133	7838.3	6702.1	–3217.6	.591657	1.27105	6.046	.24	.29	.27	.18
T92+G+I	131	7841.6	6722.6	–3229.8	.595902	1.40764	5.747	.27	.27	.23	.23
HKY+G	132	7843.2	6715.6	–3225.3	n/a	0.18198	5.704	.24	.29	.27	.18
T92+G	130	7844.2	6733.7	–3236.4	n/a	0.18238	5.502	.27	.27	.23	.23
TN93+G+I	134	7844.8	6700.1	–3215.6	.600298	1.44673	5.898	.24	.29	.27	.18
TN93+G	133	7847.4	6711.3	–3222.2	n/a	0.17768	5.625	.24	.29	.27	.18
K2+G	129	7847.9	6745.9	–3243.5	n/a	0.18463	5.340	.25	.25	.25	.25
K2+G+I	130	7848.1	6737.6	–3238.3	.590279	1.37493	5.536	.25	.25	.25	.25
T92+I	130	7868.7	6758.2	–3248.6	.648535	n/a	4.961	.27	.27	.23	.23
GTR+G	136	7868.8	6707.1	–3217	n/a	0.18053	5.935	.24	.29	.27	.18
HKY+I	132	7870.2	6742.6	–3238.8	.648806	n/a	5.028	.24	.29	.27	.18
K2+I	129	7876.9	6774.9	–3258	.647701	n/a	4.818	.25	.25	.25	.25
GTR+G+I	137	7879.3	6709.1	–3217	0	0.18053	5.935	.24	.29	.27	.18
GTR+I	136	7895.5	6733.7	–3230.4	.648192	n/a	5.339	.24	.29	.27	.18

Note. Table headings were slightly changed for convenience. Some data is omitted from the original output table for brevity. Models with the lowest BIC scores (Bayesian Information Criterion) are considered to describe the substitution pattern the best. For each model, AIC value (Acaike Information Criterion, corrected), Maximum Likelihood value (lnL) are also provided. Many other parameters are also included in the output, including transition/transvertion bias (R). GTR, General Time Reversible; HKY, Hasegawa-Kishino-Yano; TN93, Tamura-Nei; T92, Tamura 3-parameter; K2, Kimura 2-parameter; JC, Jukes-Cantor.

15.5 TREE BUILDING USING MEGA-5 PROGRAM PACKAGE AND ANNOTATION ON PROGRAMS PAUP, MRBAYES AND SOME OTHERS

Options or model parameters and models themselves for the computation of a gene phylogenetic tree are provided by different programs, PAUP* (Swofford 2000), MEGA-4 (Kumar et al. 1993), MEGA-5 (Tamura et al. 2011) and others. User friendly interface and diverse statistical properties, including sequence analysis, is provided by ARLEQUIN (Schneider et al. 2000). There are very convenient descriptions for molecular phylogenetic analysis in Hall's (2001; 2011) books. The latest edition of the book (Hall 2011) is focused on MEGA-5 basically, and contains many descriptions on general principles of phylogenetic analysis of DNA and Polypeptide sequences as well. Beyond this, the book has many examples and much advice on the usage of CLUSTAL-W, MUSCLE, MrBayes, etc. For phylogenetic analysis, we will basically exemplify only MEGA-5 below, which is the best software in many ways nowadays.

The beginning of the main analytical job in MEGA-5 may be done immediately after alignment has been finished (see Section 15.3). For this we should first close the file of the Alignment Explorer (it has extension .mas, both in MEGA-4 and MEGA-5). When attempting to execute MEGA-5, the program uses the .mas file directly for computation. Prior to the tree building it is recommended to investigate sequences further using **Sequence Data Explorer** (Fig. 15.5.1). In MEGA-4 and MEGA-5, working windows of this explorer look identical. The Sequence Data Explorer allows us to compare nucleotide content, estimate nucleotide bias, evaluate

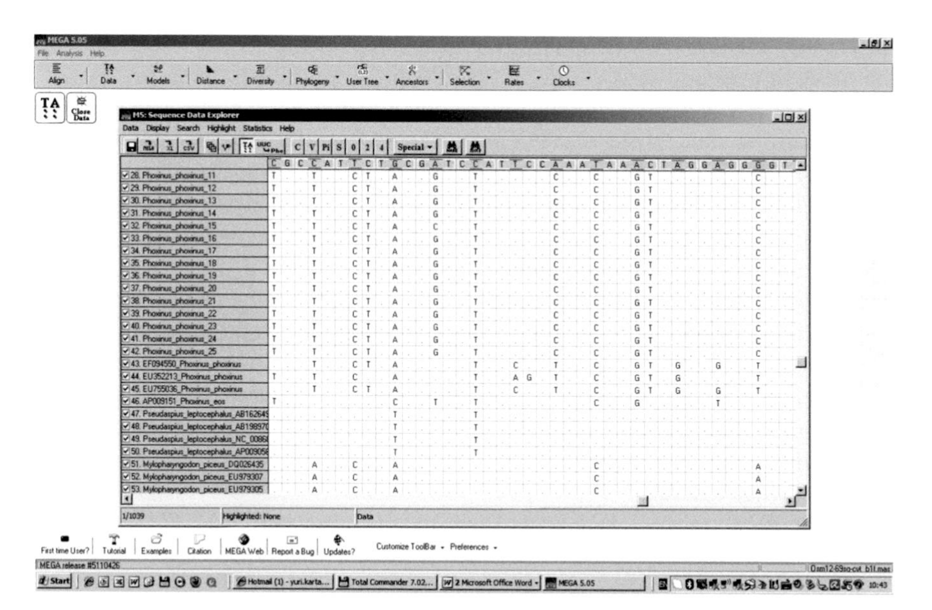

Fig. 15.5.1. A window of the MEGA-5 with opened sequence processor, **Sequence Data Explorer**. Identical nucleotides are replaced with dots.

codon usage and analyse other information that may be useful. In MEGA-5 there are several new options, which extend the volume of dynamic memory, and increase the speed of computations and the length of nucleotide vectors-sequences. More details on MEGA-5 usage, user manual and the program itself can be downloaded from the PP site (http://www.megasoftware.net/).

With the Sequence Data Explorer the user can perform additional jobs with the sequence set when necessary, e.g., turn off some sequences from the analysis, translate nucleotide sequences into protein sequences, analyse nucleotide content, etc. The last function is done from an option in the Sequence Data Explorer by clicking: **Statistics** and **Nucleotide Composition.** The result of this computation is a table that is convenient for use in a paper after slight editing (The table may be converted to MS Excel file format).

Closing the Sequence Data Explorer allows access to the PP main menu and MEGA-5 toolbar. Let us first view the main menu of MEGA-5. It contains three options: **File**, **Analysis** and **Help**. We will consider only the second option in detail. The **File** option is quite typical for OS Windows, and allows operations with a file, like **Open a File/Session**, **Open Recently Used File**, etc. From the available possibilities the most frequently used is the option **Convert File Format to MEGA**, which allows conversion of .meg file format to others, say .nex. From this dialog box one can also call the **Text Editor**, and finally may terminate the program with **Exit.**

The **Analysis** option is a key option of the program, and it is replicated with options in the icon toolbar below the main menu. In this toolbar there is also the icon (**Align**), for doing an alignment, and icon (**Data**), allowing various actions with the data set. The work with the **Align** option was introduced before (Section 15.4), while the **Data** option that is absent in the main menu must be described before discussing the **Analysis** option.

The **Data** icon options provide several opportunities for working with the data: **Open a File/ Session, Explore Active Data, Export Data, Save Data Session to File, Select Genetic Code Table, Select Genes and Domains, Select Taxa and Groups** (Fig. 15.5.2). The first option allows opening a file for the analysis. One could also run **Data Explorer** again, which was briefly introduced above, but has a variety of other useful functions. Using different options of this explorer it is possible to: show (underline) certain sequences (specimens), show properties of gene organization or domains in polypeptides (**Show & Edit Gene/Domain**), trace nucleotides by colour, show conservative sites, show variable sites, show singleton sites, translate nucleotide sequence, call the **Statistics** option, which was already mentioned, etc. In the option **Data**, some of these functions are repeated for convenience. Thus running **Select Taxa and Groups** runs an interactive utility that allows making groups for comparison within the whole dataset.

Under the **Data** icon, the **Export Data** option allows converting file format for using this file in other PP. Another option, **Select Genetic Code Table**, allows selecting for the dataset the appropriate coding, such as vertebrate mitochondrion, etc.

Now we can move to the description of the **Analysis** option. The opened window with the list of options in it is shown below (Fig. 15.5.3). The following functions are seen: **Phylogeny, Distances, Sequence Diversity, Sequence Models, Substitution**

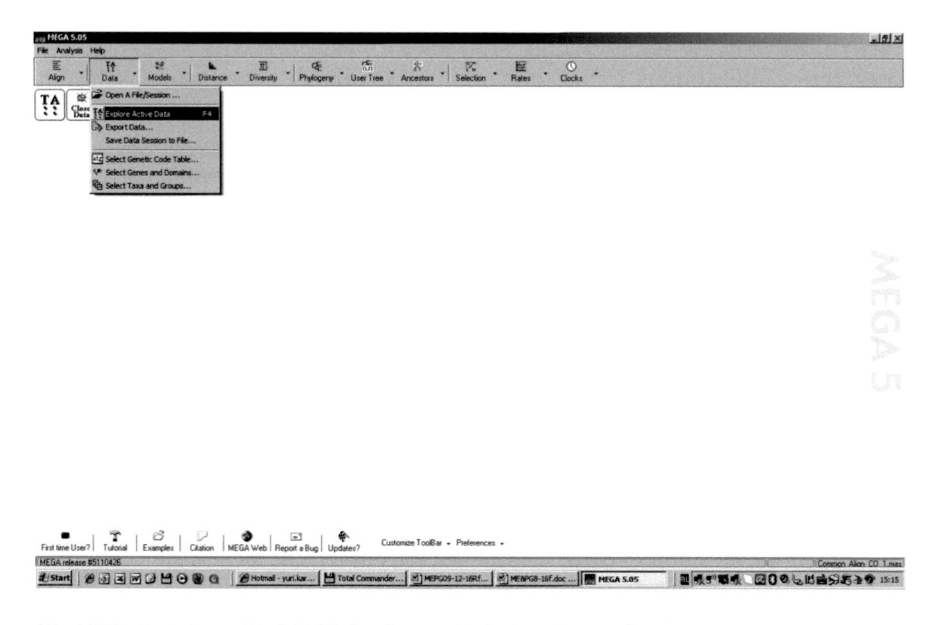

Fig. 15.5.2. A window of the MEGA-5 software with the icon **Data** options.
A dialog window for **Data** option provides several functions that are explained in the text.

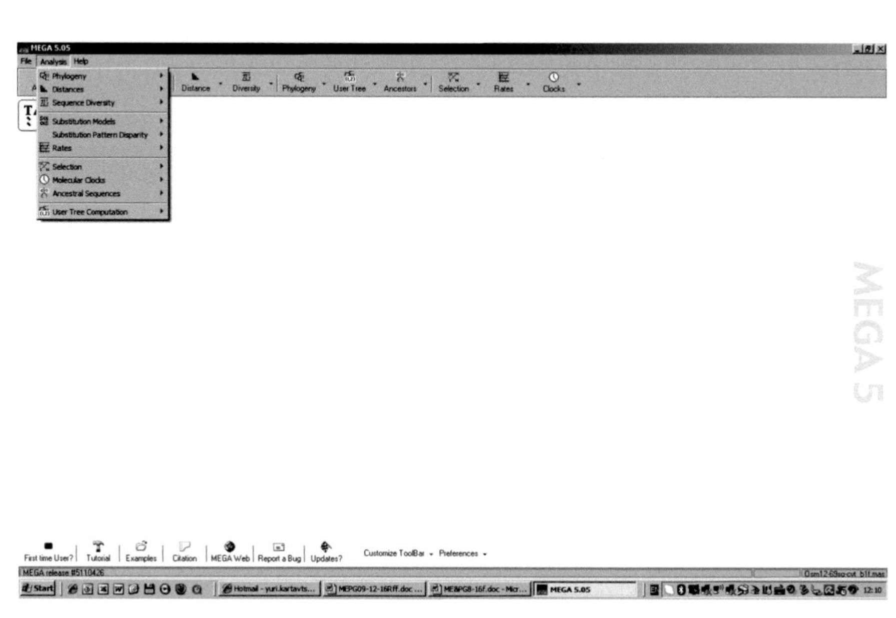

Fig. 15.5.3. A window of the MEGA-5 software with main menu options and the toolbar.
A dialog window for file (**Analysis**) option with several functions is shown.
All options in main menu are repeatedly represented as toolbar icons. In the bottom there are some other facilities of MEGA.

Pattern Disparity, Rates, Selection, Molecular Clocks, Ancestral Sequences and **User Tree Computation**. Let us start with the **Distances** option, and finish with **Phylogeny**.

The **Distances** option allows a variety of functions for computations (Fig. 15.5.4). Choosing **Compute Pairwise Distances** in the dialog box and completing the next dialog box provides a matrix of distances. In this case the distance model may be simply necessary. So, for this computation one may use *p*-distance, *K2P*, etc. (see Chapters 7 and 13, Section 13.1). In the second opened dialog box it is possible to set such options as **Pattern among Lineages,** different substitution modes, like **Same (Homogeneous)** or **Different (Heterogeneous)** and set varying rates among sites (**Rates Among Sites**). All these setting will be especially helpful when actual trees are built. The next option in the first dialog box allows us to **Compute Overall Mean** distance for all sequences. The next three options from the **Distance** window are specifically useful when there is a need to estimate distances for certain groups within the dataset, e.g., for species, genera, families, etc. In this case we have to first define the groups themselves (using the **Data** option) and then make computations.

Among the possibilities of the next toolbar icon **Diversity**, there are three utilities to compute nucleotide diversity (π) and the coefficient of differentiation.

The next most important option of the toolbar main menu is **Phylogeny** (Fig. 15.5.5). We will conclude our examination of the **Analysis** option at this and the **Selection** options, leaving the rest for students' self-education; really, in our short assignment on MEGA-5 it is not possible to describe the entire capability of this PP. Besides, the few options that are left are not so fundamental for building a phylogenetic tree for illustrative purposes.

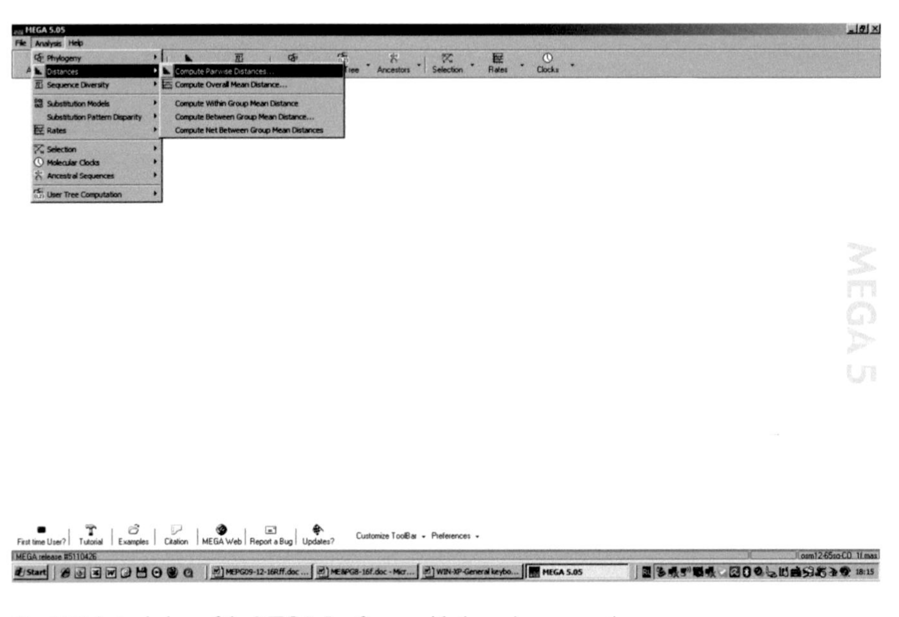

Fig. 15.5.4. A window of the MEGA-5 software with the main menu options.
A dialog window for distances (**Distances**) option with several functions that were explained in the text is shown.

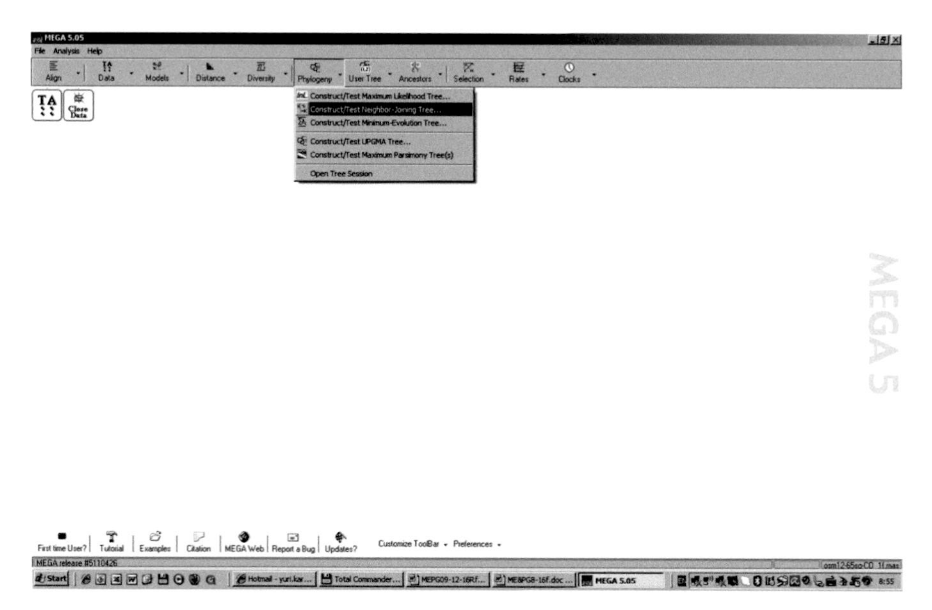

Fig. 15.5.5. A window of the MEGA-5 software with the toolbar **Phylogeny** menu options. A dialog window for phylogeny options is shown. This menu includes five main programs of tree building: (1) **Maximum likelihood (ML)**, (2) **Neighbour Joining (NJ)**, (3) **Minimal Evolution (ME)**, (4) **UPGMA** and (5) **Maximum Parsimony (MP)**. Several other functions that are explained in the text are available when the program runs.

Clicking on the **Phylogeny** icon as in the case presented (Fig. 15.5.5), or on this function at the **Analysis** option in the main menu, produces a dialog box with five toolbar choices: (1) **Construct/Test Maximum Likelihood Tree**, (2) **Construct/Test Neighbour Joining Tree**, (3) **Construct/Test Minimal Evolution Tree**, (4) **Construct/Test UPGMA Tree** and (5) **Construct/Test Maximum Parsimony Tree**. Choosing each of these five options produces a special window with dialog box for completing the tree creation (Fig. 15.5.6). You might remember that such a dialog box was seen when the **Distance** icon was explored and the best model was found (see above). For exploring major tree properties at the beginning, it is better to start with the simplest and quickest algorithms, like NJ and UPGMA. If the tree is not too large we may directly use the bootstrap option for tree testing; under this the tree is built and its support is also estimated as bootstrap support probabilities (frequencies) shown for each node. If the tree is large including, say, more than 50 sequences of 1,000–1,500 bp long, then bootstrapping is a time consuming procedure that may be avoided simply to see the basic tree nodes. During Training Course #15 there will be an example of tree building with the NJ method. Now let us continue considering the next option in the **Phylogeny** window.

In the dialog box (Fig. 15.5.6), there are several important settings we have to learn for appropriate tree reconstruction. These settings allow the specification of the number of bootstrap replications (n = 500–1,000 is typical). When the tree is built, the appropriate model (*p*-distance is no longer an appropriate measure) of substitutions is

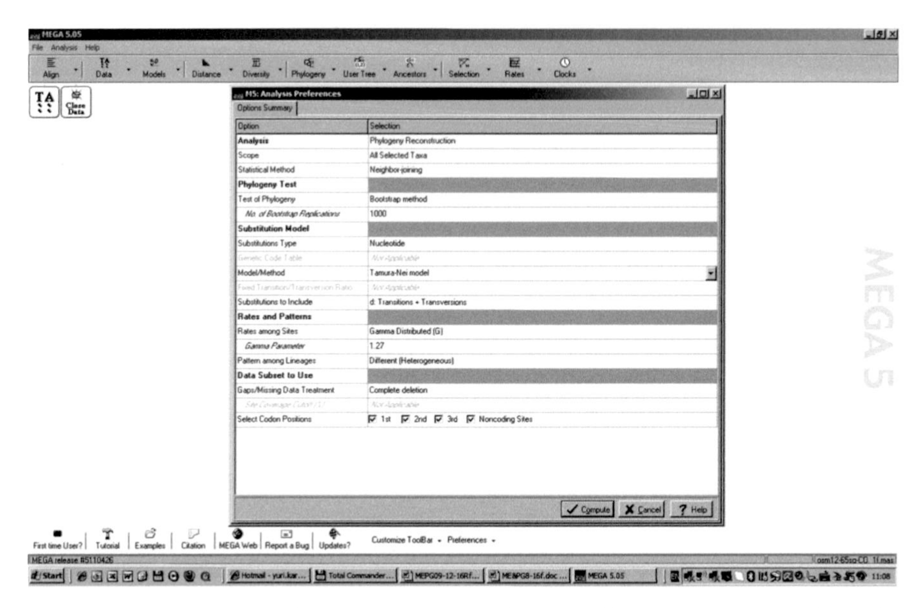

Fig. 15.5.6. A dialog box opened after running the option **Construct/Test Neighbor Joining Tree** of the MEGA-5 software with the toolbar **Phylogeny** menu options.
A dialog box with the settings for appropriate phylogenetic tree reconstruction is shown.

necessary, and thus we set it here to the TrN/TN93+G+I model, which is most suitable from the list available currently in MEGA-5 according to BIC and other parameters that were given in Table 15.4.1. At this window we can also set the option for all substitution usage, **Transition + Transversion** or other variants on command. For protein coding genes it is quite usual for the ratio between purines and pyrimidines to be biased (Kartavtsev et al. 2007a,b), so including this option is reasonable.

The last option in the icon menu bar that we will consider is **Selection** (Fig. 15.5.7). The program allows calling four functions listed in the corresponding window (Fig. 15.5.7). In the menu list the **Tajima's Test of Neutrality** function is ready to run (Fig. 15.5.7). Estimation of possible effects of natural selection since Darwin's time takes a very special place in evolutionary biology and evolutionary genetics in particular. Such research became very popular after papers by King and Jukes (1969) and Kimura (1969; 1985) about the prevalent role of genetic drift or selective neutrality in molecular evolution. The tests listed are very powerful, including the underlined **Tajima's Test of Neutrality.** But the researcher should be very careful when performing them. We should not forget about the many uncontrolled factors, including not representative sampling and heterogeneity of data, which may influence the test and thus your conclusion. The authors of MEGA are well aware of this, so many functions point out that it is the author's responsibility of getting correct results and making their own interpretations.

Beyond MEGA PP that was considered above there are many others, the best known of them being PAUP 4.0, MrBayes, PHYLIP, etc.

PAUP 4.0 (Swofford 2000) is a commercially distributed PP, and in this version it is adopted by OS Macintosh. Popular representation of PAUP 4.0, as noted, is provided

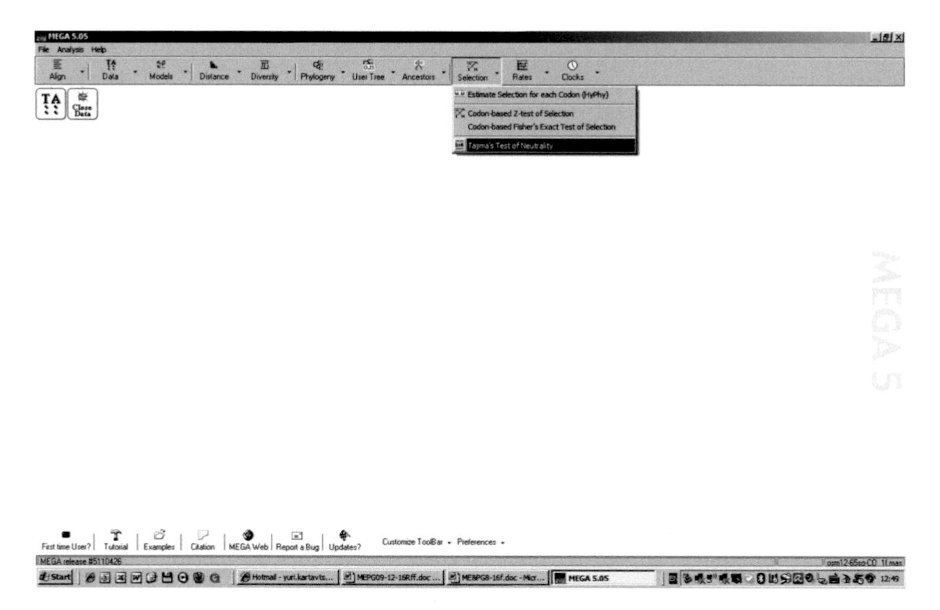

Fig. 15.5.7. A window of the MEGA-5 software with the icon toolbar menu options. A dialog window for **Selection** option with a sub-menu opened and its 3 options are shown.

by Hall (2001). PAUP 4.0 10b for OS Windows is also available, but the interface of this version is not user friendly; it has its own language to set necessary options for calculations. Thus, only an experienced user is normally able to run it seriously. Nevertheless, PAUP 4.0 repays those efforts. PAUP includes everything required for gene tree building and analysis of a molecular phylogeny with its own very useful algorithms. For instance, there are programs for ML-, NJ- and MP-tree building. There are also different options for the estimation of the tree suitability, including bootstrap support. ML-tree building requires large computer resources, and is very time consuming, especially in PAUP. Thus, one of the tree building sessions, for 67 sequences of *Cyt-b* gene (Kartavtsev et al. 2007a), continued for three weeks on quite a quick PC. Taking into account this obstacle along with demands of many science journals to present ML trees in the papers, the PAUP version for OS Linux/Unix is available, which allows runs on a multi-processor computer.

MrBayes (Huelsenbeck and Ronquist 2001; Ronquist and Huelsenbeck 2003) is a relatively small PP. However, its computing algorithm is very efficient. Thus, the 67-sequence set of the *Cyt-b* complete gene mentioned above was executed in two days on the same PC. The trees are built like ML-trees, and are probabilistic trees (see Chapter 14), which are very welcomed in quality molecular phylogenetic journals. Beyond trees, MrBayes PP provides some other computing possibilities for simulation of evolution, including morphological traits. MrBayes PP is not able to build the tree image itself, but its output files contains a record of the tree, including the very complex consensus tree, and can be used for these purposes by other programs. For example, TreeView PP (Page 1996) easily allows the drawing of the tree from the output file of MrBayes.

PHYLIP PP (Felsenstein 1995) is very useful tool for molecular phylogenetic analysis. Its large advantage, as is the case with MEGA, is that it is based on theoretical considerations, algorithms and recommendations, which are summarized in the books by Nei and Kumar (2000) and Felsenstein (2004). PHYLIP PP allows the computation of all basic tree types. However, its interface was made for OS DOS, and in this it is certainly less convenient to use.

15.6 TRAINING COURSE, #15

15.1 Submission of the gene sequence to GenBank with the Bankit utility (http://www.ncbi.nlm.nih.gov/BankIt/help.html#getting). Open Bankit window, insert any nucleotide sequence and other requested information.

15.2 Sequence alignment with CLUSTAL-W.

1) Run MEGA-5.
2) Run **Alignment Explorer** as explained in the Section 15.3, and perform alignment of any three sequences, retrieving them from GenBank (http://www.ncbi.nlm.nih.gov).

15.3 Tree building with MEGA-5 PP.

1) Run MEGA-5. Then open **File/Session** option and execute a file from the Examples folder of MEGA. Use the file Drosophila_Adh.meg.
2) In accordance with recommendations in Section 15.5, run the option **Construct/Test Neighbour Joining Tree**. At the dialog box, choose in the toolbar menu: **Bootstrap Method** and set 1000. At the **Model** option choose **Kimura 2-parameter model**, as shown in the previously opened dialog box (Fig. 15.6.1). Click **Compute**. The

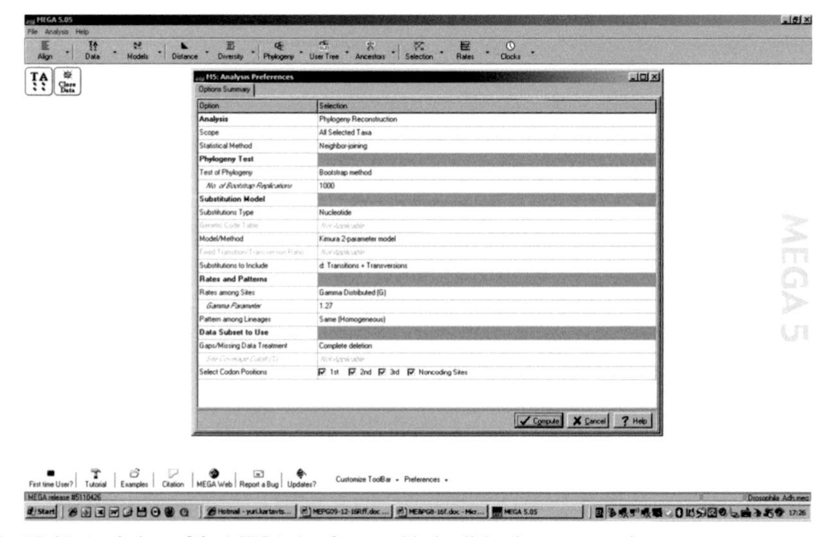

Fig. 15.6.1. A window of the MEGA-5 software with the dialog box menu options.
A dialog window for **Phylogeny** icon after running the option **Construct/Test Neighbour Joining Tree**.

constructed tree will be opened in the window of **TreeExplorer**, and will have the following view (Fig. 15.6.2).

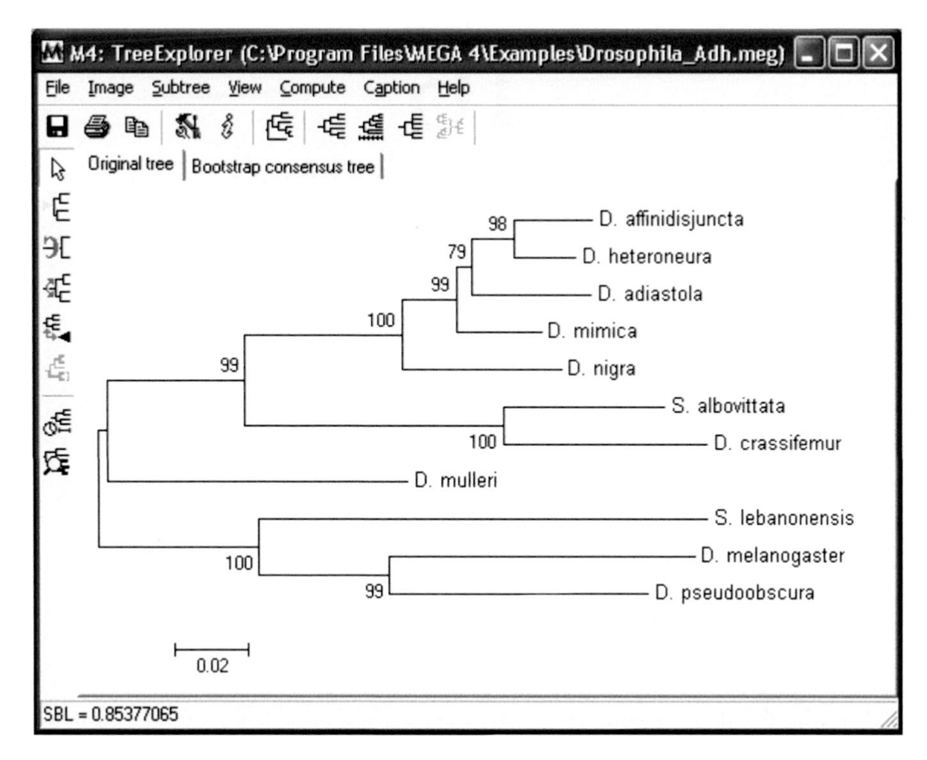

Fig. 15.6.2. A view of the MEGA-5 software in option of the **TreeExplorer**.
A file with the tree built by NJ is opened. On the tips of branches names for *Drosophila* species are given that were analysed for MDH gene sequences. The example is from MEGA distributive folder (Examples). In the left bottom corner a line represents the scale for branch length. Numbers in the tree nodes are values for bootstrap support, %.

16

GENOMIC EVOLUTION

MAIN GOALS

16.1 Origin and Change of Genomes
16.2 Evolutionary Change of Genome Size
16.3 Formation of New Genes
16.4 Repetitive DNA and Multigene Families
16.5 TRAINING COURSE, #16

SUMMARY

1. The minimal number of genes necessary for the existence of an organism as an independent, self-reproducing system was established from comparisons of *M. genitalium* and *M. pneumonia* genomes, and on mutation experiments in *M. genitalium.* Now it is estimated that living cellular organisms require a minimum of 250–350 genes.
2. Tandem duplications and genome multiplications (polyploidy) are the main sources of new genes.
3. The genome of eukaryotes contains various types of repetitive DNA. The most important are (1) satellite DNA, (2) VNTRs, minisatellites, (3) microsatellites, (4) SINEs, (5) LINEs, and (6) gene families (rRNA, tRNA, globin, immunoglobulin, etc.).
4. Since its origin and during a long history of life, the genome evolved from quite simple ring linear molecules to highly organized chromosomes and several orders of hierarchical and network-like relationships which permitted it to develop the complexity of all living higher forms.

16.1 ORIGIN AND CHANGE OF GENOMES

The first simple cellular organisms, as indicated by the fossil record, originated on the Earth about 3.5 billion years ago (Chapter 1). The genome of these organisms was probably already composed of a double-stranded DNA molecule. Long before that time a chemical evolution should have developed simple, self-reduplicating creatures like viruses or virus-like forms. Maybe some virus-like and/or proto-bacteria forms were inseminated on the Earth via a meteorite or comet attack in a time when the atmosphere was not yet a sufficient barrier to prevent life forms from entering. So, the creation of life itself may have occurred somewhere else in the Universe. The first eukaryotic fossils, resembling single-celled green algae, dated from about 1.4 billion years ago. During their life history both prokaryotes and eukaryotes constantly met variable environment, and had to change themselves to survive. In simple terms, these changes of organisms as a response to changing environments are evolution, when considered over in time. The environment is complex and comprises both the abiotic components (water, minerals, etc.) and the biotic components (other organisms and the products of their metabolism). The **Genome**, as *an interrelated unity of all functioning genes of living beings*, also had to change dynamically. That is why the causes of evolution are multifactorial. The initial causes of evolutionary changes can originate from inorganic compounds of the environment, as well as from organisms themselves and their genome. These changes, obviously as nowadays, included mutation, recombination, transposition, deletion, duplication, and others. By examining and comparing genomes of organisms of different phyla that exist today, we may imagine what mechanisms are operating and how the genome is changing, or reconstruct its evolution.

The Minimum Genome of Living Cells

How many genes are essential for a living being? We will talk about bacteria, not viruses, whose life cycles are now mostly connected with bacteria, and which may be treated as simplified parasites. It is not possible to answer precisely this question yet because the functions of most genes in the genome is unknown, even if it is fully sequenced, as reported for human by Venter et al. 2001 (Fig. 16.1.1). Still, based on genes with known functions, it is possible to estimate the minimum number of genes that are necessary to perform the basic cell functions in a single-celled organism. For this we will use sequence information from two of the smallest bacterial genomes, *Mycoplasma genitalium* and *M. pneumoniae*. These organisms are among the simplest prokaryotes known at present (see also Chapter 3). The genome of *M. genitalium* is 0.6 Mb (mega base pairs), while that of *M. pneumoniae* is 0.8 Mb. Both species belong to a group of bacteria that do not have a cell wall, and that may cause disease. There is a wide range of host organisms, including plants, insects, and humans, for which the invasion of these bacteria creates, usually, respiratory and genital infections.

The list in Fig. 16.1.1 is based on many instances of similarity to proteins of known function from translated open reading frames (ORFs) or genes. Among the most typical genes are those involved in nucleic acid metabolism (7.5% of identified genes' activities), receptors (5%), protein-kinases (2.8%) and cytoskeletal structural proteins (2.8%). From a total of 12,809 proteins predicted on the basis of sequence analysis, 5,252 (41%) have no known functions, showing how much work is still forward to define the certain function of the genes for the human only.

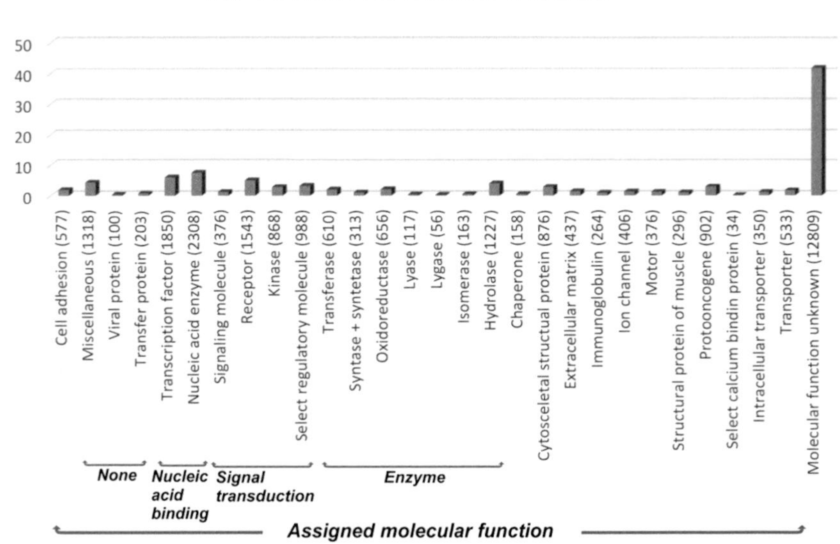

Fig. 16.1.1. A preliminary list of assigned functions for 26,588 genes in the human genome. In brackets numbers of obtained open reading frames (ORF) and on the Y-axis percentage of frequency are given for each kind of ORF obtained.

In the *M. genitalium* genome there are 467 ORFs, while in *M. pneumoniae* there are 677. All 467 genes from *M. genitalium* are found among the 677 genes of *M. pneumoniae.* For comparison, other bacteria have tenfold larger genome size: *E. coli* 4.6 Mb with 4,288 ORFs, and *Haemophilus influenzae* 1.8 Mb with 1,727 ORFs. Some functions of genes in these bacteria along with their numbers are summarized below (Table 16.1.1). Obviously, in cells among the most critical genes must be those

Table 16.1.1. Functional classes of genes in three bacterial species (Klug and Cummings 2002).

Functional Class	*E. coli*	*H. influenzae*	*M. genitalium*
Protein coding genes	4288	1727	470
DNA replication, reparation	115	87	32
Transcription	55	27	12
Translation	182	141	101
Regulatory proteins	178	64	7
Biosynthesis of amino acids	131	68	1
Biosynthesis of nucleic acids	58	53	19
Lipids metabolism	48	25	6
Energy metabolism	243	112	31
Transport and recognition proteins	427	123	34

encoding products required for DNA replication, transcription and translation, general cellular functions (anabolism, catabolism, and immunity), and at last performing the transport of macromolecules.

Let us compare the genes required for a certain function in different species. Thus, *E. coli* has 131 genes for amino acid metabolism, *H. influenzae* has 68 and *M. genitalium* has only 1. The physiological capabilities of *Haemophilus* are similar to *Mycoplasma* but it has nearly four times the number of genes; thus, even at this level there is uncertainty, which is so common for studies of genome functions. Many more genes might participate in enzyme control of function in groups with wider biotic and abiotic demands. The advance of *Haemophilus* involves its remarkable plasticity in biosynthetic capability. This more complex bacterium has 68 genes involved in amino acid biosynthesis, while *Mycoplasma*, as mentioned above, has only one such gene (Table 16.1.1). Lack of such plasticity in *Mycoplasma* leads this bacterium to depend on numerous metabolic products from its host. The above comparisons of bacteria species and mutation experiments in *M. genitalium* lead us to suppose that the minimal number of genes necessary for an organism to exist as independent, self-reproducing system is a minimum of 250–350 in living prokaryotic organisms.

Origin and Evolution of the Eukaryotic Genome

Eukaryotes are traditionally distinguished from prokaryotes by several primary features: a membrane-bound nucleus; cytoplasmic membrane systems (e.g., endoplasmic reticulum); and a cytoskeleton. New research, and genome projects in particular, give us more information on comparisons between prokaryotic and eukaryotic genomes, providing data on the origin of life and links in eukaryotic and prokaryotic genome evolution. Much evidence, including amino acid analysis of proteins, gene sequences, and metabolic pathways are relevant. Such studies indicate that the eukaryotic genome is a complex construction, and is a mosaic that has received several impacts, at least from the Archaea and the Eubacteria groups of prokaryotes. It is well known that the eukaryotic nuclear genome has no operons, and its genes include introns. These features are also seen in Archaea. The eukaryotic mtDNA (mitogenome) is very similar to that of alpha-proteobacteria. To explain these facts, it has been suggested that eukaryotes originated as a consequence of a symbiotic link between an anaerobic archebacteria host and an aerobic alpha-proteobacterium (*Rickettsia*-like form), which evolved in eukaryotes into the mitochondrion. The unity of this cell type organization in eukaryotes suggests the monophyletic origin of the whole group; it might also be suggested that this event occurred successfully only once for the life forms on the Earth. However, many less significant symbiotic associations occurred through the evolution of the genome. In an early stage that was incorporation of a sex-factor; later this evolved in a complex sex differentiation mechanism. Another such event is suggested for the genome of higher plants, harboring of chloroplast-like prokaryotes.

16.2 EVOLUTIONARY CHANGE OF GENOME SIZE

Genomic Duplications

As current data show, the most frequent mechanisms in the evolution of prokaryotic genomes are mutations, assimilations of smaller mobile elements (viruses, plasmids, etc.), and tandem duplications. Such changes and all chromosomal duplications are the main focus of this section. Polyploidy, or multiplication of the whole genome, is a peculiarity of eukaryotes; no bacteria known have more than one chromosome. For a long time most attention of geneticists was focused on duplications of single genes. The enormous significance of polyploidy for plant evolution is not debated. With animals the situation is more complicated. Ohno (1970; 1973) has argued that whole genome duplications are an important evolutionary mechanism for animals too. Analysis of nucleotide sequence data supports the idea that much of the difference in gene number in prokaryotes and eukaryotes is the result of expansions of a proto-genome. It is recognized now that major expansions in eukaryote genomic size resulted from genome duplication events. In particular, such expansions are associated with the appearance of the vertebrates, although duplications have occurred repeatedly in their evolution (Ohno 1970; 1984; Ferris and Whitt 1979).

Analysis of the yeast genome shows an ancient expansion by genomic duplication. Wolfe and Shields (1997) investigated the sequence of each yeast gene in the genome. They compared 55 duplicated regions containing 376 genes (50% of the genome). Of the 55 regions, 50 are in the same relative position on different chromosomes (Wolfe and Shields 1997). For instance, chromosomes XI and XII contain a duplicated block of genes (block 43), in which the gene order and peculiar distal orientation to the centromere has been conserved. Other chromosomes contain internal duplications, such as block 53, located on both sides of the centromere on chromosome XII. Phylogenetic analysis indicates that this genome duplication event took place 100 MYR ago (Wolfe and Shields 1997). Analysis of the human genome also shows evidence of an ancient vast duplication followed by rearrangements and gene loss in some duplicated regions (Venter et al. 2001). These duplications in humans involve segments as small as only a few genes, as well as large sections with almost the whole chromosome. Large duplications appear to date to the origin of the vertebrates, 500 MYR ago. In total, there are 1,077 blocks of duplicated regions in the human genome, including more than 10,000 genes; one such duplication is defined between chromosomes 18 and 20 (Klug and Cummings 2002). The large DNA contents of higher organisms are thought to have occurred mainly by gene duplication. Actually, there are two types of gene duplication. One is the entire chromosome duplication, and the other is the duplication of a small segment of chromosome, so called tandem duplication. The first type of chromosome duplication occurred via polyploidy, as a whole-genome duplication. As shown in Table 16.2.1, mammalian DNA is about 1,000 times as large as *E. coli* DNA. However, the increased DNA content has not always led to more a perfect organism.

In plant evolution, genome duplication via polyploidization played an important role. In the evolution of animals it also seems to play an important role. Both cytological and biochemical research showed that genome duplication has occurred several times in the evolution of fish and amphibian (Ohno 1970; Ferris and Whitt 1979). Genome

Table 16.2.1. DNA content in various organisms (Nei 1987).

Organism	Nucleotide pairs per genome	Organism	Nucleotide pairs per genome
Mammals	3.2×10^9	*Fruit fly*	0.1×10^9
Birds	1.2×10^9	*Maize*	7×10^9
Lizards	1.9×10^9	*Neurospora*	4×10^7
Frogs	6.2×10^9	*E. coli*	4×10^6
Most bony fish	0.9×10^9	*T4 phage*	2×10^5
Lungfish	111.7×10^9	*λ phage*	1×10^5
Echinoderm	0.8×10^9	*φX174 phage*	6×10^3

duplication was quite common in animal evolution before sex chromosomes were differentiated, and retain its importance in some groups of animals today (Borkin and Darevsky 1980; Altukhov 1989). When sex differentiation via sex-chromosomes became fixed in such groups as Mammals, Aves, and to lesser extant in Reptilia, polyploidy seems to become less significant. In most fish and amphibians, sex chromosomes are not well developed. So, for example, tetraploid groups are well known in fish (Ohno 1970; Frolov 2000).

Ohno (1970) has argued that genome duplication is generally more important than tandem gene duplication because the latter may duplicate only parts of the integrated genetic system, disrupting fine relationships between the structural and regulatory elements (genes) of genome. Conversely, when polyploidy duplicates the entire genetic system the harmful effect may be small, especially when autopolyploid occurs. However, molecular studies of genome organization in eukaryotes indicate that most genes do not exist as single copies in the genome, but rather in clusters of several copies. Therefore, we may conclude that tandem duplication also played a significant role in evolution. Well known examples of clustered genes are the rRNA genes and tRNA genes in a variety of organisms (Table 16.2.2). The mitochondrial genome of mammals as well as bacteria has only one set of 12S and 16S rRNA genes.

Table 16.2.2. Numbers of rRNA genes and tRNA genes per haploid genome in various organisms (Nei 1987).

Gene	Organism	Number	Genome size (base pairs)
rRNA	Human mitochondrial genome	1	16,600
	Mycoplasma capricolum	1	1×10^6
	Escherichia coil	1	4×10^6
	Saccharomyces cerevisiae	140	5×10^7
	Drosophila melanogaster	130–250	1×10^8
	Xenopus laevis	400–600	8×10^9
	Human	300	3×10^9
tRNA	Human mitochondrial genome	22	16,600
	E. coli	100	4×10^6
	S. cerevisiae	320–400	5×10^7
	D. melanogaster	750	1×10^8
	X. laevis	7,800	8×10^9
	Human	1,300	3×10^9

Tandem duplications are important not only for increasing the number of genes with the typical biochemical functions, but also for generating genes with new functions. Let us consider the β globin gene as an example. The human β globin gene family has five functional genes (Fig. 16.2.1): ε is functional in the early embryonic stage only, G_γ and A_γ produce globin polypeptides only in the fetal stage, α and β are responsible for synthesis of polypeptides after birth. However, the nucleotide sequences of these genes have very high similarity, so apparently all of them were produced by tandem duplications.

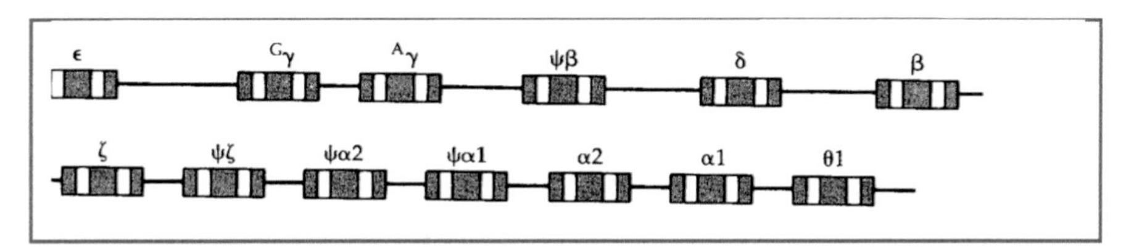

Fig. 16.2.1. Organization of human **β** globin and **α** globin gene families (From Hartl and Clark 1989 with consent from Sinauer Associates, Inc.).

16.3 FORMATION OF NEW GENES

Gene Duplication

Tandem gene duplication and consequent diversification is the easiest way of gene origin. Let us first clarify, what are the mechanisms for tandem duplications. Tandem duplications can occur by three major mechanisms: (1) **unequal crossingover**, a recombination between members of a homologous pair of chromosomes in which a DNA segment is duplicated in one of the homologs and lost in the second; (2) **unequal sister chromatid exchange,** a recombination event similar to unequal crossingover, but occurring between the two chromatids of a single chromosome; (3) **replication errors**, in which during replication of a template chain slippage can occur in this strand, causing the insertion of a short segment into the second, newly synthesized strand; this can change ORF, or give rise to a new gene. Such a gene as a rule will be non functional, but over time it may evolve functionality.

If a gene is duplicated and two functionally identical copies occur, one of them may mutate without harmful effects on viability of an individual, and so may achieve a different function (Ohno 1970). We can determine whether a gene pair has descended from a common ancestor by examining the similarity of the nucleotide sequences of the genes or the amino acid sequences in the proteins encoded by these genes. Ingram (1957) showed that the genes responsible for the three chains of human hemoglobin were produced by gene duplication. Comparison made for the three chains indicates that the proportion of amino acids shared by α and β chains is 41%, while that for β and γ chains is even higher, 73% (Table 16.3.1). Obtained similarities are so high that the probability or their random occurrence is negligible. Ingram further argued

Table 16.3.1. Levels of divergence and functional differences between proteins originated by gene duplications (Dayhoff and Barker 1972; adopted from Nei 1987).

Proteins	**Differences in amino acids (%)**	**Divergence time (MYR)**	**Chemical activities**	**Aggregation properties**	**Action sites**
Hemoglobin-myoglobin	77	1,100	—	++	+
Growth hormone - prolactin	75	200	+	—	+
Immunoglobulin, heavy and light chains	75	400	++	+	—
Immunoglobin, μ and γ chain C regions	70	350	+	+	+
Trypsin-thrombin	65	1,500	+	—	+
Lactalbumin - lysozyme	63	350	++	—	+
Immunoglobulin κ and λ chain C regions	62	300	—	—	—
Hemoglobin α and β chains	59	600	—	+	—
Hemoglobin β and γ chains	27	130	—	—	—
Protamines, salmine AI and AII	22	100	—	—	—
Chymotrypsin A and B	21	270	—	—	—
Growth hormone - lactogen	15	23	+	—	+
Hemoglobin β and δ chains	8	40	—	—	—
Alcohol dehydrogenase F and S chains	2	10	+	—	—

Note. + + Very different; + Different; — Similar. Chemical activities include differences in catalytic action and in binding to substrates, inhibitors, antigens, etc.

that myoglobin also originated from the same common ancestor as that of the three chains of hemoglobins. Later many other examples of formation of new genes by gene duplication were discovered (Table 16.3.1).

The evolutionary fate of a single duplication is as a rule loss of function and its conversion to a pseudogene, because deleterious mutations are much more frequent than advantageous ones (Chapter 6). Experimental evidence supporting this is available for Eukaryotes (Lynch and Conery 2000) and numerous Prokaryotes (Kondrashev et al. 2002). Nevertheless, the rate of duplications in vertebrates, for example, is known to be high, about one duplication fixed per 100 MYR (Lynch and Conery 2000). Because of this high rate of duplications, which roughly equals the nucleotide substitution rate, their role in evolution remains very significant. Interestingly, some genes, such as the gene coding for xanthine dehydrogenase (*Xdh*), were duplicated repeatedly and evolved by convergence to a new function, provided by the gene for aldehyde oxidase, *Ao* (Rodriguez-Trelles et al. 2003; Pereira 2004), although theoretically it should be very rare event (Pereira 2004). A review of this theme indicated that some pseudogenes are expressed (Balakirev and Ayala 2003), thus showing signs of re-functionality or residual activity.

Gene Elongation

Another way for a gene to change is increase of size, or elongation. Gene elongation may be caused by duplication of a part of gene or the entire gene. Many proteins of living organisms have internal repeats of amino acid sequences, and these repeats correspond to different structural and functional domains of the protein (see Chapter 3). We may guess that the genes coding for such proteins were formed by internal gene duplication. A popular view is that most eukaryotic genes were produced by duplication and elongation of primordial genes or minigenes, which may have existed in the early history of genes (Darnell 1978; Doolittle 1978; Ohno 1981; Blake 1985; Nei 1987; Nei and Kumar 2000).

In particular, this point of view is supported by the nucleotide sequences of the ovomucoid genes in birds. Ovomucoid is a protein present in eggs, and is responsible for the trypsin inhibitory activity in egg whites. The ovomucoid protein chain consists of three functional domains (Stein et al. 1980; Fig. 16.3.1). Each of these domains is capable of binding to one molecule of trypsin or other serine protease. The amino acid similarities between domains I–II, II–III, and I–III are 46, 30, and 33%, respectively. Such quite high degrees of similarity suggest that the ovomucoid gene was produced by triplication of a primordial domain gene (Kato et al. 1978).

Examples of gene elongation by duplication may be extensive (Li 1983; Doolittle 1985; Klug and Cummings 2002). Many cases involve one or more domain repeats. The duplication events can be inferred from nucleotide sequence data as well (see above). When it is difficult to identify functional domains, the protein folding structure may be helpful to reveal the ancient functional domains or modules of a protein (Go 1981), its tertiary and quaternary structures. It is quite possible that each such module represents a primordial gene. It is argued that primordial genes were very short, and that most current genes were produced by several rounds of duplication of these parental genes (Ohno 1984).

Hybrid Genes

Unequal crossover may occur in a DNA region including two genes. This may produce a new gene containing parts of the two genes. Hybrid genes, as stated earlier for tandem duplications, should have deleterious effects. However, if the original gene was duplicated and retained, the hybrid gene still has opportunity to evolve into a new gene. A possible example of this case is the clupeine Z gene in the herring, which probably arose through a crossover between the clupeine VI and VII genes (Fitch 1971b). Probability that these three genes arose by simple duplication and subsequent amount of amino acid substitution is small for this case (Fitch 1971b). Gilbert (1978) proposed the concept of exon shifting, and later Doolittle (1985) examined in this respect the

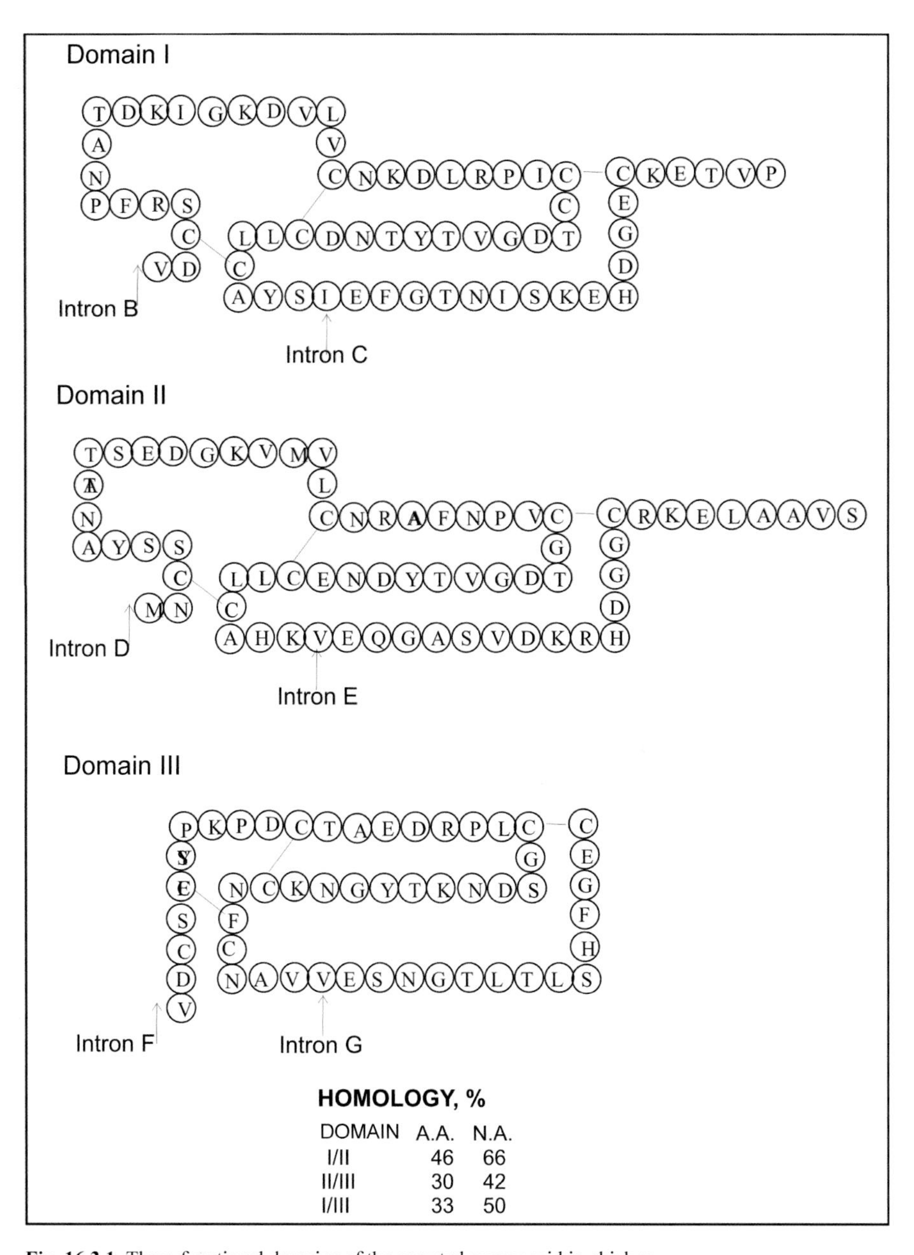

Fig. 16.3.1. Three functional domains of the secreted ovomucoid in chicken.
The homologies between domains at the amino acid (A.A.) level and the mRNA (N.A.) level are shown at the bottom of the figure (Adopted from Nei 1987 with publisher permission).

amino acid sequences of many proteins in higher organisms. These examinations found that seemingly unrelated proteins often have partial sequence homology, and thus this may have originated by exon shuffling among different genes. The cited authors regard this as an important mechanism of new genes origin. New data, for example for human low-density lipoprotein (LDL), strongly support this concept (Klug and Cummings 2002, p. 272; see also Fig. 3.3.7).

16.4 REPETITIVE DNA AND MULTIGENE FAMILIES

The genome of eukaryotes contains various types of repetitive DNA (Fig. 16.4.1), firstly summarized by Britten and Kohne (1968). Repetitive DNA includes nucleotide sequences that are repeated from several times to millions in the genome. The amount of repetitive DNA varies greatly in different organisms. Thus, highly repetitive DNA constitutes about 5% of the human genome and 10% of the mouse genome. The function of much repetitive DNA is unknown, and some types probably lack any function ("junk DNA"; Ohno 1972). However, some groups of repetitive DNA have definite functions: coding for tRNA, rRNA, multigene families, centromeric repeats, etc. (Fig. 16.4.1).

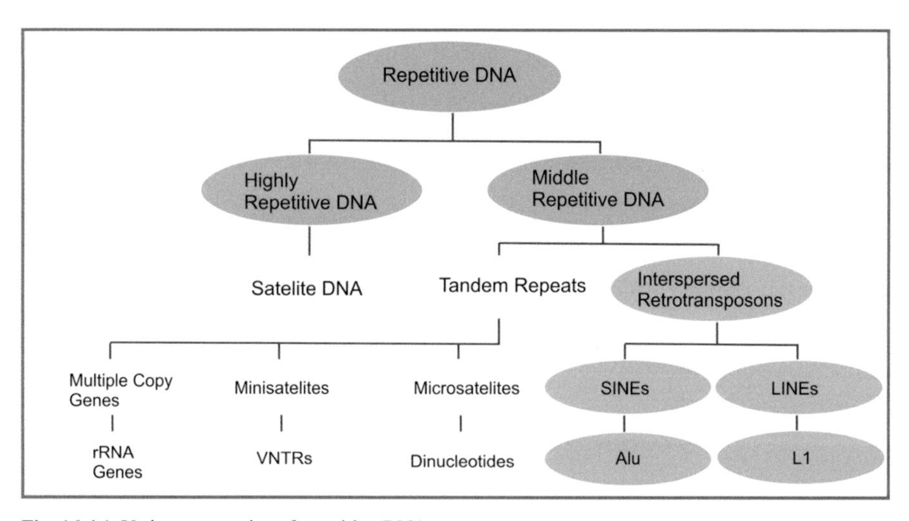

Fig. 16.4.1. Various categories of repetitive DNA.
Explanation of abbreviations and DNA repeats role are in the text.

The nucleotide composition of DNA (e.g., the percentage of G – C to A – T pairs) determines its density in a solution and could be measured with ultra centrifugation. If analyze eukaryotic DNA by this technique, the bulk of its molecules comprise one main peak (band) of density and one or more additional peaks, which representing DNA slightly different in density. This second component of total DNA, called satellite DNA. This DNA represents a variable proportion of the total DNA, depending on the species. As an example of a typical profile of DNA sedimentation, the mouse DNA analysis is given (Fig. 16.4.2). Prokaryotic genome contains only one DNA peak. The role of satellite DNA was analyzed in mid-1960s, by Britten and Kohne (1967) by invented technique of reassociation kinetics of single stranded DNA into normal two strands. They demonstrated that certain portions of DNA reannealed more rapidly than others, and concluded that rapid reassociation was due to multiple DNA repeats.

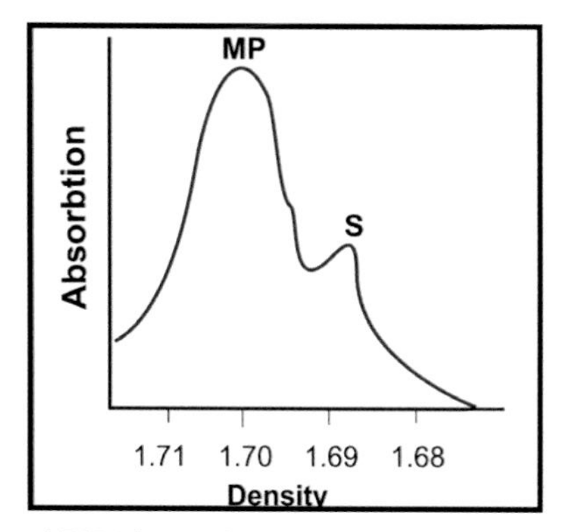

Fig. 16.4.2. Separation of total DNA into main peak (MP) and satellite (S) DNA from the mouse by ultracentrifugation in a CsCl gradient. Absorbtion measured at optical density in ultraviolet light at 260 nm.

Middle Repetitive DNA Sequences

As depicted (Fig. 16.4.1) beyond satellite DNA there is another prominent category of repetitive DNA that is important in understanding the structure and evolution of the eukaryotic genome. This is a category of **middle** or **moderately repetitive DNA**. These DNAs are quite well studied. In humans middle repetitive DNA most frequently constitutes either tandemly repeated or interspersed sequences. No function is known for this repetitive DNA in the genome. Among those characterized, there is a type of DNA called **variable number tandem repeats** (VNTRs). VNTRs are usually 15–100 bp long, and are found within and between genes. VNTRs, often referred to as **minisatellites**, are dispersed throughout the genome. Another group of tandemly repeated sequences is the dinucleotides, also referred to as **microsatellites**. Similarly to VNTRs, they are dispersed throughout the genome and vary among individuals in the number of repeats present at different sites. In humans, the most common microsatellite is the dinucleotide CA, with the repeats number up to 5–50. The microsatellites are now widely used in forensic investigations, and serve as very suitable molecular markers for population genetic analysis.

Repetitive Transposed DNA Sequences

DNA also consists of multiple sequences that are interspersed throughout the genome, but which not tandemly repeated. These sequences can be variable in length, and many of them are so-called **transposable sequences** or **transposons**, mobile and **insertion sequences** (IS); they are consequently able to change their locations in the genome or even jump to another genome. On their similarity with retroviruses they are called interspersed retrotransposons (Fig. 16.4.1). A large portion of eukaryotic genomes is composed of such sequences; they are also found in bacteria DNA and signs of them have been observed even in mtDNA (see below). Transposable elements were first discovered by McClintock in corn. One of these elements, **short interspersed elements** (SINEs) are smaller than 500 bp. The frequency of SINEs sometimes exceeds 500,000 copies in the genome, including the human genome. Human SINEs are a set of closely related sequences known as the ***Alu* family** (named after the restriction endonuclease *AluI*). SINEs of the *Alu* family are also found in other mammalian taxa. In mammals they are 200–300 bp long and dispersed uniformly throughout the genome, incorporated inside and outside the genes. SINEs of the *Alu* family are sometimes transcribed. There is no well understood role yet for this transcribed RNA, but in part it should be related to their mobility in the genome. Origin of the *Alu* sequences is thought to have been from an RNA element, whose DNA complement was formerly dispersed throughout the genome, caused by the reverse transcriptase activity of a transposase enzyme (Tpase).

The other transposon group, **long interspersed elements** (LINEs), is also repetitive DNA sequences. In humans, the best known LINE is a family designated L1. These LINE sequences are approximately 6,400 bp long, and there are up to 100,000 copies in the genome. The 5'-end of LINEs is highly variable; no definite role is yet defined for either L1 or other LINEs. The basis for transposition of L1 elements is know, however. The LINE L1 sequence is first transcribed into an RNA molecule. Then the RNA is uses as the template for the synthesis of the DNA complement via the Tpase enzyme. This enzyme is encoded among others by the L1 sequence. Finally, the new L1 copy inserts into the chromosome's DNA, usually at another site. Because of the similarity of this mechanism of transposition with that used by retroviruses, LINEs are also called **retrotransposons**. Jointly SINEs and LINEs form a large fraction of the human genome. SINEs and LINEs share the same organizational feature: comprising a mixture of sequences, some 70% unique and 30% repetitive DNAs in each group. SINEs and LINEs sequences occupy approximately 10% of the genome.

One more group of mobile elements, ISs, was mentioned above. That is very wide group of transposons having normally small size, 740–7,900 bp (Bacterial Database of ISs: http://www-is.biotoul.fr/is.html). Recent genome projects revealed the existence of numerous transposable elements and simpler mobile elements, ISs, both in eukaryotes (Pritham 2009) and in prokaryotes (Craig et al. 2002; Wagner and Chaux 2008), extending our knowledge of the complexity of life forms, even such relatively simple forms as bacteria. Organization of a typical IS is quite simple. The IS is represented as a modified ORF, in which the terminal inverted repeats (IRs) are found, usually labeled as IRL (left inverted repeat) and IRR (right inverted repeat). A single ORF encoding the Tpase normally includes the entire length of the IS, and extending into the IRR sequence. Also within IS may be short directly repeated sequences (DRs) generated in the target DNA as a consequence of insertion. The Tpase enzyme promoter, p, which is partially localized in IRL, occurs together with a couple of domains, Domains I and II, that represent the base pairs necessary for sequence-specific recognition and binding by the Tpase. More details of IS structure and other features may be found in the relevant literature (Mahilion and Chandler 1998). In the case of bacterial ISs, only approximately 50 had been analyzed at the nucleotide level by 1989 (Galas and Chandler 1989), compared to over 3,000 nowadays (Wagner and Chaux 2008). This is equally true of eukaryotic "ISs" such as *mariner*, derivatives of which have been found in over 240 insect species alone, as well as many in fungi, mammals, fish, and plants (Robertson and Lampe 1995), and related elements such as Tc*1* have also been observed. This enormous ISs diversity and distribution is astonishing, and requires further analysis, classification and understanding the degree of their homology, which remain unresolved in many instances at least between prokaryotes and eukaryotes in general and among organelle and nuclear genomes in particular. There is an interesting question, are ISs can also harbor such simplest genome as mitogenome is? Our analysis (Kartavtsev 2011c) of 74 complete mitogenomes in bird and fish species, with a focus on the presence of the IS in mtDNA, revealed that mitogenomes hold no full-length ISs, but include probably many of their sections. In our survey of 74 complete mitogenomes, from 4 to 15 short IS elements per genome were detected with IS-Finder software. The IS elements found are probably inverted repeats of real ISs. Both gender-dependent and horizontal transmission routs of these IS segments were found among 74 representatives of vertebrate species. The horizontal transmission through food chains like fish → bird, although that may be relatively rare for distant lineages, is potentially possible for human consumers of a raw marine food like sushi and sashimi. This important point requires further consideration.

Middle Repetitive Multiple Copy Genes

In some cases, middle repetitive DNA includes functional genes present tandemly in multiple copies, referred to as multigene families (Fig. 16.4.4). For example, many copies exist of the genes encoding not only rRNA, but tRNA too, as already discussed in Section 16.2 (see Table 16.2.2).

Centromeric and Telomeric DNA Sequences

These chromosome elements also include repetitive sequences. The precision of chromosome division during mitosis and meiosis and their transmission throughout generations is critical for cellular organisms. It is very high, but sometimes mistakes still occur. The evaluation of such errors during mitosis shows their rate is indeed very low: 1×10^{-5} to 1×10^{-6}; so, only 1 error could occur at up to 1 mln cell divisions. The molecular mechanism of this high precision was recovered after sequencing of the centromere region (CEN). Its function and structure are now quite clear. The structure of the CEN region is two-fold: (i) it binds proteins forming the centromere, and (ii) it provides a place for the kinetechore (used during cell division). The wide-scale analysis of the CEN regions of yeast *S. cerevisiae* chromosomes provided the basis for the original model (Carbon 1984). All CENs have very similar organization because identity of a centromere function. For example, the CEN of yeast chromosomes consists of about 225 bp; each CEN can be divided into three regions (Fig. 16.4.3). The regions I and III are short conserved sequences and comprised of 8 and 26 bp in length, respectively. Region II is larger (78–85 bp), AT-rich (up to 94–95% of all nucleotides) and contains variable sequences for different chromosomes.

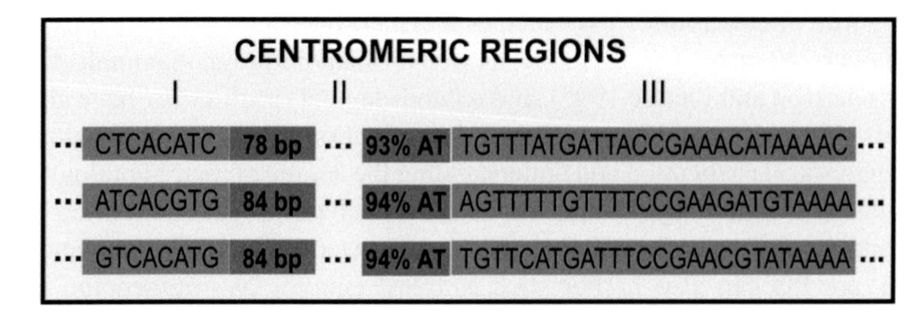

Fig. 16.4.3. Nucleotide sequences obtained from DNA of the three major centromere regions of three yeast *S. cerevisiae* chromosomes, 4, 6, 11: CEN 4, CEN 6, and CEN 11 (from top to bottom).

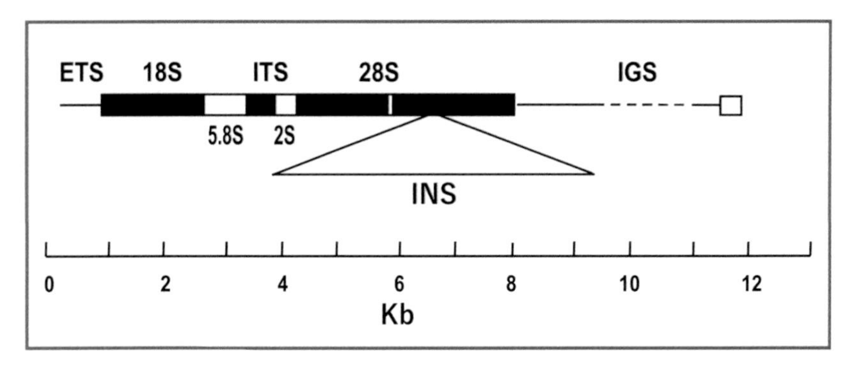

Fig. 16.4.4. Organization of the rRNA genes in *Drosophila melanogaster*, a tandemly repeated multigene family.
In addition to the 18S, 5.8S, 2S, and 28S rRNA genes, the repeated unit contains an external transcribed spacer (ETS), an internal transcribed spacer (ITS), and an intergenic sequence. Some copies also contain insertion sequences (INS) (Modified from Hartl and Clark 1989 with consent from Sinauer Associates, Inc.).

The telomere regions are also very significant for a chromosome's integrity and stability in Eukaryotes. There are two types of telomere sequences (TS). One TS type, called telomeric DNA sequences, consists of short tandem repeats. In the ciliate *Tetrahymena,* over 50 tandem repeats of the hexanucleotide sequence GGGGTT are found. In humans, there is similar highly repeating sequence, GGGATT. The analysis of TS DNA has shown that they are strongly conserved throughout evolution; this most probably reflects the critical role the TS play in maintaining the integrity of linear chromosomes in higher organisms. Another TS type is telomere-associated sequences (TAS). They are repetitive sequences too. TAS is found located both in TS neighborhood and within the telomere. TAS vary among organisms, but their function role is unknown yet.

Evolution in Multigene Families

Repeated sequences sometimes presented by multigene families. Within individuals of the species, the copies of the multigene family usually have a greater similarity among each other than they exhibit to other gene families present in the genomes of related taxa (see Fig. 16.2.1, Table 16.3.1). These genes may increase their number through duplication, the main mechanisms of which considered in next section. Complementarily to them, RNA-mediated transposition is possible, as we saw above, by gene amplification and genetic exchange (Maeda and Smithies 1986). Using techniques of tree construction, analogous to phylogenetic tree building (see Chapter 14), we can establish similarity and the ancestry of individual members of gene families. For instance for the myoglobin family as widely known case (Klug and Cummings 2002, p. 381) the reconstruction of the family evolution in time is shown (Fig. 16.4.5).

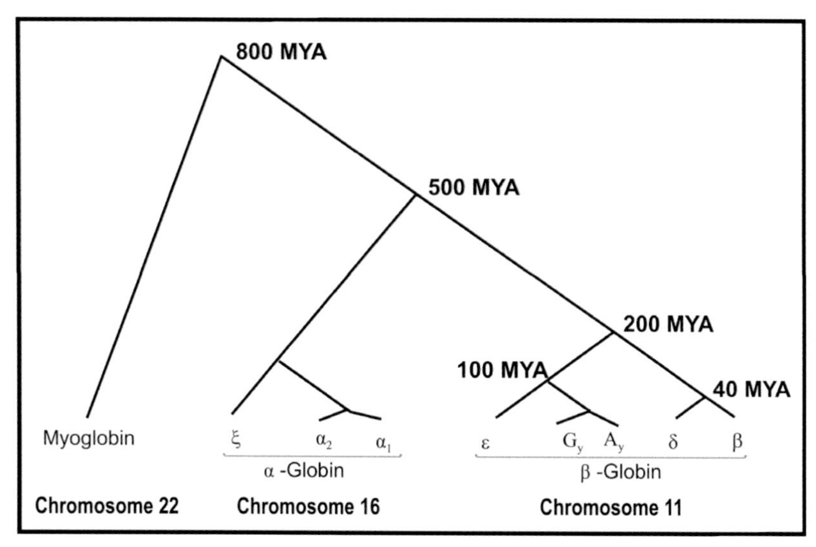

Fig. 16.4.5. Evolutionary history of the myoglobin gene family.
About 800 million years ago (MYA), a duplication event in an ancestral gene gave rise to two lineages. One led to the myoglobin gene, which in humans is located on chromosome 22. One of lineages then goes through new duplication event about 500 MYA and gave rise the descendants of the alpha and beta subfamilies. Duplications about 200 MYA produced the gamma and beta globin subfamilies. Location of all genes after rearrangements on chromosomes in humans at present is shown.

Mechanisms of Concerted Evolution

Members of multigene families tend to retain more similarity than may be expected under random drift. They tend to evolve together (in concert) because of mechanisms that operate to homogenize the sequences within the family. This tendency of a homogenization is known as **concerted evolution.** Two mechanisms of concerted evolution are suggested (Fig. 16.4.6). The first is a **gene conversion,** a process in which nucleotide pairing between two sufficiently homologous genes is accompanied by the loss of all or part of the nucleotide sequence in one gene and its replacement by a replica of the nucleotide sequence from the other gene (Fig. 16.4.6A). Formally, the result is that the sequence in one gene "converts" the sequence in the other gene to be exactly like itself. The second is concerted evolution by means of **unequal crossing over** (Fig. 16.4.6B).

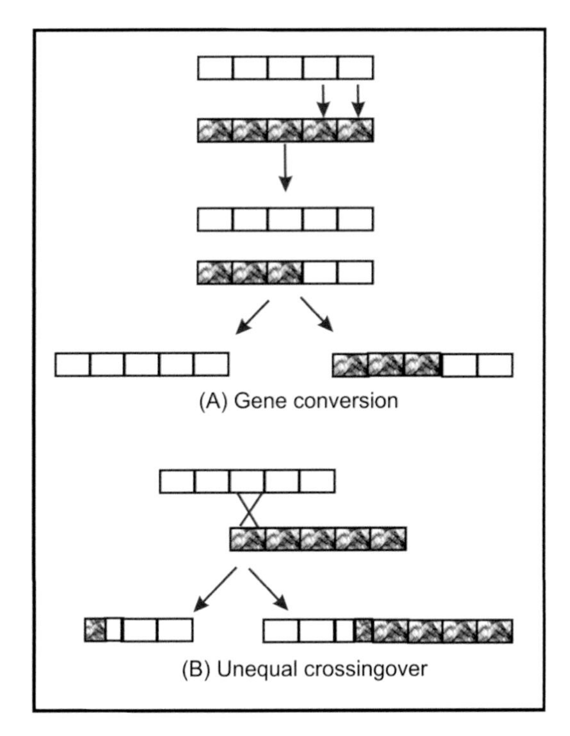

Fig. 16.4.6. Two mechanisms of concerted evolution that can homogenize the sequences of members of multigene families.
(A) Gene conversion, in which genetic information in one gene is replicated and replaces genetic information in its homolog. (B) Unequal crossing over. The increase in the number of members of the gene family because of unequal crossing over can be corrected by a subsequent unequal crossing over, resulting in identical DNA sequences among several adjacent members of the multigene family (Modified from Hartl and Clark 1989 used with permission from Sinauer Associates, Inc.).

In tandem multigene families, misaligned pairing of the genes during meiosis can occur, and crossing over results in gametes that can have either an increase or a decrease in copy number. Unequal crossing over by itself does not result in a steady increase in copy number; rather it results in an increased variance in copy number. Thus, a subsequent crossing over can either correct the copy number or increase it further. Formally, the final result is genetically equivalent to gene conversion, because the nucleotide sequences change in the same way.

The role of unequal crossover in increasing the number of genes in the genome was postulated long ago (Bridges 1936). However, only when molecular genetic data became numerous, was it realized that it plays a very significant role in increasing or decreasing DNA content and, consequently, gene numbers. Particularly in multigene families such as immunoglobulin genes and ribosomal RNA genes, unequal crossover may be the main factor (Hood et al. 1975). Gene conversion is also suggested as a factor for increasing genome size (Radding 1982). Slightom et al. (1980) noted that part of the second intron of the human A_γ globin gene is very similar to that of the nonhomologous gene G_γ from the same individual. They hypothesized that the part of the A_γ gene had been "converted" by an intergenic exchange to become more like the G_γ gene. An example of gene conversion was obtained in common mussel (Rawson et al. 1996; see Chapter 10).

Gene conversion and unequal crossing over maintain similarity between copies of genes in multigene families. Theoretical studies of the processes have been carried out by Ohta (1982) and by Nagylaki (1984).

Pseudogenes

The evolutionary change of pseudogenes is of special interest because they can be used for testing the neutral mutation theory. As for nonfunctional DNA, the rate of nucleotide substitution for pseudogenes must be higher than that for structural genes. As we discussed, pseudogenes indeed evolve faster (Chapter 6, Table 6.5.1). Once a gene becomes nonfunctional, it is generally difficult to revive it. This is because there are many ways to silence a gene, while a back mutation that may revive a pseudogene must occur at exactly the same site where the first forward mutation occurred. Obviously, the chance of such event occurring is very small. Moreover, when a gene becomes nonfunctional, many more deleterious mutations may occur in it, and thus the probability of reviving a pseudogene decreases with evolutionary time. Nevertheless, as we discussed some inverse processes are also found (see Section 16.3).

Endogenous Retroviruses in Mammals

Different organisms, as we emphasized above, including mammals, normally contain numerous ISs and several kinds of retroviruses. DNA sequences that are homologous to retroviruses and that have become fixed in the genome of a species are known as endogenous retroviral sequences, or virogenes. Several distinct types of endogenous sequences occur in the vertebrate genome. Virogenes usually are represented by 10–100 copies per genome, and originated in reverse transcription and reinsertion in host DNA. Under suitable conditions in cells, endogenous viruses can give rise to mature retrovirus particles. Sometimes they may develop diseases in cells of their host and release particles, which are infectious in cells of related species. The released viruses are typical retroviruses, in that a small number of reverse transcripts of the viral RNA become incorporated again into the host DNA. These observations and others suggest that sequences of endogenous retroviruses are advantageous to the host in conferring some degree of immunity to viral infection (Benveniste 1985). Perhaps, the same reasoning could be applied to all mobile elements: transposons, ISs, etc.

Endogenous viruses are retained in the genome for a long time. For example, the Type C endogenous retrovirus of baboons is found in all species of Old World monkeys but is absent in New World monkeys. This kind of retrovirus therefore invaded the lineage of all monkey descendants after faunal separation due to continental drift, and it remained within the Old World lineage during numerous speciation events in the group. The Type C endogenous retrovirus of baboons also illustrates the rapid rate of evolution typical for the virogenes (Fig. 16.4.7). The data summarize DNA hybridization experiments and obtained with two types of probe DNA from the baboon,

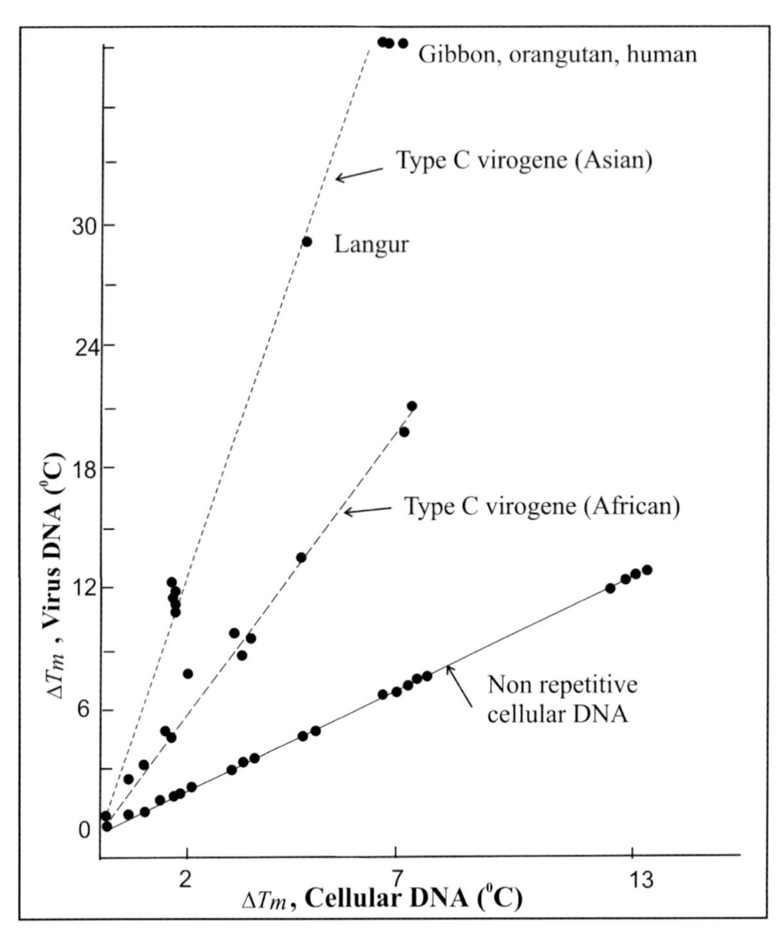

Fig. 16.4.7. Evolutionary rate of baboon Type C endogenous retrovirus (dashed lines) as compared with nonrepetitive cellular DNA (solid).
In African primates, these virogenes have evolved almost threefold faster than nonrepetitive cellular DNA, and in Asian primates they have evolved about sixfold faster (after Benveniste 1985; adopted from Hartl and Clark 1989).

Papio cynocephalus, when hybridized with cellular DNA from other primate species (Fig. 16.4.7). The first probe represents baboon cellular DNA, which gave the solid line plotted with equal values on both axes. Note that the DNA from all species falls on the line, representing a fine example of a molecular clock with a constant rate of nucleotide substitution; it is reflected in the plotted ΔTm, when averaged for the whole functional genome, i.e., excluding repetitive DNA (Fig. 16.4.7).

We must finish this endless theme on genome evolution as endless living forms are. We obviously recognize that life and the genome as its most important part evolve from simplicity to complexity. Simple ring DNA in prokaryotes coding for 250–350 genes changed through evolution into chromosomes in eukaryotes, coding more than 26,500 genes, as in humans. Activity of genes in prokaryotes and eukaryotes has many differences, and gene functions again evolved, as a rule, towards more complexity. This functional complexity recruits through evolution the multiple regulatory mechanisms within the eukaryotic genome at the levels of cells, individuals, and populations; these processes are organized in a hierarchical, cascade, homeostatic, and network-like fashion. Evolution of many parts of genes is caused by random or stochastic processes, but major gene functions, and genome expression in general, are highly conserved, maintained by purifying selection. Other forms of selection also can be detected, as we have seen during our course in different chapters; directional selection retains its dominant role in a creation of genetic novelties with their new functions, while others like normalizing selection may play a key role too. More common in nature is a dynamic combination of random and directional forces.

16.5 TRAINING COURSE, #16

16.1 Read Appendix II and Appendix IV. Ask any questions on these two and other appendices.

16.2 Consultation on all themes of Training Courses.

16.3 Learn few new terms:

- **Tandem duplication** can originate by three major mechanisms: (1) **unequal crossing over**, a recombination event between members of a homologous pair of chromosomes in which a DNA segment is duplicated in one of the recombination products; (2) **unequal sister chromatid exchange,** a recombination event similar to unequal crossing over that occurs between the two chromatids of a single chromosome; (3) **replication errors**; during replication of a template molecule, slippage can cause the insertion of a short segment into the newly synthesized strand.
- **Gene Elongation** is another way for gene change. Gene elongation caused by duplication of a gene or its part.
- **Multigene family.** Family of homologous genes with related functions is called a multigene family, the members of which are often tandemly arranged within the chromosome.
- **Concerted evolution.** The tendency of a homogenization or nonrandom substitution rate in gene families is known as concerted evolution. It is realized due to gene conversion and unequal cross over.
- **Transposons** are one several types of mobile elements of the genome, like SINEs, LINEs, and ISs.

APPENDIX I

METHODICAL INSTRUCTION FOR SELF-EDUCATION OF STUDENTS IN THE MASTER COURSE

A goal of methodical instruction is to deepen the basics, and understand the most important aspects of population evolutionary and ecological genetics of fish and shellfish. To achieve these goals it is recommended to learn and develop one of the themes and provide the abstract in the written form and give oral presentation. The main questions of the theme, specified below, must be developed with the usage of the literature sources given for each theme.

The Themes of the Abstracts:

1. Structure of a Chromatin. Mobility of a Genome.
2. Chromosomes: The Morphology and Evolution. Speciation Modes.
3. Population Genetic Structure in Fish and Shellfish Species as Compared to Land Animals.
4. Methods and Principles of Biochemical Genetics (BG). The BG and Biology and Ecology.
5. Application of Molecular Genetic Markers (MGM) in Fishery and Biology in General. The Advances and Perspectives in MGM Usage.
6. The Genetic Aspects of Animal Speciation: Examples from Favourable Animal Groups.
7. Genetics in Aquaculture.
8. Phylogenetics.

Main Topics to be Developed in the Abstract and the Literature

THEME 1. The Structure of a Chromatin. Mobility of a Genome.

Goals that must be reached:

1. A chromatin's condensation and its decondensation. 2. A regulation of replication, transcription and translation. 3. A gene structure in Eukaryotes. 4. The role of the mobile genetic elements in development and evolution.

Literature:

1. Klug, W.S. and M.R. Cummings. 2002. Essential Genetics, 4th Edition, Prentice Hall, Inc., Upper Saddle River, NJ 07458.
2. Hartl, Daniel L. and Elizabeth W. Jones. 2002. Essential Genetics, 3rd Edition. Jones and Bartlett Publishers, Sudburry, Massachusetts, Boston-Toronto-London-Singapore.
3. Lewin, B. 1983. Genes. Ed. Cell. John Wiley and Sons. New York, Chichester, Brisbane, Toronto, Singapore.

THEME 2. Chromosomes: The Morphology and Evolution. Speciation Modes.

Goals that must be reached:

1. Methods of chromosome staining. A chromosome morphology and karyograms.
2. Major ideas on the chromosome numbers and chromosome evolution in fish and shellfish. 3. Karyotypic differences and their role in the development of Reproductive Isolation Barriers (RIB). A polyploidy and hybrid speciation. A problem of parapatric (stasipatric) speciation in fish.

Literature:

1. Ohno, S. 1970. Evolution by Gene Duplication. Springer-Verlag, Berlin.
2. King, Max. 1993. Species Evolution: The Role of Chromosome Change. Published by the Press Syndicate of the University of Cambridge. The Pitt Building, Trumpington Street, Cambridge CB2 1RP 40 West 20th Street, New York, NY 10011-4211, USA 10 Stamford Road, Oakleigh, Melbourne 3166, Australia.
3. Ryman, N. and F. Utter (eds.). 1989. Population Genetics & Fishery Management. Washington Sea Grant Program. Distributed by University of Washington Press, Seattle and London.
4. Altukhov, Yu.P. 1990. Population Genetics: Diversity and Stability. Harwood Academic Publishers, London (Post Office Box 197, London WC2E 9PX, UK or correct in Internet).

THEME 3. Population Genetic Structure in Fish and Shellfish Species as Compared with Land Animals.

Goals that must be reached:

1. Basic intraspecies groups in the animal realm. 2. The specifics of population genetic structure in fish and shellfish species. 3. Models of population genetic organization. The factors of population genetics, dynamic and stationary. Genetic basis for sustainable fishery.

Literature:

1. Ryman, N. and F. Utter (eds.). 1989. Population Genetics & Fishery Management. Washington Sea Grant Program. Distributed by University of Washington Press, Seattle and London.
2. Altukhov, Yu.P. 1990. Population Genetics: Diversity and Stability. Harwood Academic Publishers, London (Post Office Box 197, London WC2E 9PX, UK or correct in Internet).
3. Altukhov, Yu.P., E.A. Salmenkova and V.T. Omeltchenko. 2000. Salmonid Fishes: Population Biology, Genetics and Management. Blackwell Science Ltd., Oxford, 368 p. Look in www.fishknowledge.com.
4. Nei, M. 1987. Molecular Evolutionary Genetics. Columbia University Press, N.Y., 512 p.
5. Li, C.C. 1976. First Course in Population Genetics. The Boxwood Press, Pacific Grove, California.
6. Hartl, D.L. and A.G. Clarke. 1989. Principles of Population Genetics, 2nd Edition. Sinauer Assoc. Inc. Publishers, Sunderland, Massachusetts.
7. Soule, M.E. (ed.). 1987. Viable Populations for Conservation. Cambridge University Press, Cambridge. NY, New Rochelle, Melbourne, Sydney.

THEME 4. Methods and Principles of Biochemical Genetics (BG). The BG and Biology and Ecology.

Goals that must be reached:

1. Electrophoretic methods of research of proteins as gene markers. 2. Genetics of enzymes: isozymes, allozymes. Principles of genetic interpretation of the iso- and allozyme variability. 3. Heterozygosity and genetic distance measurements and their application in biology and ecology.

Literature:

1. Lewontin, R.C. 1974. The Genetic Basis of Evolutionary Change. Columbia University Press. New York and London.
2. Nei, M. and R.K. Koehn (eds.). 1983. Evolution of Genes and Proteins. Sinauer Assoc., Sunderland (Mass.).
3. Ryman, N. and F. Utter (eds.). 1989. Population Genetics & Fishery Management. Washington Sea Grant Program. Distributed by University of Washington Press, Seattle and London.

4. Nei, M. 1987. Molecular Evolutionary Genetics. Columbia University, Press, N.Y., 512 p.
5. Altukhov, Yu.P. 1990. Population Genetics: Diversity and Stability. Harwood Academic Publishers, London (Post Office Box 197, London WC2E 9PX, UK or correct in Internet).

THEME 5. Application of Molecular Genetic Markers (MGM) in Fishery and Biology in General. The Advances and Perspectives in MGM Usage.

Goals that must be reached:

1. Methods of research of DNA-length variation as gene markers. 2. DNA polymorphism within and between populations; number of polymorphic restriction sites, nucleotide diversity. 3. The advances and perspectives in MGM usage.

Literature:

1. Nei, M. 1987. Molecular Evolutionary Genetics. Columbia University Press, N.Y. 512 p.
2. Park, L.K. and P. Moran. 1994. Developments in molecular genetic techniques in fisheries. Rev. Fish. Biol. 4: 272–279.
3. Ryman, N. and F. Utter (eds.). 1989. Population Genetics & Fishery Management. Washington Sea Grant Program. Distributed by University of Washington Press, Seattle and London.

THEME 6. The Genetic Aspects of Animal Speciation: Examples from Favourable Animal Groups.

Goals that must be reached:

1. Main modes of speciation. 2. Heterozygosity and genetic distance. Phylogenetic trees. 3. Genetic divergence in taxa of different rank. 4. Genetic aspects of species origin. Examples from most studied groups.

Literature:

1. King, Max. 1993. Species Evolution: The Role of Chromosome Change. Published by the Press Syndicate of the University of Cambridge. The Pitt Building, Trumpington Street, Cambridge CB2 1RP 40 West 20th Street, New York, NY 10011-4211, USA 10 Stamford Road, Oakleigh, Melbourne 3166, Australia.
2. Nei, M. 1987. Molecular Evolutionary Genetics. Columbia University Press, N.Y., 512 p.
3. Ryman, N. and F. Utter (eds.). 1989. Population Genetics & Fishery Management. Washington Sea Grant Program. Distributed by University of Washington Press, Seattle and London.
4. Lewontin, R.C. 1974. The Genetic Basis of Evolutionary Change. Columbia University Press, New York and London.

5. Kartavtsev, Y.P., V.V. Sviridov, T. Sasaki and N. Hanzawa. 2002. Genetic divergence of far eastern dace of Tribolodon genus (PISCES, CYPRINIDAE) and close taxa. Genetica 38(9): 1518–1531 (English translation from Russian).

THEME 7. Genetics in Aquaculture.

Goals that must be reached:

1. Half-Cyclic Systems of Rearing (HCSR) and Full-Cyclic Systems of Rearing (FCSR). 2. Major genetic tasks and their differences in the HCSR and FCSR. 3. The genetic basis of salmon ranching; the rehabilitation and introduction from the genetic point of view. 4. The genetic basis of fish breeding under FCSR; inheritance and selection modes.

Literature:

1. Calaprice, I.R. 1976. Mariculture—Ecological and genetic aspects of production. J. Fish. Research Board of Canada 1(4).
2. Krueger, C.C. et al. 1981. Genetic aspects of fisheries rehabilitation. Canadian J. Fish. a. Aquat. Sci. 38(12): 1877–1881.
3. Ryman, N. and F. Utter (eds.). 1989. Population Genetics & Fishery Management. Washington Sea Grant Program. Distributed by University of Washington Press, Seattle and London.
4. Altukhov, Yu.P., E.A. Salmenkova and V.T. Omelchenko. 2000. Salmonid Fishes: Population Biology, Genetics and Management. Blackwell Science Ltd., Oxford, 368 p. Look in www.fishknowledge.com.

THEME 8. Phylogenetics.

Goals that must be reached:

The subject of phylogenetics as a scientific field. Origin and Change of Genome. Types of Phylogenetic Trees. Method of Distance Matrix. Parsimony and Maximal Likelihood Methods. Role of Population Genetic Theory in a Tree Construction. Tree building software.

Literature:

1. Nei, M. 1987. Molecular Evolutionary Genetics. Columbia University Press, N.Y., 512 p.
2. Hartl, D.L. and A.G. Clarke. 1989. Principles of Population Genetics, Second edition. Sinauer Assoc. Inc. Publishers, Sunderland, Massachusetts.
3. Nei, M. and S. Kumar. 2000. Molecular Evolution and Phylogenetics. Oxford University Press, N.Y., 333 p.
4. Swofford, D.L., G.J. Olsen, P.J. Waddel and D.M. Hillis. 1996. Phylogenetic inference. pp. 407–514. *In*: D.M. Hillis, C. Moritz and B. Mable (eds.). Molecular Systematics. Sinauer. Assoc. Inc., Sunderland, Massachusetts, USA.

APPENDIX II
EXAM QUESTIONS

Section 1: Define the Concept

1. Main Stages in Earth Geological Evolution.
2. Paleontology Dating of Evolution.
3. Molecular Biological Evolutionary Dating.
4. Chromosome Structure.
5. Meiosis and Gamete Formation.
6. Mendel's or Transmission Genetics. Mono-, Di- and Trihybrid Crosses. Mendel's Laws.
7. Eukaryote Gene Structure. Mutations and Modifications.
8. Components that are Critical for Protein Synthesis.
9. The process of Translation: From RNA to Polypeptide.
10. Heredity, Proteins and Function.
11. Introduction to Immunogenetics.
12. Inheritance of Immunogenetic Traits and Population Variability.
13. Immunogenetic Dating of Evolution including Amino Acid Sequence Analysis.
14. Introduction to Methods of Detection of Protein Variability.
15. Interpreting of Protein Variability.
16. Genetics of Isozymes.
17. Main Modes of Speciation. Genetic Aspects of Speciation.
18. Population Structure in Bisexual Species.
19. Modes of Intraspecies Structure.
20. Methods and Principles in Studies of Population Genetic Structure.
21. Main Outcomes of Population Genetic Research.
22. Main Genetic Parameters of Population Variability.
23. Quantitative Measures of Differentiation.
24. Models of Population Genetic Structure.

25. Empirical Estimates of Parameters.
26. What are Hybrids and Hybridization?
27. Methods of Detecting Hybridization.
28. Genetic Interpretation of Hybridized Populations.
29. Empirical Investigations of the Hybrid Zones.
30. Quantitative Variation and its Laws.
31. The Multiple Factor Hypothesis.
32. Main Tasks of Genetic Research of Quantitative Traits (QT): σ^2_{PH}, σ^2_{G}, Cv, h^2 and other Variables.
33. Heritability.
34. Mapping of Quantitative Trait Loci.
35. Concept of Genetic Homeostasis.
36. Heterosis and Heterozygosity.
37. Empirical Results on QT Heterozygosity Association and Interpreting.
38. Introduction to Mutation Study.
39. Structure and Function of Genes.
40. Type of Mutations on the DNA Level.
41. Mutations and Amino Acid Substitutions.
42. Mutation Rate.
43. Genetic Approach to Speciation: Advances and Limitations.
44. What Data are Necessary to Describe Evolution? What is the Data Base?
45. Species Concept. Explain Data on Heterozygosities/Diversities and Distances.
46. Main Speciation Modes. Notes on Population Genetic View.
47. DNA Sequence Polymorphism. Nucleotide Diversity.
48. DNA Polymorphism Estimated from Restriction Site Data.
49. DNA Length Polymorphism. Theory and Observations.
50. Divergence of Populations at DNA Level.
51. Types of Phylogenetic Trees.
52. Distance Matrix Methods for Tree Building.
53. Parsimony and Probability Methods for Tree Inference.
54. Population Genetic Theory and Tree Construction.
55. Edition of Nucleotide Sequences and their Submission to a Gene Bank.
56. Data Formats and Gene Banks Available.
57. Alignment of Sequences.
58. Searching for an Optimal Nucleotide Substitution Model.
59. Tree Building Using MEGA-5 Program Package and Annotation on Programs PAUP, MrBayes and Some Others.

60. Origin and Change of Genome.
61. Evolutionary Change of Genome Size.
62. Formation of New Genes.
63. Repetitive DNA and Multigene Families.

Section 2: Show the Skills of the Training Course

1. Genetic Vocabulary: Give definitions for following notations—*AA, Aa, BbCc, M, m, P, F1, Fa.* Explain Terms: *Locus, Gene, Allele, Cistron.*
2. Chi-Square Test of Independence in Frequency Distribution of Qualitative Traits. X2 = [(f' - f) / f'], f' and f are expected and observed numbers, d.f. = n - 1, where n—number of phenotypic classes.
   Chi-Square Test of Independence in Frequency Distribution of Qualitative Traits. Find solution for one numerical example.

P		F_2		n	d.f.
C_1C_2	C_2C_2	C_1C_2	C_2C_2		
-	-	105	95	200	1
-	-	64	77	141	1
-	-	63	86	149	1
-	-	81	69	150	1
-	-	313	327	640	1

   $X^2 = \Sigma\ [(f\text{–}f')^2/f']$; $X^2_{.95}\ (1) = 3.84$; Locus GPD, pink salmon (Kartavtsev et al. 1985).
3. Allele Frequency and its Standard Error. Heterozygosity.
   Calculate allele frequency, SE and H exp.: AA = 100, Aa = 70, aa = 30.
4. Chi-Square Test of Independence in Frequency Distribution at Qualitative Traits. Observed and Expected Numbers in Natural Population Matings. Check H-W expectations.
5. SPECSTAT-Software. Input Data Table. Output Table 1. To solve one Numerical Example on Literature Data and Two on Your Own Samples.
6. Chi-Square Test of Heterogeneity.
7. Estimates of Genetic Distances and Constructing of UPGMA Dendrograms.
8. Software BIOSYS. Solve Package Examples for Distances or Similarities and making Dendrograms on their Basis.
9. Software MEGA-5. Solve Package Examples for p-Distance Calculation. Build NJ Tree on appropriate Distance Basis.

Section 3: Chose an Appropriate Answer

1. Genetic Vocabulary: Which is the correct answer of the two following notations?
   i) *AA, Aa, BbCc. All 3 genotypes are all heterozygotes.*
   ii) *AA, Aa, BbCc. AA is homozygote and Aa, BbCc are heterozygotes.*
2. Genetic Vocabulary: What is the appropriate answer from the following notations? Tick: correct (i), (ii), both.
   i) *P, F1, Fa are notations for parental, first and backward crosses.*
   ii) *P, F1, Fa are notations for parental, first and analytical crosses.*
3. Which of the two (i, ii or both) precisely explain the terms: *Locus* and *Gene.*
   i) **Locus** is a place on a chromosome where the gene is located.
   ii) **Locus** is a section of a gene and is located on a chromosome.
   i) **Gene** is a noncoding sequence of DNA, which determined a regulation of a function of an organism (like polypeptide synthesis).
   ii) **Gene** is a coding sequence of DNA, which determined an elementary function of an organism (like polypeptide synthesis).
4. Which of the two (i, ii or both) better suit the terms: *Allele* and *Cistron.*
   i) **Allele** is a variant of a gene with certain peculiarities of a function (e.g., differences in a charge of proteins).
   ii) **Allele** is a byproduct of a gene with certain similarities of a function (like proteins heat resistance).
   i) **Cistron** is a unit of a genetic ontogenetic regulation, which includes a DNA coding sequence and regulatory elements for turnover of molecular synthesis.
   ii) **Cistron** is a unit of a genetic function, which includes a DNA coding sequence and regulatory elements for macromolecular synthesis.
5. What is ***linkage***? The joint inheritance of a pair or more genes which code for independent traits.
   Basic researches in the beginning of the 20th were made:
   a) by Morgan, Stertevant and Bridges.
   b) by De-Frieze, Koltsov and Weissman.
   Which is true, (a) or (b)?
6. Do the two definitions below belong to the **Biological Species Concept** (BSC)? Do both or does one of them belong to the BSC?
   i) **Species is** a ***biological unity, which is reproductively isolated from other unities and consists from one to several more or less stable populations of sexually reproducing individuals that occupy a certain area in nature***.
   ii) ***A species is a reproductive community of populations (reproductively isolated from others) that occupies a specific niche in nature***.
7. Who are the authors of these two definitions: Altukhov (1974) and (Dobzhansky 1951; 1970)? Who for which?
   i) A ***local stock (local population)*** *is a system of sub-populations, which is theoretically equivalent to a* **subdivided population**, *and is the least stable reproductive unit of a species*. (ii) *A* **population** *is a reproductive unit of organisms which share a common gene pool.*

8. Classification of the modes of intraspecies structures by using population genetic tools.
   Two different principal modes are possible to define, if based on population genetic tools:
   i) Panmictic population.
   ii) Subdivided population.
   Is it correct?
9. Allele frequency, p.
   Empirically under codominance or incomplete dominance, $\boldsymbol{p}$ can be estimated by the direct count of genotypes: Which formula is correct (1) or (2)?
   1) $\boldsymbol{p_{A1} = (2D\, A_1A_1 + H\, A_1A_2)/2n = (D\, A_1A_1 + 1/2_H A_1A_2)/n,}$
   2) $\boldsymbol{(p + q)^2 = p^2 + 2pq + q^2.}$ Is the general notation of the Hardy-Weinberg Law suits for multiple alleles:
      $\boldsymbol{(\Sigma\, p_k A_{ij})^2 = \Sigma\, p^2_k A_i A_i + \Sigma\, 2p_k\, (1 - p_k)\, A_i A_j.}$
10. The Hardy-Weinberg law can be expressed as follows below.
    Which expression is correct: (1), (2), (3), (1) and (2), all?
    1) $\boldsymbol{(p + q)^2 = p^2 + 2pq + q^2.}$
    2) $\boldsymbol{(\Sigma\, p_k A_{ij})^2 = \Sigma\, p^2_k A_i A_i + \Sigma\, 2p_k (1 - p_k)\, A_i A_j.}$
    3) $\boldsymbol{(p + q + r)^2 = p^2 + q^2 + r^2 + 2pq + 2pr + 2qr.}$

APPENDIX III

MAIN LITERATURE AND WEB SUPPORT

1. Nei, M. 1987. Molecular Evolutionary Genetics. Columbia Univ. Press, N.Y., 512 p.
2. Ryman, N. and F. Utter (eds.). 1989. Population Genetics & Fishery Management. Washington Sea Grant Program. Distributed by University of Washington Press, Seattle and London.
3. Hartl, D.L. and A.G. Clarke. 1989. Principles of Population Genetics, Second edition. Sinauer Assoc. Inc. Publishers, Sunderland, Massachusetts.
4. Li, C.C. 1976. First Course in Population Genetics. The Boxwood Press, Pacific Grove, California.
5. Soule, M.E. (ed.). 1987. Viable Populations for Conservation. Cambridge University Press, Cambridge. NY, New Rochelle, Melbourne, Sydney.
6. King, M. 1993. Species Evolution. The Role of Chromosome Change. Cambridge University Press, N.Y.
7. Lewontin, R.C. 1974. The Genetic Basis of Evolutionary Change. Columbia University Press. New York and London.
8. Altukhov, Yu.P. 1990. Population Genetics: Diversity and Stability. Harwood Academic Publishers, London (Post Office Box 197, London WC2E 9PX, UK or correct in Internet).
9. POPULUS 1.4. Alstad, D. Department of Ecology, Evolution & Behavior University of Minnesota 318 Church St. SE Minneapolis, MN, USA 55455-0302/ Tel: 612-625-0488. E-mail: DNA@UMNACVX.bitnet_.
10. BYOSIS. Swofford, D.L. and R.B. Selander. 1981. Biosys-1: A Fortran program for the comprehensive analysis of electrophoretic data in population genetics and systematics. J. Hered. 72(4): 281–283.
11. NTSYS. Rholf, F.J. 1988. NTSYS-pc: Numerical taxonomy and multivariate analysis system. Exter Publ., Ltd., N.Y.
12. SPECSTAT. Kartavtsev, Y.P. and A.A. Soloviev. 1993. SPECSTAT: PC-software for statistical data analysis in the field of population genetics. Genetica 88: 79–82.

13. Tamura, K., D. Peterson, N. Peterson, G. Stecher, M. Nei and S. Kumar. 2011. MEGA5: molecular evolutionary genetics analysis using maximum likelihood, evolutionary distance, and maximum parsimony methods. Molecular Biology and Evolution 28: 2731–2739.
14. STATISTICA 6.0. StatSoft, Inc. 1999. STATISTICA for Windows [Computer program manual]. Tulsa, OK: StatSoft, Inc., 2300 East 14th Street, Tulsa, OK 74104, phone: (918) 749-1119, fax: (918) 749-2217, email: info@statsoft.com, WEB: http://www.statsoft.com.
    http://www.jbpub.com/genetics http://csep10.phys.utk.edu/genetics_demo. http://www.gene.com.achttp://gened.emc.maricopa.edu/bio/BIO181/BIOBK/BioBookTOC.html. http://www.columbia.edu/cu/biology/courses/c3032/.

Appendix IV

EXPLANATION OF BASIC TERMS

Chromosome is a cell particle which can be seen during metaphase in an optical microscope, and consists of DNA and proteins. Chromosome is the main genes holder.

Gene is a unit of function in the organism. Locus ≈ Gene ≈ Allele ≈ Cistron these terms are among the most important in genetics and students must know their meaning and differences.

- **Locus** is a place on a chromosome, where the gene is located.
- **Gene** is a coding sequence of DNA, which determines an elementary function of an organism (like polypeptide synthesis).
- **Allele** is a variant of a gene with the certain peculiarities of a function (e.g., differences in a charge of proteins).
- **Cistron** is a unit of genetic function, which includes a DNA coding sequence and regulatory elements for macromolecular synthesis.
- **Dominant** allele controls a trait, which expressed in the presence of any other allele; such an alternative allele of a pair in a genotype is called **Recessive.**
- **Complementation** is a case of gene interaction, when alleles of different loci give a new phenotype that is not present in both parents.
- **Epistasis** is a case of intergene action, when alleles of one locus inhibited the action of alleles at another locus.

Gene Structure. In accordance with the modern concept, a gene structure is comprised by the following parts: I, a site of initiation of transcription, R, **Regulator** (**CAAT**) and O, **Operator** sites of a **Promoter**, **Exon**, coding part of the gene, **Intron**, non-coding part of the gene, T, a site of termination of transcription. Other elements may also be pointed out such as an enhancing site of the Promoter (in some viruses) and complex pathways both for regulation of transcription and translation.

Translation is the biological polymerization of amino acids into a polypeptide chain on a mRNA matrix.

Protein is a linear nonbranched polymer of amino acids.

Primary structure of a protein is defined by sequence order of amino acids in molecule. Secondary, tertiary and quaternary structures of protein molecule are also recognized.

Normal/Immune Sera are the blood sera that developed in an organism without/with the immunization and introduction of an antigen.

Multiple Alleles are several alleles that segregate at the same genetic locus in a population.

Null Allele is an allele of the locus that gives no active gene product (produces no antigen).

Hardy-Weinberg Law. In a random mating population there is an equilibrium binomial ratio between gametic and genotypic frequencies, which stay infinitely stable in space and in time in the absence of evolutionary forces' action.

Multiple Molecular Forms of Enzymes (MMFE) represented by isozymes and allozymes, are genetically determined molecular forms as well as posttranslational modifications.

Allozymes are isozymes that are caused by allele segregation at the same genetic locus. Electrophoresis in biochemistry and biochemical genetics is a process of separation of biological macromolecules in a gel by their electric charges.

Isoloci are loci that do not differ in the electrophoretic mobility of their consequent enzymes.
In terms of their function, genes can be classified into two groups: protein-coding or **structural genes** and RNA-coding, **regulatory genes**.

Spontaneous mutations are those that occasionally happen in nature. No specific agents are associated with their occurrence. In contrast to such spontaneous events, those that result from the influence of any artificial factor are considered to be **induced mutations.**

Nutritional or **biochemical mutations** are mutations that alter variation of corresponding traits.
Mutations that result in synonymous codons are called **synonymous** or **silent mutations**, whereas others are called **nonsynonymous** or **amino acid altering mutations**.

Viability mutations, may be classified as **lethal mutations, neutral, advanced** or **deleterious.**

Nucleotide substitutions. They can be divided into two different classes: **transition** and **transvertion**. **Transition** is a substitution of a purine for another purine (A⇔G) or the substitution of a pyrimidine for another pyrimidine (T⇔C). **Transvertions** are inverse types of nucleotide substitutions, when a purine is substituted by a pyrimidine and vice versa (G⇔T, A⇔C, C⇔G or A⇔T).

Population is a reproductive unit of organisms, which share a common gene pool. **Local populations** or **local stocks** are usually represented by subdivided populations, which consist of **demes** or **subpopulations**.

Main genetic parameters which determine dynamics of populations are (1) ***p*, allele frequency**, (2) *Ne*, **effective size of population**, (3) ***m*, coefficient of migration** defining a gene flow and (4) *s*, **coefficient of natural selection**.

Island Model of population structure is one of the models of a subdivided population and most popular among researchers.

Fst (and **Gst)** is standardized allele frequency variance and the most important measure of genetic differentiation of the populations within species. **Gst (Fst)** is used as a tool for hierarchical population analysis. **Gst** is called a coefficient of gene differentiation as well.

Hybrid is a mixture, an offspring of cross(es) between genetically different organisms. Hybrids may be treated as well as individuals having a mixed ancestry. In a certain sense, a heterozygote at one or more loci could be named a hybrid individual. However, normally the term hybrid is attributed to more distant cross, e.g., between different species.

Hybridization is a process through which hybrids occur.

Reproductive isolation between biological species in a pure sense means absence of any gene flow.

Hybrid zone is an area in nature where hybrids occur between supposed two or more parental forms.

Cline is a gradual or abrupt allele frequency change in a mode: Species 1 (deme 1) → hybrids → Species 2 (deme 2) and that is maintained by balance between dispersal and selection against hybrids.

Gametic disequilibrium refers to the nonrandom association of alleles between loci.

Polygenic trait is a trait that is controlled by several genes.

Heritability is the ratio of the genetic variance to the total or phenotypic variance.

Additive allele. Each locus may be occupied by either an **additive allele,** which add an impact to the phenotype, or by a **nonadditive allele,** which does not contribute cumulatively to the phenotype.

Clone is a unity of genotypically identical individuals.

Variance

Genetic variance is a component of variability at a QT that can be attributed to genotypic control. It has a complex nature.

Environmental variance is a component of variability at a QT that cannot be attributed to genotypic control.

Normal distribution is a distribution of density of probability with two parameters, which are **mathematical expectation** (μ) and **variance** (σ^2). Estimates of these parameters under sampling procedures are **mean** and **standard deviation**.

Genetic homeostasis can be defined as the property of the population to equilibrate its genetic composition and to resist sudden changes.

Heterosis. Hybrid vigour or increased "power" of F_1 offspring as compared to both parental forms.

Inbred depression hypothesis suggests that the loci itself are selectively neutral and they are only markers of genome heterozygosity.

Balance hypothesis or hypothesis of overdominance predicts that the loci itself or closely linked markers influence H_0 & QT association.

Nucleotide diversity is the average number of nucleotide differences per site between two sequences.
The entire DNA segment under investigation is a **nucleon** and a particular DNA sequence (or restriction site sequence) for the segment studied is a **haplotype** or a **nucleomorph**.

RFLP, Restriction Fragment Length Polymorphism is a technique based on the reaction of specific endonucleases to cut (restrict) DNA in a certain place and are used as molecular genetic markers.

RAPD, Random Amplified Polymorphic DNAs are molecular polymorphisms identified at random and used as molecular markers to detect variation and for **genetic mapping**.

AFLP, Amplified Fragment Length Polymorphism. AFLP is a variant of RAPD.

Silent polymorphism is DNA polymorphism that is not expressed at the amino acid level.

Phenogram is a rooted tree in which only phenotypic resemblance is accounted for primarily without reference to kinship.

Cladogram is a phylogenetic tree, which is constructed by considering genetic kinship and various possible ways of evolution, following certain rules and choosing the best possible tree.

A **nucleotide site** is **informative** only when there are at least two different kinds of nucleotides and one represented at least two times. The nucleotides that exist uniquely in an OTU (**singular nucleotides**) are not informative.

Tandem duplication can originate by three major mechanisms: (1) **unequal crossing over**, a recombination event between members of a homologous pair of chromosomes in which a DNA segment is duplicated in one of the recombination products; (2) **unequal sister chromatid exchange,** a recombination event similar to unequal crossing over that occurs between the two chromatids of a single chromosome; (3)

replication errors; during replication of a template molecule, slippage can cause the insertion of a short segment into the newly synthesized strand.

Gene Elongation is another way for gene change. Gene elongation is caused by the duplication of a gene or its part.

Multigene family. Family of homologous genes with related functions is called a multigene family, the members of which are often tandemly arranged within the chromosome.

Concerted evolution. The tendency of a homogenization or nonrandom substitution rate in gene families is known as concerted evolution. It is realized due to gene conversion and unequal crossover.

Transposons are several types of mobile elements of the genome, like SINEs, LINEs and ISs.

APPENDIX V

COMPLETE LIST OF REFERENCES

Afifi, A.A. and S.P. Azen. 1982. Statistical Analysis: A Computer Oriented Approach. Mir. Publ., Moscow, 488 p. (Russian Edition).

Air, J.M. 1981. Sequence relationships among of the hemagglutinin genes of 12 subtypes of influenza A virus. Proc. Nat. Acad. Sci. U.S.A. 78: 7639–7643.

Alikhanyan, S.I., A.P. Akifiev and L.S. Chernin. 1985. General Genetics. Vishaya Shkola, Moscow, 448 p.

Allendorf, F.W. 1977. Electromorphs or alleles? Genetics 87: 821–822.

Allendorf, F.W. and G.H. Thorgaard. 1984. Tetraploidy and the evolution of salmonid fishes. pp. 1–53. *In*: B. Turner (ed.). Evolutionary Genetics of Fishes. Plenum Press, N.Y.

Allendorf, F., F. Utter and N. Ryman. 1987. Genetics and fishery management: Past, present, and future. pp. 1–20. *In*: N. Ryman and F. Utter (eds.). Population Genetics and Fishery Management. Washington Sea Grant Programs, Univ. Washington Press, Seattle–London.

Aleoshin, V.V. 2013. Invertebrate phylogeny in light of molecular data: the prospect of phylogenetics as a finished scientific discipline. Issues of Zoological Institute of RAS. Supl. 2: 9–38.

Altukhov, Yu.P. 1974. Populyatsionnaya genetika ryb (Fish Population Genetics). Pishchevaya Promyshlennost, Moscow.

Altukhov, Yu.P. 1983. Geneticheskie protsessy v populyatsiyakh (Genetic Processes in Populations). 1st ed. Nauka Publ., Moscow.

Altukhov, Yu.P. 1989. Genetic Processes in Populations, 2nd ed. Nauka Publ., Moscow.

Altukhov, Yu.P. 1997. Species and speciation. Soros Obrazovat. Zh. (Soros' Educational J.) 4: 2–10.

Altukhov, Yu.P. and Y.G. Rychkov. 1970. Population systems and their structural components. Genetic stability and variability. Rus. J. General Biol. 31(5): 507–526.

Altukhov, Yu.P. and Y.G. Rychkov. 1972. Genetic monomorphism of species and its possible biological significance. Rus. J. General Biol. 33: 281–301.

Altukhov, Yu.P. and G.M. Livshits. 1978. Factors of integration and differentiation of gene pool in isolated population of mollusk Chondrus bidens. Reports Rus. Acad. Sci. 238(24): 955–958.

Altukhov, Yu.P. and N.V. Varnavskaya. 1983. Varnavskaya Adaptive genetic structure and its relation with intraspecies differentiation at sex, age, growth rate in pacific salmon *Oncorhynchus nerka* (Walb.). Rus. J. Genetics 19(5): 796–807.

Altukhov, Yu.P., A.I. Pudovkin, E.A. Salmenkova and S.M. Konovalov. 1975. Steady state of allele frequency distribution at genes lactate dehydrogenase and phosphoglucomutase at the system of subpopulations of fish stock (an example for Oncorhynchus nerka (Walb.)). 2. Random gene drift, migration, and selection as factors of steady state. Rus. J. Genetics 11(4): 54–62.

Altukhov, Yu.P., V.A. Dukharev and L.A. Zhyvotovsky. 1983. Selection against rare electrophoretic variants of proteins and rates of spontaneous mutagenesis in populations. Rus. J. Genetics 19(2): 264–276.

Altukhov, Yu.P. and E.A. Salmenkova. 1987. Transplantation, system organization and sustainable usage of natural populations of fish. pp. 333–342. *In*: N. Ryman and F. Utter (eds.). Population Genetics and Fishery Management. Washington University Press, Seattle–London.

Altukhov, Yu.P., E.A. Salmenkova and Y.P. Kartavtsev. 1991. Relationships of allozyme heterozygosity with viability and growth rate of pink salmon. Ukr. J. Cytology a. Genet. 25(1): 47–51.

Altukhov, Yu.P., E.A. Salmenkova and V.T. Omeltchenko. 1997. Population Genetics of Salmon Species. Nauka Publ., Moscow, 288 p.

Anderson, P.K. 1970. Ecological structure and gene flow in small mammals. Symp. Zool. Soc. London 26: 299–325.

Aquadro, C.F. and J.C. Avise. 1981. Genetic divergence between rodent species assessed by using two-dimensional electrophoresis. Proc. Nat. Acad. Sci. U.S.A. 78: 3784–3788.

Arnason, U., X. Xu, A. Gullberg and D. Graur. 1996. The "Phoca Standard": An external molecular reference for calibrating recent evolutionary divergences. J. Mol. Evol. 43: 41–45.

Arnold, M.L. and S.K. Emms. 1998. Paradigm lost: Natural hybridization and evolutionary innovation. pp. 379–389. *In*: D.J. Howard and S.H. Berlocher (eds.). Endless Forms: Species and Speciation. Oxford University Press, Oxford-N.Y.

Aronshtam, A.A., L.Y. Borkin and A.I. Pudovkin. 1977. Isozymes in Population and Evolutionary Genetics. pp. 199–249. *In*: Genetika Izofermentov (Genetics of Isozymes). Nauka Publ., Moscow.

Asmussen, M.A., J. Arnold and J.S. Avise. 1987. Definition and properties of disequilibrium statistics for associations between nuclear and cytoplasmic genotypes. Genetics 115: 755–768.

Atchley, W.R., J.J. Rutledge and D.E. Cowley. 1982. A multivariate statistical analysis of direct and correlated response to selection in the rat. Evolution 36: 677–698.

Avise, J.C. 1976. Genetic differentiation during speciation. pp. 106–122. *In*: F.J. Ayala (ed.). Molecular Evolution. Sinauer Associates, Inc. Sunderland, Massachusetts.

Avise, J.C. 1994. Molecular Markers, Natural History and Evolution. Chapman & Hall, New-York.

Avise, J.C. 2000. Phylogeography: The History and Formation of Species. Harvard University Press, Cambridge.

Avise, J.C. 2001. Cytonuclear genetic signatures of hybridization phenomena: Rationale utility and empirical examples from fishes and other aquatic animals. Rev. Fish Biol. Fisheries 10: 253–263.

Avise, J.C. and M.H. Smith. 1974. Biochemical Genetics of Sunfish. I. Geographic Variation and Subspecific Intergradation in the Bluegill, *Lepomis macrochirus*. Evolution 28: 42–56.

Avise, J.C. and C.F. Aquadro. 1982. A comparative summary of genetic distances in the vertebrates: Pattern and correlations. Evol. Biol. 15: 151–185.

Avise, J.C. and M.J. Van Den Avyle. 1984. Genetic analysis of reproduction of hybrid white bass striped bass in the Savannah River. Trans. Amer. Fish Soc. 113: 563–570.

Avise, J.C. and N.C. Saunders. 1984. Hybridization and introgression among species of sunfish (*Lepomis*): Analysis by mitochondrial DNA and allozyme markers. Genetics 108: 237–250.

Avise, J.C. and K. Wollenberg. 1997. Phylogenetics and origin of species. Proc. Nat. Acad. Sci. U.S.A. 94: 7748–7755.

Avise, J.C. and D. Walker. 1999. Species realities and numbers in sexual vertebrates: Perspectives from an asexually transmitted genome. Evolution 9(3): 992–995.

Avise, J.C., J.C. Patton and C.F. Aquadro. 1980. Evolutionary genetics of birds. III. Comparative molecular evolution in New World warblers and rodents. Heredity 71: 303–310.

Avise, J.C., J.F. Shapira, S.W. Daniel, C.F. Aquadro and R.A.Q. Lansman. 1983. Mitochondrial DNA differentiation during the speciation process in *Peromyscus*. Mol. Biol. Evol. 1: 38–56.

Avise, J.C., E. Bermingham, L.G. Kessler and N.C. Saunders. 1984. Characterization of mitochondrial DNA variability in a hybrid swam between subspecies bluegill sunfish (*Lepomis macrochirus*). Evolution 38: 931–941.

Awramik, S.M., J.W. Schopf and M.R. Walter. 1983. Filamentous fossil bacteria from the Arhean Western Australia. Precambr. Res. 20: 357–374.

Ayala, F.J. 1975. Scientific hypotheses, natural selection and neutrality theory of protein evolution. pp. 19–42. *In*: F.M. Salzano (ed.). The Role of Natural Selection in Human Evolution. North-Holland Publ. Co. Amsterdam.

Ayala, F.J. 1984. Introduction in Population Genetics. Mir. Publ., Moscow, 232 p.

Ayala, F.J., M.L. Tracey, D. Hedgecock and R.C. Richmond. 1974. Richmond Genetic differentiation during the speciation process in *Drosophila*. Evolution 28: 576–592.

Bacterial Database of ISs: http://www-is.biotoul.fr/is.html.

Bailey, N. 1970. Mathematics in Biology and Medicine. Mir. Publ., Moscow, 326 p.

Baker, C.S., A. Perry, G.K. Chambers and P.J. Smith. 1995. Population variation in the mitochondrial cytochrome-b gene of the Orange Roughy Hoplostethus atlanticus and the Hoki Macruronus novaezelandiae. Marine Biol. 122(4): 503–509.

Baker, R.L., A.B. Chapman and R.T. Wardell. 1975. Direct response to selection for post weaning weight gain in the rat. Genetics 80: 171–189.

Balakirev, E.S. and F.J. Ayala. 2003. Pseudogenes: Are they "junk" or functional DNA? Annual Rev. Genetics 37: 123–151.
Balloux, F., W. Amos and T. Coulson. 2004. Does heterozygosity estimate inbreeding in real populations? Mol. Ecol. 13: 3021–3031.
Bams, R.A. 1976. Survival and propensity for homing as affected by presence or absence of locally adapted paternal genes in two transplanted populations of pink salmon (*Oncorhynchus gorbuscha*). J. Fish. Res. Bd. Can. 33: 2716–2725.
Barghoorn, E.S. and J.W. Schopf. 1966. Microorganisms three billion years old from the Precambrian of South Africa. Science 152: 758–763.
Barns, M.R. 2003. Predictive functional analysis of polymorphisms: An overview. pp. 249–271. *In*: M.R. Barnes and I.C. Gray (eds.). Bioinformatics for Geneticists. Wiley, Chichester.
Barton, N.H. 1979. The dynamics of hybrid zones. Heredity, London 43: 341–359.
Barton, N.H. 1983. Multilocus clines. Evolution 37: 454–471.
Barton, N.H. and G.M. 1985. Hewitt Analysis of hybrid zones. Ann. Rev. Ecol. Syst. 16: 113–148.
Beacham, T.D. and R.E. 1985. Withler Heterozygosity and morphological variability of the pink salmon (*Oncorhynchus gorbuscha*) from the Southern British Columbia and Puget Sound. Can. J. Genet. Cytol. 27: 571–579.
Beaumont, A.R., E.M. Gosling, C.M. Beveridge, M.D. Budd and G.M. Burnell. 1985. Studies of heterozygosity and size in the scallop Pecten maximus (L.). pp. 443–455. *In*: R.E. Gibbs (ed.). Proc. 19-th European Mar. Biol. Symp., Cambridge University Press, Cambridge, U.K.
Beet, E.A. 1949. The genetics of sickle cell trait in a bantus tribe. Ann. Eugenics 14: 279–284.
Begon, M. 1977. The effective size of natural *Drosophila subobscura* populations. Heredity 38: 13–18.
Beland, K.F., F.L. Roberts, R.L. Sanders. 1981. Evidence of *Salmo salar* × *Salmo trutta* hybridization in a North American river. Can. J. Fish. Aquat. Sci. 38: 552–554.
Bell, J.I., M.J. Selby and W.J. Butter. 1982. The highly polymorphic region near the human insulin gene is composed of simple randomly repeated sequences. Nature 295: 31–35.
Bennet, J.H. 1954. On the theory of random mating. Annals of Eugenics 11: 311–317.
Benveniste, R.E. 1985. The contribution of retroviruses to the study of mammalian evolution. pp. 359–417. *In*: R.J. McIntire (ed.). Molecular Evolutionary Genetics. Plenum, N.Y.
Bertsch, A., H. Schweer and H. Tanaka. 2005. Male Labial Gland Secretions and Mitochondrial DNA Markers Support Species Status of *Bombus criptarum* and *B. magnus* (Hymenoptera, Apidae). Insect. Soc. 52: 45–54.
Beverley, S.M. and A.C. Wilson. 1984. Molecular evolution in Drosophila and higher Diptera. II. A time scale for the evolution. J. Mol. Evol. 21: 1–13.
Beverley, S.M. and A.C. Wilson. 1985. Ancient origin of Hawaiian Drosophilinae inferred from protein comparison. Proc. Nat. Acad. Sci. U.S.A. 82: 4753–4757.
Bezrukov, V.F. 1989. Algebraic regularities of the relationships of heterozygosity with mean and variance of quantitative trait values. Rus. J. Genetics (Moscow) 25(7): 1310–1319.
Billington, N. and R.M. 1995. Strange Mitochondrial DNA analysis confirms the existence of a genetically divergent Walleye population in Northeastern Mississippi. Trans. Am. Fish. Soc. 124(5): 770–776.
Blake, C.C.F. 1985. Exons and the evolution of proteins. Intl. Rev. Cytol. 93: 149–185.
Blanken, R.L., L.C. Klotz and A.G. Hinnebusch. 1982. Computer comparison of new and existing criteria for constructing evolutionary trees from sequence data. J. Mol. Evol. 19: 9–19.
Borkin, L.Y. and I.S. Darevsky. 1980. Network (hybrid) speciation. Rus. J. General Biol. 41(4): 485–505.
Bodmer, M. and M. Ashburner. 1984. Conservation and change in the DNA sequences coding for alcohol dehydrogenase in sibling species of *Drosophila*. Nature 309: 425–430.
Brannon, E.L. 1982. Orientation mechanisms of homing salmonids. pp. 219–227. *In*: E.L. Brannon and E.O. Salo (eds.). Proceedings of the Salmon and Trout Migratory Behavior Symposium. School Fish., University of Washington, Seattle.
Brenner, S., F. Jacob and M. Meselson. 1961. An unstable intermediate carrying information from genes to ribosomes for protein synthesis. Nature 190: 575–580.
Brereton, J. 1962. Evolved regulatory mechanisms of population control. pp. 81–93. *In*: G.W. Leeper (ed.). Evolution of Living Organisms. Melbourne Univ. Press., Melbourne, Australia.
Bridges, C.B. 1936. Genes and chromosomes. Teaching Biol. 11: 17–23.
Britten, R.J. and D.E. Kohne. 1968. Repeated Sequences in DNA. Science 161(3841): 529–540.
Brower, A.V.Z. 1999. Delimitation of phylogenetic species with DNA sequences: A critique of Davis and Nixon's population aggregation analysis. Systematic Biology 48: 199–213.

Brown, W.M. 1983. Evolution of animal mitochondrial DNA. pp. 62–88. *In*: M. Nei and R.K. Koehn (eds.). Evolution of Genes and Proteins. Sinnauer Assoc., Sunderland, Mass.

Brown, W.M., M. George and A.C. Wilson. 1979. Rapid evolution of animal mitochondrial DNA. Proc. Nat. Acad. Sci. U.S.A. 76: 1967–1971.

Brown, W.M., E.M. Pragner, A. Wang and A.C. Wilson. 1982. Mitochondrial DNA sequences of primates: Tempo and mode of evolution. J. Mol. Evol. 18: 225–239.

Brown, C.J., C.F. Aquadro and W.W. Anderson. 1990. DNA sequence evolution of the amylase multigene family in Drosophila pseudoobscura. Genetics 126: 131–138.

Buri, P. 1956. Gene frequency in small populations of mutant *Drosophila*. Evolution 10: 367–402.

Buroker, N.E. 1984. Gene flow in mainland and insular populations of *Crassostrea* (Mollusca). Biol. Bull. 166: 550–557.

Busack, C.A. and G.A.E. Gall. 1981. Introgressive hybridization in populations of Paiute cutthroat trout (*Salmo clarki seleniris*). Can. J. Fish. Aquat. Sci. 38: 939–951.

Busack, C.A., J.H. Torgaard, M.P. Bannon and G.A.E. Gall. 1980. An elelctrophoretic karyotypic and meristic characterization of the Eagle Lake trout, *Salmo gairdnery aquilarum*. Copeia 3: 418–424.

Bush, G.L. 1975. Modes of animal speciation. Ann. Rev. Ecol. Syst. 6: 339–364.

Bush, R.M., P.E. Smouse and F.T. 1987. Ledig The fitness consequences of multiple–locus heterozygosity: the relationships between heterozygosity and growth rate in pitch pine (*Pinus rigida* Mill.). Evolution 41: 787–798.

Butlin, R. 1998. What do hybrid zones in general and the *Chortippus paralellus* zone in particular, tell us about speciation. pp. 367–378. *In*: D.J. Howard and S.H. Berlocher (eds.). Endless Forms: Species and Speciation. Oxford University Press, Oxford – N.Y.

Bucklin, A. and P.H. Wiebe. 1998. Low mitochondrial diversity and small effective population sizes of the copepods Calanus finmarchicus and Nannocalanus minor: Possible impact of climatic variation during recent glaciation. J. Hered. 89(5): 383–392.

Calder, N. 1983. Timescale: An Atlas of the Fourth Dimension. Viking Press, New York.

Campton, D.E. 1987. Natural hybridization and introgression in fishes. Method of detection and genetic interpretation. pp. 161–192. *In*: N. Ryman and F. Utter (eds.). Population Genetics & Fishery Management. Washington Sea Grant Program. Distributed by University of Washington Press, Seattle and London.

Campton, D.E. and F.M. Utter. 1985. Natural hybridization between steelhead trout (*Salmo gairdneri*) and coastal cutthroat trout (*Salmo clarki clarki*) in two Puget Sound streams. Can. J. Fish. Aquat. Sci. 42(1): 110–119.

Campton, D.E. and J.M. Johnston. 1985. Electrophoretic evidence for a genetic admixture of native and non-native trout in Yakima River, Washington. Transaction of Amer. Fish. Soc. 114: 782–793.

Cann, R.L. 1982. The evolution of human mitochondrial DNA. Ph.D. Thesis. Berkeley, Univ. California.

Cann, R.L., W.M. Brown and A.C. Wilson. 1982. Evolution of human mitochondrial DNA: A preliminary report. pp. 157–165. *In*: B. Bonne-Tamir (ed.). Human Genetics. Part A: The Unfolding Genome. Alan R. Liss, N.Y.

Cannon, W.B. 1932. The Wisdom of the Body. Norton, N.Y., 312 p.

Carbon, J. 1984. Yeast centromeres: Structure and function. Cell 37: 352–353.

Carson, H.L. 1975. Genetics of speciation at the diploid level. Amer. Nat. 109: 83–92.

Cavalli-Sforza, L.L. and A.W.F. Edwards. 1967. Phylogenetic analysis. Models and estimation procedures. Amer J. Hum. Genet. 19: 233–257.

Cavalli-Sforza, L.L. and W.F. Bodmer. 1971. The Genetics of Human Populations. W.H. Freeman and Co., San Francisco, 965 p.

Cavalli-Sforza, L.L., J. Barrai and A.W.F. Edwards. 1964. Analysis of human evolution under random genetic drift. Cold Spring Harbor Sym. P. Quant. Biol. 29: 9–20.

Cavender, J.A. 1981. Test of phylogenetic hypotheses under generalized model. Math. Biosci. 54: 217–229.

Chakraborty, R. 1977. Estimation of time of divergence from phylogenetic studies. Can. J. Genet. Cytol. 19: 217–223.

Chakraborty, R. 1981. The distribution of the number of heterozygous loci in an individual in natural populations. Genetics 98: 461–466.

Chakraborty, R. and N. Ryman. 1983. Relationships of mean and variance of genotypic values with heterozygosity per individual in a natural population. Genetics 103: 149–152.

Chakraborty, R. and O. Leimar. 1987. Genetic variation within a subdivided population. pp. 89–120. *In*: N. Ryman and F. Utter (eds.). Population Genetics & Fishery Management. Washington Sea Grant Program. Distributed by University of Washington Press, Seattle and London.

Chakraborty, R., P.A. Fuerst and M. Nei. 1978. Statistical studies on protein polymorphism in natural populations. II. Gene differentiation between populations. Genetics 88: 367–390.

Chakraborty, R., M. Haag, N. Ryman and G. Stahl. 1982. Hierarchical gene diversity analysis and its application to brown trout population data. Hereditas 97: 17–21.

Chakravarti, A., K.H. Buetow, S.E. Antonarakis, P.G. Waber, C.D. Boehm and H.H. Kazazian. 1984. Nonuniform recombination within a human β-globin gene cluster. Amer. J. Hum. Genet. 36: 1239–1258.

Champion, A.B., E.M. Prager, D. Wachter and A.C. Wilson. 1974. Microcomplement fixation. pp. 397–416. *In*: C.A. Wright (ed.). Biochemical and Immunological Taxonomy of Animals. Academic Press, London.

Chapman, B.S., K.A. Vincent and A.C. Wilson. 1986. Persistence or rapid generation of DNA length polymorphism at the zeta locus of humans. Genetics 112: 79–92.

Chapman, J.R., S. Nakagawa, D.W. Coltman, J. Slates and B.C. Sheldon. 2009. A quantitative review of heterozygosity-fitness correlations in animal populations. Molecular Ecology 18: 2746–2763.

Chichvarkhin, A.Yu., Y.P. Kartavtsev and A.I. Kafanov. 2000. Genetic relationships among some species of mussels *Mytilidae (Mollusca: Bivalvia)* of North Pacific Ocean. Russian J. Genetics 36(9): 1206–1220 (In Russian, Translated in English).

Chow, S. and Y. Fujio. 1987. Comparison of intraspecific genetic diversity levels among local populations in Decapod Crustacean species; with some references of phenotypic diversity. Nippon Suisan Gakkaishi 53(5): 691–693.

Chromas-pro. 2007. http://www.flu.org.cn/en; http://www.technelysium.com.au/chromas.html.

Clark, A.G. 1984. Natural selection with nuclear and cytoplasmic transmission: I. A deterministic model. Genetics 107: 679–701.

Cockerham, C.C. and B.S. Weir. 1977. Digenic descent measures for finite populations. Genetic Res. 30: 121–127.

Coen, E.S., J.M. Thoday and G. Dover. 1982. Rate of turnover of structural variants in the rDNA gene family of *Drosophila melanogaster.* Nature 295: 564–568.

Cohn, V., M. Thompson and G. Moore. 1984. Nucleotide sequence comparison of the *Adh* gene in three drosophilids. J. Mol. Evol. 2: 31–37.

Collier, G.E. and S.J. O'Brien. 1985. A molecular phylogeny of the Felidae: Immunological distance. Evolution 39: 473–487.

Coltman, D.W. and J. Slate. 2003. Microsatellite measures of inbreeding: a meta-analysis. Evolution 57: 971–983.

Comesana, A.S., J.E. Toro, D.J. Innes and R.J. Thompson. 1999. A molecular approach in the ecology of a mussel (*Mytilus edulis–M. trossulus*) hybrid zone on the east coast of Newfoundland, Canada. Mar. Biol. 133: 213–221.

Coyne, J.A. 1982. Gel electrophoresis and cryptic protein variation. Isozymes 5: 1–32.

Coyne, J.A. 1992. Genetics and speciation. Nature 355: 511–515.

Cracraft, J. 1983. Species concepts and speciation analysis. Current Ornithology 1: 159–187.

Craig, N., R. Craigie, M. Gellert and A. Lambovitch (eds.). 2002. Mobile DNA II. ASM Press, Washington DC.

Creer, S., A. Malhotra and R.S. Thorpe. 2003. Assessing the phylogenetic utility of four mitochondrial genes and a nuclear intron in the Asian pit viper genus, *Trimeresurus*: Separate, simultaneous, and conditional data combination analyses. Mol. Biol. Evol. 20(8): 1240–1251.

Crick, F.H.S. 1958. On protein synthesis. Symposia of the Society for Experimental Biology 12: 138–163. editors names missing?

Crick, F.H.S. 1966. Codon-anticodon paring: The wobble hypothesis. J. Mol. Biol. 19: 548–555.

Crow, J.F. 1954. Breeding structure of populations. 2. Effective population number. pp. 543–556. *In*: Gowen J.W. and J.L. Lush (eds.). Statistics and Mathematics in Biology. Iowa State College Press, Ames.

Crow, J.F. and M. Kimura. 1970. An Introduction to Population Genetics Theory. Harper & Row, N.Y.

Danzman, R.G., M.M. Ferguson and F.W. Allendorf. 1989. Genetic variability and components of fitness in hatchery strains of rainbow trout. J. Fish Biol. 35(Suppl. A): 313–319.

Darnell, D.W. and K.M. Klotz. 1975. Subunit constitution of proteins: A table. Arch. Biochem and Biophys. 166Z: 651–682.

Darnell, J.E. 1978. Implication of RNA–RNA splicing in evolution of eukaryotic cells. Science 202: 1257–1260.
Davis, J.I. and K.C. Nixon. 1992. Populations, genetic-variation, and the delimitation of phylogenetic species. Systematic Biology 41: 421–435.
David, P., B. Delay and P. Berthou. 1995. Jarne Alternative models for allozyme-associated heterosis in the marine bivalve Spisula ovalis. Genetics 139: 1719–1726.
David, P. 1997. Modeling the genetic basis of heterosis: tests of alternative hypothesis. Evolution 51(4): 1049–1051.
Dayhoff, M.O. (ed.). 1969. Atlas of Protein Sequence and Structure. National Biomedical Research Foundation, Silver Springs, Md., 4.
Dayhoff, M.O. 1972. Atlas of Protein Sequence and Structure. National Biomedical Research Foundation, Silver Springs, Md., 5.
Dayhoff, M.O. and C.M. Park. 1969. Cytochrome c: Building a phylogenetic tree. pp. 7–16. *In*: M.O. Dayhoff (ed.). Atlas of Protein Sequence and Structure. National Biomedical Research Foundation, Silver Springs, Md., V. 4.
Dayhoff, M.O. and W.S. Barker. 1972. Mechanisms in molecular evolution: Examples. *In*: M.O. Dayhoff (ed.). Atlas of Protein Sequence and Structure. National Biomedical Research Foundation, Silver Springs, Md. 5: 41–45.
Dayhoff, M.O. 1978. Survey of new data and computer methods of analysis. *In*: M.O. Dayhoff (ed.). Atlas of Protein Sequence and Structure. National Biomedical Research Foundation, Silver Springs, Md. 5(2): 2–8.
Delaney, M.E. and S.E. Bloom. 1984. Replication banding patterns in the chromosomes of rainbow trout. Journal of Heredity 75: 431–434.
Deng, H.W. and Y.X. Fu. 2000. Counting mutations by parsimony and estimation of mutation rate variation across nucleotide sites—A simulation study. Mathematical and Computer Modelling 32(1-2): 83–95.
De Qeiros, K. 1998. The general lineage concept of species, species criteria and process of speciation: A conceptual unification and terminological recommendations. pp. 57–75. *In*: D.J. Howard and S.H. Berlocher (eds.). Endless Forms: Species and Speciation. Oxford University Press, Oxford–N.Y.
Dickerson, R.E. 1971. The structure of cytochrome c and the rates of chromosome evolution. J. Mol. Evol. 1: 26–45.
Diehl, W.J. and R.K. Koehn. 1985. Multiple locus heterozygosity, mortality, and growth in a cohort of *Mytilus edulis*. Marine Biol. 88: 265–271.
Diehl, W.J., P.M. Gaffney and R.K. Koehn. 1986. Physiological and genetical aspects of growth in the mussel *Mytilus edulis*. 1. Oxygen consumption, growth, and weight loss. Physiol. Zool. 59(2): 201–211.
Dobzhansky, T. 1937. Genetics and the Origin of Species. Columbia Univ. Press, N.Y.
Dobzhansky, T. 1943. Genetics of natural populations. 9. Temporal changes in the composition of populations of *Drosophila pseudoobscura*. Genetics (US) 28: 162–186.
Dobzhansky, T. 1951. Genetics and the Origin of Species, 3rd Edn. Columbia Univ. Press, New York.
Dobzhansky, Th. 1955. Evolution, Genetics and Man. John Wiley & Sons, Inc.–N.Y., Chapman & Hall, Limited – London, 398 p.
Dobzhansky, Th. 1970. Genetics of the Evolutionary Process. Columbia Univ. Press, N.Y., 505 p.
Dobzhansky, Th. 1972. Species of *Drosophila*. Science 177: 664–669.
Donaldson, L.R. and G.H. Allen. 1958. Return of silver salmon, *Oncorhynchus kisutch* (Walbaum) to point of release. Trans. Amer. Fish. Soc. 87: 13–22.
Doolittle, R.F. 1985. The genealogy of some recently evolved vertebrate proteins. Trends in Biochem. Sci. 10: 233–237.
Doolittle, W.F. 1978. Genes in pieces: Were they ever together? Nature 272: 581–582.
Dowling, T.E. and W.S. Moore. 1984. Level of reproductive isolation between two cyprinid fishes, *Notropis cornutus* and *N. crysocephalus*. Copeia: 617–628.
Dowling, T.E. and W.S. Moore. 1985. Evidence for selection against hybrids in the family Cyprinidae (genus *Notropis*). Evolution 39: 152–158.
Dowling, T.E. and W.M. Brown. 1993. Population structure of the bottle-nosed dolphin (*Tursiops truncatus*) as determined by restriction endonuclease analysis of mitochondrial DNA. Marine Mamm. Sci. 9(2): 138–155.
Drake, J.W. 1966. Spontaneous mutations accumulating in bacteriophage T4 in the complete absence of DNA replication. Proc. Nat. Acad. Sci. U.S.A. 55: 738–743.

Drozdov, A.L. and O.G. Kussakin. 2002. State of the art and problems of megasystematics. pp. 18–29. *In*: A.P. Kryukov and L.V. Yakimenko (eds.). Problems of Evolution. Dalnauka, Vladivostok.
Dubrova, Yu.E., T.M. Karafet, R.I. Sukernik and T.V. Gol'tsova. 1990. Analysis of the relationships between heterozygosity and fertility parameters in forest Nenets and Nganasans. Rus. J. Genetics (Moscow) 26(1): 122–129.
Dubrova, Yu.E., E.A. Salmenkova, Yu.P. Altukhov, Y.P. Kartavtsev et al. 1994. The influence of parental heterozygosity on interfamily variation of the progeny body length in Pink Salmon. Russ. J. Genet. (Moscow) 30(3): 365–371.
Dubrova, Y.E., E.A. Salmenkova, Y.P. Altukhov, Y.P. Kartavtsev, E.V. Kalkova and V.T. Omeltchenko. 1995. Family heterozygosity and progeny body length in pink salmon *Oncorhynchus gorbuscha* (Walbaum). Heredity 75: 281–289.
Duftner, N. and S. Koblmuller, C. Sturmbauer. 2005. Evolutionary relationships of the Limnochromini, a tribe of benthic deepwater Cichlid fish endemic to Lake Tanganyika, East Africa. J. Mol. Evol. 60(3): 277–289.
Durand, P. and F. Blanc. 1986. Divergence genetique chezun bivalve marine tropical: *Pinctada margaritifera*. Coll. Nat. CNRS "Biologie des Populations", Lyon: 323–330.
East, E.M. 1936. Heterosis. Genetics 21: 376–397.
Eck, R.V. and M.O. Dayhoff. 1966. Atlas of Protein Sequence and Structure. National Biomedical Research Foundation Natl. Biomed. Res. Found, Silver Springs, Md.
Edgar, R.C. Muscle. 2004. Multiple sequence alignment with high accuracy and high throughput. Nucleic Acids Res. 32: 1793–1797.
Edgel, M.H., S.C. Hardies, B. Brown, C. Voliva, A. Hill, S. Phillips, M. Comer, F. Burton, S. Weaver and C.A. Hutchison III. 1983. Evolution of the mouse β-globin complex locus. pp. 1–13. *In*: M. Nei and R.K. Koehn (eds.). Evolution of Genes and Proteins. Sinauer Assoc., Sunderland, Mass.
Ehrlich, P.R. 1965. The population biology of the butterfly, Euphydryas editha. 2. Structure of the Jasper Ridge colony. Evolution 19: 327–336.
Eldredge, N. and J. Cracraft. 1980. Phylogenetic Patterns and the Evolutionary Process. Columbia Univ. Press, N.Y.
Elston, R.C. 1971. The estimation of admixture in racial hybrids. Annals of Human Genetics 35: 9–17.
Engels, W.R. 1981. Estimating genetic divergence and genetic variability with restriction endonucleases. Proc. Nat. Acad. Sci. U.S.A. 78: 6329–6333.
Faith, J.P. 1985. Distance methods and the approximation of most parsimonious trees. Syst. Zool. 34: 312–325.
Farris, J.S. 1972. Estimating phylogenetic trees from distance matrices. American Naturalist 106: 645–668.
Farris, J.S. 1977. On the phenetic approach to vertebrate classification. pp. 823–850. *In*: M.K. Hecht, P.C. Goody and B.M. Hecht (eds.). Major Patterns of Vertebrate Evolution. Plenum Press, New York.
Felsenstein, J. 1973. Maximum likelihood and minimum-steps methods for estimating evolutionary trees from data on discrete characters. Syst. Zool. 22: 240–249.
Felsenstein, J. 1981. Evolutionary trees from DNA sequences: A maximum likelihood approach. J. Mol. Evol. 17: 368–376.
Felsenstein, J. 1982. Numerical methods for inferring evolutionary trees. Quart. Rev. Biol. 57: 379–404.
Felsenstein, J. 1985. Confidence limits on phylogenies with a molecular clock. Syst. Zool. 34: 152–161.
Felsenstein, J. 1995. PHYLIP (Phylogeny Inference Package) Version 3.57c. University Washington.
Felsenstein, J. 2004. Inferring phylogenies. Sinauer Associates Inc., Sanderland, Massachusetts, 664 p.
Ferguson, J.W.H. 2002. On the use of genetic divergence for identifying species. Biol. J. Linn. Soc. 75(4): 509–516.
Ferris, S.D. and G.S. Whitt. 1979. Evolution of the differential regulation of duplicate genes after polyploidization. J. Mol. Evol. 12(3): 267–317.
Ferris, S.D., R.D. Sage, C.-M. Huang, J.T. Nielsen, U. Ritte and A.C. Wilson. 1983. Flow of mitochondrial DNA across a species boundary. Proc. National Acad. Sci. U.S.A. 80: 2290–2294.
Fitch, W.M. 1967. Evidence suggesting a non-random character to nucleotide replacement in naturally occurring mutations. J. Mol. Biol. 26: 499–507.
Fitch, W.M. 1971a. Toward defining the course of evolution: Minimum change for a species tree topology. Syst. Zool. 20: 406–416.
Fitch, W.M. 1971b. Evolution of clupeine Z, a probable crossover product. Nature New Biol. 229: 245–247.
Fitch, W.M. 1977. On the problem of discovering of most parsimonious tree. Amer. Natur. 3: 223–257.
Fitch, W.M. and E. Margoliash. 1967a. Construction of phylogenetic trees. Science 155: 279–284.

Fitch, W.M. and E. Margoliash. 1967b. A method for estimating the number of invariant amino acid coding positions in a gene using cytochrome c as a model case. Biochem. Genet. 1: 65–71.

Fitch, W.M., J.M.E. Leiter, X. Li and P. Palese. 1991. Positive Darwinian evolution in human influenza A viruses. Proc. National Acad. Sci. U.S.A. 88: 4270–4274.

Fox, G.E., L.J. Magrum, W.E. Balch, R.S. Wolf and C.R. Woese. 1977. Classification of methanogenic bacteria by16S ribosomal RNA characterization. Proc. Nat. Acad. Sci. U.S.A. 74: 4537–4541.

Fox, L.R. and P.A. Morrow. 1981. Speciation: species property or local phenomen phenomenon? Science 211: 887–893.

Fox, S. and W. Dose. 1975. Molecular Evolution and Origin of Life. Mir. Publ., Moscow, 374 p.

Frolov, S.V. 2000. Variability and Evolution of Karyotype of Salmon Fish. Dalnauka Publ., Vladivostok, 229 p.

Gaffney, P.M. 1990. Enzyme heterozygosity, growth rate, and viability in *Mytilus edulis*: another look. Evolution 44(1): 204–210.

Gaffney, P.M., T.M. Scott, R.K. Koehn and W.J. Diehl. 1990. Interrelationships of heterozygosity, growth rate and heterozygote deficiencies in the coot clam, *Mulinia lateralis*. Genetics 124(3): 687–699.

Galas, D.J. and M. Chandler. 1989. Bacterial insertion sequences. pp. 109–162. *In*: D.E. Berg and M.M. Howe (eds.). Mobile DNA. American Society for Microbiology, Washington, D.C.

Garcia-Machado, E., P.P. Chevalier Monteagudo and M. Solignac. 2004. Lack of mtDNA Differentiation among Hamlets (*Hypoplectrus*, Serranidae). Marine Biol. 144: 147–152.

Gardner, J.P.H. and D.O.F. Skibinski. 1988. Historical and size-dependent genetic variation in hybrid mussel populations. Heredity 61: 93–105.

Garton, D.W. 1984. Relationship between multiple locus heterozygosity and physiological energetics of growth in estuarine gastropod *Thais hemastoma*. Physiol. Zool. 57(5): 530–543.

Garton, D.W., R.K. Koehn and T.M. Scott. 1984. Multiple locus heterozygosity and physiological energetics of growth of the coot clam, *Mulinia lateralis*, from a natural population. Genetics 108(2): 445–455.

Geist, V. 1971. Mountain Sheep, a Study in Behavior and Evolution. University Chicago Press, Chicago. 383 p.

GenBank. 2006. http://www.ncbi.nlm.nih.gov.

Gentini, M.R. and A.R. Beaumont. 1988. Environmental stress, heterozygosity and growth rate in *Mytilus edulis*. J. Exper. Mar. Biol. Ecol. 120: 145–153.

Gerber, A.S., C.A. Tibbets and T.E. Dowling. 2001. The role of introgressive hybridization in the evolution of the Gila robusta complex (Teleostei: Cyprinidae). Evolution 55(10): 2028–2039.

Gilbert, W. 1978. Why genes in pieces? Nature 271: 501.

Gillespie, J.H. 1989. Lineage effects and the index of dispersion of molecular evolution. Mol. Biol. Evol. 6: 636–647.

Glass, B. 1955. On the unlikelihood of significant admixture of genes from the North American Indians in the present composition of the Negroes of the United States. Amer. J. Hum. Genet. 7: 368–385.

Glass, B. and C.C. Li. 1953. The dynamics of racial intermixture—an analysis based on American Negro. Amer. J. Hum. Genet. 5: 1–20.

Go, M. 1981. Correlation of DNA exonic regions with protein structural units in hemoglobin. Nature 291: 90–92.

Gojobori, T. and S. Yokoyama. 1985. Rates of evolution of retroviral oncogene of Moloney murine sarcoma virus and of its cellular homologues. Proc. Nat. Acad. Sci. U.S.A. 82: 4198–4201.

Gojobori, T., W.-H. Li and D. Graur. 1982. Patterns of nucleotide substitution in pseudogenes and functional genes. J. Mol. Evol. 18: 360–369.

Goldman, D., P.R. Giri and S.J. O'Brien. 1987. A molecular phylogeny of the hominid primates as indicated by two-dimensional protein electrophoresis. Proc. Nat. Acad. Sci. U.S.A. 84: 3307–3311.

Gonzalez-Willasenor, L.I. and D.A. Powers. 1990. Mitochondrial DNA restriction site polymorphisms in the teleost *Fundulus heteroclitus* support secondary intergradation. Evolution 44: 27–37.

Goodman, M., G.W. Moore and G. Matsuda. 1975. Darwinian evolution in the genealogy of hemoglobin. Nature 253: 603–608.

Goodman, M., A.E. Romerro-Herrara, H. Dene, J. Czelusniak and R.E. Tashian. 1982. Amino acid sequence evidence on the phylogeny of primates and other eutherians. pp. 115–191. *In*: M. Goodman (ed.). Macromolecular Sequences in Systematic and Evolutionary Biology. Plenum Press, N.Y.

Gosling, E.M. 1989. Genetic heterozygosity and growth rate in a cohort of *Mytilus edulis* from the Irish coast. Marine Biol. 100: 211–215.

Grant, V. 1985. The Evolutionary Process: A Critical Review of Evolutionary Theory. Columbia Univ. Press, N.Y.

Graur, D. and W.H. Li. 1999. Fundamentals of Molecular Evolution. Sinauer Ass., Sunderland.

Green, R.H., S.M. Sing, B. Hicks and J.M. McCuaig. 1983. An arctic intertidal population of *Macoma baltica* (Mollusca pelecypoda): genotypic and phenotypic components of population Structure. Can. J. Fish. Aquat. Sci. 40(9): 1360–1371.

Greenfield, D.W. and T. Greenfield. 1972. Introgressive hybridization between *Gila orcutti* and *Hesperoleucus symmetricus* (Pisces: Cyprinidae) in the Guyama River Basin, California. I. Meristics, morphometrics, and breeding. Copeia: 849–859.

Greenfield, D.W., F. Abdel-Hameed, G.D. Deckert and R.R. Finn. 1973. Hybridization between *Chrosomus erythrogaster* and *Notropis cornutus* (Pisces: Cyprinidae). Copeia: 54–60.

Greenwood, J.J.D. 1975. Effective population number in the snail *Cepaea nemoralis*. Evolution 28: 513–526.

Greenwood, J.J.D. 1976. Effective population number in *Cepaea*: a modification. Ibid 30: 186.

Hahn, M.W., J.G. Mesey, D.J. Begun, J.H. Gillespie, A.D. Kern, C.H. Langley and L.C. Moyle. 2005. Evolutionary genomics: Codon bias and selection on single genomes. Nature 433: E5–E6 (brief communication 10.1038/nature03221).

Hall, B. 2001. Phylogenetic Trees Made Easy: A How-To Manual for Molecular Biologists, 1st ed. Synauer Assoc. Inc., Sunderland, Massachusetts, 179 p.

Hall, B. 2011. Phylogenetic Trees Made Easy: A How-To Manual for Molecular Biologists, 4th ed. Sinauer Assoc. Inc., Sunderland, Massachusetts, 282 p.

Haldane, J.B.S. 1932. The causes of evolution. London: Longmans and Green.

Hansson, B. and L. Westerberg. 2002. On the correlation between heterozygosity and fitness in natural populations. Molecular Ecology 11: 2467–2474.

Harris, H., D.A. Hopkinson and E.B. Robson. 1974. The incidence of rare alleles determining electrophoretic variants: Data on 43 enzyme loci in man. Ann. Hum. Genet. 37: 237–253.

Harris, H. and D.A. Hopkinson. 1976. Handbook of Enzyme Electrophoresis in Human Genetics. North Holland Publ. Co., Amsterdam.

Harrison, R.G. 1998. Linking evolutionary pattern and process: The relevance of species concepts for the study of speciation. pp. 19–31. *In*: D.J. Howard and S.H. Berlocher (eds.). Endless Forms: Species and Speciation. Oxford University Press, Oxford–N.Y.

Hartl, D.L. 1980. Genetic dissection of segregation distortion. III. Unequal recovery of reciprocal recombinants. Genetics 96: 986–996.

Hartl, D.L. and A.G. Clarke. 1989. Principles of Population Genetics. Sinauer Assoc. Inc. Publishers, Sunderland, Massachusetts (Second edition).

Hartl, D.L. and A.G. Clarke. 1997. Principles of Population Genetics. Sinauer Assoc. Inc. Publishers, Sunderland, Massachusetts (Third edition).

Hartl, D.L. and E.W. Jones. 2002. Essential Genetics, 3rd Edition. Boston–Toronto–London–Singapore: Jones and Bartlett Publishers, Sudbury Sudburry, Massachusetts.

Hartmann, W.L. and R.V. Raleigh. 1964. Tributary homing of sockeye salmon at Brooks and Karluk Lakes Alaska. J. Fish. Res. Bd. Can. 21: 485–504.

Hays, J.D., J. Imbrie and N.J. Shackleton. 1976. Variation in earth's orbit: pacemaker of the ice ages. Science 194: 1121–1132.

Heath, D.A., P.D. Rawson and T.J. Hilbish. 1995. PCR-based nuclear markers identify alien blue mussel (*Mytilus* spp.) genotypes on the west coast of Canada. Can. J. Fish. Aquat. Sci. 52: 2621–2627.

Hebert, P.D.N., A. Givinska and S.L. Ball. 2002a. Biological identification through DNA barcodes. Proc. R. Soc. London, B. 270(1512): 02PB0653.1–02PB0653.9.

Hebert, P.D.N., S. Ratnasingham and J.A. deWaard. 2002b. Barcoding Animal Life: Cytochrome c Oxidase Subunit 1 Divergences among Closely Related Species. Proc. R. Soc. London, B. 270(1512): 03BL0066.S1–03BL0066.S4.

Hedgecock, D. and C. Nelson. 1981. Genetic variation of enzymes and adaptive strategies in crustaceans. pp. 105–129. *In*: Genetika i razmnozhenie morskikh zhivotnykh (Genetics and Reproduction of Marine Animals). Vladivostok: Dal'nevost. Nauchn. Tsentr Akad. Nauk SSSR.

Hedgecock, D. 1986. Population Genetic Bases for Improving Cultured Crustaceans, FIFAC/FAO Symp. on Selection, Hybridization and Genetic Engineering in Aquaculture of Fish and Shellfish for Consumption and Restocking. Bordeaux.

Hedrick, P.W., S. Jain and L. Holden. 1978. Multilocus systems in evolution. pp. 101–184. *In*: M.K. Hecht, W.C. Steere and B. Wallace (eds.). Evolutionary Biol., V. 11. Plenum Press, N.Y.

Hedrick, P.W. 1983. Genetics of Populations. Science Books International, Boston, Mass, 629 p.
Hershenson, S.M. 1983. Basics of Modern Genetics, 2nd Ed., Naukova Dumka, Kiev, 560 p.
Hill, W.G. 1974. Estimation of linkage disequilibrium in randomly mating populations. Heredity 33: 229–239.
Hillis, D.M., B.K. Mable and C. Moritz. 1996. Application of molecular systematics: The state of the field and a look to the future. pp. 515–543. *In*: D.M. Hillis, C. Moritz and B. Mable (eds.). Molecular Systematics. Sinauer. Assoc. Inc., Sunderland, Massachusetts.
Hines, N.O. 1976. Fish of Rare Breeding. Salmon and Trout of Donaldson Strains. Smitsonian Inst. Press, City of Washington, 167 p.
Hochachka, P. and J. Somero. 1988. Biochemical Adaptation. Mir. Publ., Moscow, 568 p.
Hoffmann, H.J. and J.W. Schopf. 1983. Early proterozoic microfossils. pp. 329–360. *In*: J.W. Schopf (ed.). Earth's Ealiest Biosphere: Its Origin and Evolution. Princeston University Press, Princeston, N.J.
Houle, D. 1989. Allozyme-associated heterosis in *Drosophila melanogaster*. Genetics 123: 789–801.
Holland, J., K. Spindler, F. Horodyski, E. Grabau, S. Nichol and S. VandePol. 1982. Rapid evolution of RNA genomes. Science 215: 1577–1585.
Hood, L., J.H. Campbell and S.C. Elgin. 1975. The organization, expression, and evolution of antibody genes and other multigene families. An. Rev. Genet. 9: 305–353.
Hori, H. and S. Osawa. 1979. Evolutionary change in 5S RNA secondary structure and a phylogenetic tree 54 5S RNA species. Proc. Nat. Acad. Sci. U.S.A. 76: 381–385.
Houle, D. 1992. Comparing evolvability of quantitative traits. Genetics 130: 195–204.
Howard, D.J. and G.L. Warning. 1991. Topographic diversity, zone width, and the strength of reproductive isolation in the zone of overlap and hybridization. Evolution 45(5): 1120–1135.
Howard, D.J. 1998. Unanswered questions and future directions in the study of speciation. pp. 439–448. *In*: D.J. Howard and S.H. Berlocher (eds.). Endless Forms: Species and Speciation. Oxford University Press, Oxford–N.Y.
Hubbs, C., R.A. Kuehne and J.C. Ball. 1953. The fishes of upper Guadelupe River, Texas. Texas Journal of Science 5: 216–244.
Hubbs, C.L. 1955. Hybridization between fish species in nature. Syst. Zoology 4: 1–20.
Hubbs, C.L. and K. Lagler. 1970. Fishes of the Great Lake Region. University of Michigan Press, Ann Arbor, Michigan, 213 p.
Hubby, J.L. and R.C. Lewontin. 1966. A molecular approach to the study of genetic heterozygosity in natural populations. I. The number of alleles at different loci in *Drosophila pseudoobscura*. Genetics 54: 577–594.
Huelsenbeck, J.P. and F. Ronquist. 2001. Mr. BAYES: Bayesian inference of phylogeny. Bioinformatics 17: 754–755.
Hunt, J.A., T.J. Hall and R.J. Britten. 1981. Evolutionary distances in Hawaian *Drosophila* measured by DNA association. J. Mol. Evol. 17: 361–367.
Hunter, R.L. and C.L. Markert. 1957. Histochemical demonstration of enzymes separated by zone electrophoresis in starch gels. Science 125: 1294–1295.
Huxley, J.S. 1954. The evolutionary process. pp. 156–180. *In:* J. Huxley, A.C. Hurdy and E.B. Ford (eds.). Evolution as a process. Georg Allen and Unwin, London.
Iliin, V.E., S.M. Konovalov and A.G. Shevlyakov. 1983. Coefficient of migration and special structure of pacific salmons. pp. 9–13. *In:* Biological basis of salmon culture development in waters of USSR. Nauka Publ., Moscow.
Ingram, V.M. 1957. Gene mutations in human hemoglobin: The chemical difference between normal and sickle hemoglobin. Nature 180(4581): 326–328.
Ingram, V.M. 1963. The Hemoglobins in Genetics and Evolution. Columbia Univ. Press, N.Y.
Ingman, M. and U. Gyllensten. 2007. A recent genetic link between Sami and the Volga-Ural region of Russia. European Journal of Human Genetics 15: 115–120.
Ivankov, V.N. 1998. Variability and Microevolution of Fish. Far Estern Univ. Publ., Vladivostok, 124 p.
Jimulev, F.V. 2002. General and Molecular Genetics. Novosibirsk Univ. Publ., Novosibirsk, 459 p.
Johns, G.C. and J.C. Avise. 1998. A comparative summary of genetic distances in the vertebrates from the mitochondrial cytochrome b gene. Mol. Biol. Evol. 15(11): 1481–1490.
Johnson, M.J., D.C. Wallace, C.D. Farris, M.C. Rattazzi and L.L. Cavalli-Sforza. 1983. Radiation of human mitochondria DNA types analyzed by restriction endonuclease cleavage patterns. J. Mol. Evol. 19: 255–271.

Kahler, A.L., R.W. Allard and R.D. 1984. Miller Mutation rates for enzyme and morphological loci in barley (*Hordeum vulgare* L.). Genetics 106: 729–734.
Kaplan, N. and C.H. Langley. 1979. A new estimate of sequence divergence of mitochondrial DNA using restriction endonuclease mapping. J. Mol. Evol. 13: 295–304.
Kartavtsev, Y.Ph. 1976. Polymorphism of shell stain in gastropod mollusk *Littorina brevicula* (Philippi). pp. 93–98. *In*: V.L. Kasyanov (ed.). Biological Research of Vostok Bay. Far Eastern Branch of Rus. Acad. Sci. Publ., Vladivostok.
Kartavtsev, Y.Ph. 1979. On possible determination of balance polymorphism at loci coding for isozymes. pp. 36–40. *In*: V.S. Kirpichnikov (ed.). Biochemical and Population Genetics of Fish. Inst. of Cytology Publ., Leningrad.
Kartavtsev, Y.Ph. 1990. Allozyme heterozygosity and morphological homeostasis in pink salmon *Oncorhynchus gorbuscha* (Pisces: Cyprinidae). Rus. J. Genetics 26(8): 1399–1407.
Kartavtsev, Y.P. 1991. In space and in time allele frequency variability among the pink salmon *Oncorhynchus gorbuscha* (Walbaum) populations. Rus. J. Ichtiology (Voprosy Ichtiologii) 31(3): 487–495 (In Russian).
Kartavtsev, Y.P. 1992. Allozyme heterozygosity and morphological homeostasis in pink salmon fry *Oncorhynchus gorbuscha* (Pisces: Salmonidae): evidences from the family analysis. J. Fish Biol. 40(1): 17–24.
Kartavtsev, Yu. Ph. 1994. Wide-scale genetic differentiation among the pink shrimp *Pandalus borealis* (Crustacea: Pandalidae) populations. pp. 41–52. *In*: A.R. Beaumont (ed.). Genetics and Evolution of Aquatic Organisms Aquatic Organisms, Chapman & Hall, London L.
Kartavtsev , Y.Ph. 1995. Genetic variability and diversity in populations of marine animals. Dissertation of Doctor of Biol. Sciences. S-Petersburg: S-Petersburg State Univ. 530 p. (In Russian. Manuscript; the abstract is available in press).
Kartavtsev, Yu. Ph. 1996. Allozyme variability and differentiation sources in pink shrimp *Pandalus borealis*. Isozyme Bull. 29: 30.
Kartavtsev, Y.Ph. 1998. Analysis of association of allozyme heterozygosity and fitness traits in pink salmon *Oncorhynchus gorbuscha* (Pisces, Salmonidae) fry. Rus. J. Mar. Biol. (Biologia Morya) 24(1): 34–37.
Kartavtsev, Y. 2000. Genetic Aspects of Speciation, Species Differentiation and Biodiversity. pp. 27. *In:* Proc. Int. Meet. of Biodiversity in Asia 2000, September 2000, Tokyo.
Kartavtsev, Y. 2003. Association between heterozygosity level and quantitative trait score: Intralocus interaction and multiple loci averaging. pp. 17. *In*: International Conference "Marine Environment: Nature, Communication, and Business", Vladivostok, 2003. Korea Maritime University, Busan (Abstract).
Kartavtsev, Y. 2005a. Association between the level of heterozygosity and quantitative trait scores: Intra-locus interaction and multiple loci averaging. Rus. J. Genetics 41(1): 100–111 (In Russian, Translated in English).
Kartavtsev, Y.P. 2005b. Molecular Evolution and Population Genetics. Far Eastern State Univ. Publ., Vladivostok, 234 p. (Textbook in Russian).
Kartavtsev, Y.Ph. 2009a. Analysis of sequence diversity at mitochondrial genes on different taxonomic levels. pp. 1–50. *In*: C.L. Mahoney and D.A. Springer (eds.). Genetic Diversity. Nova Science Publishers, Inc., New York.
Kartavtsev, Y.P. 2009b. Molecular evolution and population genetics. Far Eastern State Univ. Publ., 2-nd Ed. Vladivostok, 280 p. (Textbook in Russian, Content, Table, Figure captions are in English).
Kartavtsev, Y.Ph. 2011a. Sequence divergence at mitochondrial genes in animals: Applicability of DNA data in genetics of speciation and molecular phylogenetics. Marine Genomics 49: 71–81.
Kartavtsev, Y.Ph. 2011b. Sequence divergence at Co-1 and Cyt-b mtDNA on different taxonomic levels and genetics of speciation in animals. Mitochondrial DNA 2(3): 55–65.
Kartavtsev, Y.Ph. 2011c. Insertion Sequences in mtDNA of birds and fish: No full length but some short sequences detected for which are obvious maternal and signs of horizontal transmission. Intern. Res. J. of Biochem. and Bioinformatics 6: 139–153.
Kartavtsev, Yu.P. and N.I. Zaslavskaya. 1983. Allozyme polymorphism in a population of the common mussel *Mytilus edulis* L. (Mytilidae) from the Sea of Japan. Marine Biol. Let. 4: 163–172.
Kartavtsev, Yu.F. and A.M. Mamontov. 1983. Electrophoretic estimation of protein variability and similarity of omul, two forms of whitefish (Coregonidae) and grayling (Thymallidae) of Lake Baikal. Rus. J. Genetics 19(11): 1895–1902 (In Russian).

Kartavtsev, Y.P. and S.M. Nikiforov. 1993. Evaluation of the effective size of pacific mussel *Mytilus trossulus* populations from Peter the Great Bay Sea of Japan basing at allele frequency variability among allozyme loci. Rus. J. Genetics 29(3): 476–489 (In Russian, Translated in English).

Kartavtsev, Y.P. and I.G. Rybnikova. 1999. Genetic and morphobiological research of pacific herring populations *Clupea pallasi* from Japan and Okhotsk Seas. Rus. J. Genetics 35(8): 1093–1103 (In Russian, Translated in English).

Kartavtsev, Y.P. and O.V. Svinyna. 2003. Allozyme markers and morphometric variability in gastropod mollusk *Nucella heyseana* (Mollusca, Gastropoda) and their association with environmental change. Korean J. Genetics 25(4): 1–12.

Kartavtsev, Y.P. and J.-S. Lee. 2006. Analysis of nucleotide diversity at genes *Cyt-b* and *Co-1* on population, species, and genera levels. Russian J. Genetics 42(4): 341–362 (In Russian, Translated in English).

Kartavtsev, Y.Ph. and O.L. Zhdanova. 2012. Numerical simulations of the heterozygosity and quantitative trait relationship: intralocus interactions and multiple-locus averaging. Res. J. Biol. Biomed. 2(5): 301–314.

Kartavtsev, Yu.P., M.K. Glubokovsky and I.A. Chereshnev. 1983. Genetic variability and differentiation of two sympatric trout species (*Salvelinus*, Salmonidae) from the Chukotka. Rus. J. Genetics 19(4): 584–593 (In Russian).

Kartavtsev, Yu.P., I.V. Kartavtseva and N.N. Vorontsov. 1984. Population genetics and gene geography of wild mammals. 5. Genetic distances between representatives of different genera of Palearctic hamsters (Rodentia, Cricetini). Rus. J. Genetics 20(6): 961–967 (In Russian).

Kartavtsev, Yu.P., N.E. Polyakova and A.I. Karpenko. 1985. Inheritance, qualitative and quantitative variability of allozymes of malate dehydrogenase, malic enzyme and lactate dehydrogenase in chum salmon, *Oncorhynchus keta*. Rus. J. Genetics 21(5): 845–853 (In Russian).

Kartavtsev, Y.P., E.A. Salmenkova, G.A. Rubtsova and K.A. Afanasiev. 1990. Family analysis of allozyme variability and its relationship to body length and viability of progeny in pink salmon *Oncorhynchus gorbuscha* (Walb.). Rus. J. Genetics 26(9): 1610–1619 (In Russian, Translated into English).

Kartavtsev, Y.P., B.I. Berenboim and K.A. Zgurovsky. 1991. Population genetic differentiation in the pink shrimp *Pandalus borealis* Kroyer from the Barents and Bering Seas. J. Shellfish Res. 10(2): 333–339.

Kartavtsev, Yu.P., K.A. Zgurovsky and Z.M. Fedina. 1992. Allozyme variability and differentiation of the pink shrimp *Pandalus borealis* from the three Far-Eastern seas. Rus. J. Genetics 28(2): 110–122 (In Russian, Translated into English).

Kartavtsev, Y.P., K.A. Zgurovsky and Z.M. Fedina. 1993. Spatial structure of the pink shrimp *Pandalus borealis* from the Far Eastern seas as it proved by methods of population genetics and morphometrics. J. Shellfish Res. 12(1): 81–87.

Kartavtsev, Y.P., A.V. Sytnikov, S.M. Nikiforov and A.Y. Chichvarhin. 1998. Allozyme and morphometric variability in the predatory gastropod mollusk *Nucella heyseana* (MOLLUSCA, GASTROPODA) in polluted and normal environment. Rus. J. Genetics 34(10): 1425–1433 (In Russian, Translated in English).

Kartavtsev, Yu.P., I.G. Rybnikova, A.V. Sitnikov, E.Yu. Amachaeva and O.V. Svinyna. 2000. Genetic and morphometric variability in the gastropod mollusk *Nucella heyseana* (Mollusca, Gastropoda) in environmental optimum and pessimum. Rus. J. Genetics 36(1): 1340–1347 (In Russian, Translated in English).

Kartavtsev, Y.P., V.V. Sviridov, T. Sasaki and N. Hanzawa. 2002. Genetic divergence of far eastern dace belonging to the genus *Tribolodon* (Pisces, Cyprinidae) and closely related taxa: some insights in taxonomy and speciation. Rus. J. Genetics 38(11): 1518–1531 (In Russian, translated in English).

Kartavtsev, Y.Ph., N.I. Zaslavskaya, O.V. Svinyna and A. Kijima. 2006. Allozyme and morphometric variability in the whelk *Nucella heyseana* (Mollusca, Gastropoda) from Russian and Japanese waters: evidence for a single species under different names. Invertebrate Systematics 20: 771–782.

Kartavtsev, Y.P., Y.-M. Lee, S.-O. Jung, H.-K. Byeon and J-S. Lee. 2007a. Complete mitochondrial genome in the bullhead torrent catfish, *Liobagrus obesus* (Siluriformes, Amblycipididae) and phylogenetic considerations. Gene 396: 13–27.

Kartavtsev, Y.P., T.-J. Park, K.A. Vinnikov, V.N. Ivankov, S.N. Sharina and J.-S. Lee. 2007b. Cytochrome *b* (*Cyt-b*) gene sequences analysis in six flatfish species (Pisces, Pleuronectidae), with phylogenetic and taxonomic insights. Marine Biol. 152(4): 757–773.

Kartavtsev, Y.Ph., S.N. Sharina, T. Goto, A.Y. Chichvarkhin, A.A. Balanov, K.A. Vinnikov, V.N. Ivankov and N. Hanzawa. 2008. Cytochrome oxidase 1 (*Co-1*) gene sequence analysis in six flatfish species (Teleostei, Pleuronectidae) of Russia Far East with inferences in phylogeny and taxonomy. Mitochondrial DNA 19(6): 479–489.

Kartavtsev, Y.Ph., S.N. Sharina, T. Goto, A.A. Balanov and N. Hanzawa. 2009. Sequence diversity at cytochrome oxidase 1 (*Co-1*) gene among sculpins (Scorpaeniformes, Cottidae) and some other scorpionfish of Russia Far East with phylogenetic and taxonomic insights. Genes and Genomics 31(2): 191–205.

Kato, I., W.J. Kohr and M.J. Laskowski. 1978. Evolution of aves ovomucoids. pp. 197–206. *In*: S. Magnuson, M. Ottesen, B. Taltman, K. Dano and H. Neurath (eds.). Proc. 11th Feder. Eur. Biol. Sci. Pergamon Press, N.Y.

Kerster, H.W. 1964. Neighborhood size in the rusty lizard, *Sceloporus olivacens*. Evolution 18: 445–457.

Kimura, M. 1953. "Stepping–stone" model of population. Ann. ReP. Nat. Inst. Genet. Mishima 3: 63–65.

Kimura, M. 1965. A stochastic model concerning the maintenance of genetic variability in quantitative characters. Proc. Nat. Acad. Sci. U.S.A. 54: 731–736.

Kimura, M. 1968. Genetic variability maintained in a finite population due to mutational production of neutral or nearly neutral isoalleles. Genet. Res. 11: 247–269.

Kimura, M. 1969. The number of heterozygous nucleotide sites maintained in a finite population due to steady flux of mutations. Genetics 61: 893–903.

Kimura, M. 1971. Theoretical foundation of population genetics at the molecular level. Theor. Popul. Biol. 2: 174–208.

Kimura, M. 1980. A simple method for estimating evolutionary rates of base substitutions through comparative studies of nucleotide sequences. J. Mol. Evol. 16(2): 111–120.

Kimura, M. 1983a. The Neutral Theory of Molecular Evolution. Cambridge Univ. Press, Cambridge.

Kimura, M. 1983b. The neutral theory of molecular evolution. pp. 208–233. *In*: M. Nei and R.K. Koehn (eds.). Evolution of Genes and Proteins. Sinauer Assoc., Sunderland, Mass.

Kimura, M. 1983c. Rare variant alleles in the light of neutral theory. Mol. Biol. Evol. 1: 84–93.

Kimura, M. 1985. Molecular Evolution: The Theory of Neutrality. Mir Publ., Moscow, 398 p. (Rus. Edition).

Kimura, M. and J.F. Crow. 1963. The measurement of effective population number. Evolution 17: 279–288.

Kimura, M. and G.H. Weiss. 1964. The stepping-stone model of population structure and the decrease of genetic correlation with distance. Genetics 49: 561–576.

Kimura, M. and T. Ohta. 1971. Theoretical Aspects of Population Genetics. Princeton Univ. Press, Princeton, New Jersey, 219 p.

Kimura, M. and T. Ohta. 1973. Eukaryotes-prokaryotes divergence estimated by 5S ribosomal RNA sequences. Nature New Biol. 243: 199–200.

King, J.L. and T.H. Jukes. 1969. Non-Darwian evolution. Science 164: 788–798.

King, J. and T. Ohta. 1975. Polyallelic mutation equilibria. Genetics 79: 681–691.

King, M. 1993. Species Evolution: The Role of Chromosome Change. Cambridge University Press, Cambridge, 336 p.

Kirpichnikov, V.S. 1979. Genetic Basis of Fish Selection. Nauka Publ., Leningrad, 392 p.

Kirpichnikov, V.S. 1987. Genetic and Fish Selection. Nauka Publ., Leningrad, 520 p.

Kiseleva, E.V. 1989. Secretory protein synthesis in Chironomus salivary gland cells is not coupled with protein translocation across endoplasmic reticulum membranes. Electron microscopic evidence. FEBS Lett. 257(2): 251–253.

Klug, W.S. and M.R. Cummings. 2002. Essential Genetics, 4th Edition. Prentice Hall, Inc., Upper Saddle River, NJ.

Kochzius, M. and D. Blohm. 2005. Genetic population structure of the lionfish *Pterois miles* (Scorpaenidae, Pteroinae) in the Gulf of Aqaba and Northern Red Sea. Gene 347(2): 295–301.

Koehn, R.K. 1978. Physiology and biochemistry of an enzyme variation: The interface of ecology and population genetics. pp. 51–72. *In*: P. Brussard (ed.). Ecological Genetics: The Interface. Springer-Verlag, New York.

Koehn, R.K. and P.M. 1984. Gaffney Genetic heterozygosity and growth rate in *Mytilus edulis*. Marine Biol. 82: 1–7.

Koehn, R.K., W.J. Diehl and T.M. Scott. 1988. The differential contribution by individual enzymes of glycolysis and protein catabolism to the relationship between heterozygosity and growth rate in the coot clam, *Mulinia lateralis*. Genetics 118: 121–130.

Kolmogorov, A.N. 1935. Deviations from Hardy-Weinberg formulae under partial isolation. Proc. Rus. Acad. Sci. 3(3): 129–132.

Kondo, S. 1977. Evolutionary considerations of DNA repair and mutagenesis. pp. 313–331. *In*: M. Kimura (ed.). Molecular Evolution and Polymorphism. National Institute of Genetics, Mishima, Japan.

Kondrashov, A.S., L.Y. Yampolsky and S.A. Shabalina. 1998. On the sympatric origin of species by means of natural selection. pp. 90–98. *In*: D.J. Howarth and S.H. Berlocher (eds.). Endless Forms: Species and Speciation. Oxford University Press, Oxford.

Konovalov, S.M. 1980. Population Biology of Pacific Salmon. Nauka Publ., Leningrad, 237 p.

Kontula, T., S.V. Kirilchik and R. Vainola. 2003. Endemic diversification of the monophyletic cottoid fish species flock in the Lake Baikal explored with mtDNA sequencing. Mol. Phylogenet. Evol. 27: 143–155.

Koonin, E.V. 2014. The logic of chance. The nature and origin of biological evolution. Authors translation from English into Russian from Pearson Education Inc., 2012. Moscow: ZAO Centropolygraf Publ., 527 p. ISBN 978-5-227-04982-7.

Korochkin, L.I., O.L. Serov and G.P. Manchenko. 1977. Notion on isozymes. pp. 5–17. *In*: L.I. Korochkin (ed.). Genetics of Isozymes. Nauka Publ., Moscow.

Korochkin, L.I. 1999. An Introduction in Genetics of Development. Nauka Publ., Moscow, 252 p.

Krasilov, V.A. 1977. Evolution and Biostratigraphy. Nauka Publ., Moscow, 256 p.

Kreitman, M. 1983. Nucleotide polymorphism at the alcohol dehydrogenase locus of *Drosophila melanogaster*. Nature 304: 412–417.

Krueger, C.C., A.J. Gharrett, J.R. Dehring and F.W. Allendorf. 1981. Genetic aspects of fisheries rehabilitation programs. Can. J. Fish. Aquat. Sci. 38: 1877–1881.

Kubitschek, H.E. 1970. Introduction to Research with Continuous Cultures. Prentice-Hall, Engellwood Cliffs, N.J.

Kuby, J. 1994. Immunology, 2nd. ed. W.H. Freeman, N.Y., 660 p.

Kumar, S., K. Tamura and M. Nei. 1993. MEGA: Molecular Evolutionary Genetics Analysis (With a 130-page Printed Manual). University Park: Pennsylvania Univ. Web-base β-version, 2005 update.

Kusakin, O.G. and A.L. Drozdov. 1998. The phylem of organic world. Part 2. Prokaryota, Eukaryota, Microsporobiontes, Archeomonadobointes, Euglenobiontes, Mixobiontes, Rhodobiontes, Alveolates, Heterokontes. Nauka Publ., S-Petersburg, 359 p.

Lamotte, M. 1951. Recherches sur la structure genetique des populations naturelles de *Cepaea nemoralis* (L.). Bull. Biol. France et Belg. Supl. 35: 1–239.

Lande, R. 1975. The maintenance of genetic variability by mutation in a polygenic character with linked loci. Genet. Res. 26: 221–234.

Lande, R. 1977. The influence of the mating system on the maintenance of genetic variability in polygenic characters. Genetics 86: 485–498.

Lande, R. 1979. Quantitative genetic analysis of multivariate evolution, applied to brain: body size allometry. Evolution 33: 402–416.

Lande, R. 1980. The genetic covariance between characters maintained by pleiotropic mutations. Genetics 94: 203–215.

Lande, R. and J.F. Barrowclaf. 1989. Effective size of population, genetic variability and their usage for the management of populations. pp. 116–156. *In*: M. Suley (ed.). Viability of Populations: Nature Conservation Aspects. Mir Publ., Moscow (Rus. Edition).

Lane, S., A.J. McGregor, S.G. Taylor and A.J. Gharrett. 1990. Genetic marking of an Alaskan pink salmon population, with an evaluation of the mark and the marking process. Amer. Fish. Soc. Symp. 7: 395–406.

Langley, C.H. and W.M. Fitch. 1974. An examination of the constancy of the rate of molecular evolution. J. Mol. Evol. 3: 161–177.

Langley, C.H., E.A. Montgomery and W.F. Qattlebaum. 1982. Restriction map variation in the Adh region of *Drosophila*. Proc. Nat. Acad. Sci. U.S.A. 79: 5631–5635.

Laurie-Ahlberg, C.C., G. Maroni, G.C. Buley, J.C. Lucchesi and B.S. Weir. 1982. Qauntative genetic variation of enzyme activities in natural populations of *Drosophila melanogaster*. Proc. Nat. Acad. Sci. U.S.A. 77: 1073–1077.

Leary, R.F., F.W. Allendorf and K.L. Knudsen. 1983. Developmental stability and enzyme heterozygosity in rainbow trout. Nature 301: 71–72.

Leary, R.F., F.W. Allendorf, K.L. Knudsen and G.H. Thorgaard. 1985. Heterozygosity and developmental stability in gynogenetic diploid and triploid rainbow trout. Heredity 54: 219–225.

Leary, R.F., F.W. Allendorf and K.L. Knudsen. 1987. Differences in inbreeding coefficients do not explain the association between heterozygosity at allozyme loci and developmental stability in rainbow trout. Evolution 41(6): 1413–1415.
Lebedev, N.V. 1967. Elementary Populations of Fish. Food Ind. Publ., Moscow, 211 p.
Lerner, I.M. 1958. The Genetic Basis of Selection. John Wiley and Son, N.Y.
Lerner, M. 1970. Genetic Homeostasis. N.Y.: Dover Publ. Inc., 134 p.
Leslie, J.F. 1982. Linkage analysis of seventeen loci in poecilid fish (genus *Poecilopsis*). J. Hered. 73(1): 19–23.
Levinton, J.S. and R.K. Koehn. 1976. Population genetics of mussels. pp. 357–384. *In*: B.L. Bayne (ed.). Marine Mussels: Their Ecology and Physiology. Cambridge University Press, London and New York.
Lewin, B. Genes and Georgiev G.P. (ed.). 1987. Mir Publ., Moscow, 544 p.
Lewontin, R.C. 1964. The interaction of selection and linkage. I. General considerations. Heterotic models. Genetics 49: 49–67.
Lewontin, R.C. 1974. The Genetic Basis of Evolutionary Change. Columbia University Press, New York and London.
Lewontin, R. 1978. Genetic Basis of Evolution. Mir Publ., Moscow, 351 p.
Lewontin, R.C. and K. Kojima. 1960. The evolutionary dynamics of complex polymorphism. Evolution 14: 458–472.
Li, C.C. 1955. Population Genetics. Univ. Chicago Press, Chicago, 346 p.
Li, C.C. 1978. An Introduction to Population Genetics. Mir Publ., Moscow, 555 p.
Li, W.-H. 1983. Evolution of duplicate genes and pseudogenes. pp. 14–37. *In*: M. Nei and R. Koehn (eds.). Evolution of Genes and Proteins. Sinauer Assoc., Sunderland, Mass.
Li, W.-H. 1986. Evolutionary change of restriction cleavage sites and phylogenetic inference. Genetics 113: 187–213.
Li, W.H. 1997. Molecular Evolution. Sinauer Ass., Sunderland.
Li, W.H. and A. Zarkhih. 1995. Statistical tests of DNA phylogenies. Syst. Biol. 44: 49–63.
Li, W.-H., T. Gojobory and M. Nei. 1981. Pseudogenes as a paradigm of neutral evolution. Nature 292: 237–239.
Li, W.-H., C.-I. Wu and C.-C. Luo. 1984. Nonrandomness of pint mutation as reflected in nucleotide substitutions in pseudogenes and its evolutionary implications. J. Mol. Evol. 21: 58–71.
Lindberg, G.U. 1948. On the impact of transgression and regression phase change on the evolution of fish and shellfish. Proc. of Rus. Acad. Sci. 63(1): 93–95.
Livshits, G. and E. Kobiliansky. 1985. Lerner's concept of developmental homeostasis and the problem of heterozygosity level in natural populations. Heredity 55: 341–353.
Lukashov, V.V. 2009. Molecular Evolution and Phylogenetic Analysis. Binom Publ. Lab of Knowledge, Moscow, 256 p.
Lynch, J.D. and J.S. Conery. 2000. The evolutionary fate and consequences of duplicate genes. Science 290: 1151–1155.
Lynch, M.B. Walsh. 1998. Genetics and Analysis of Quantitative Traits. Sinauer & Associates, Inc., Sunderland, Massachusetts.
Mahillon, J. and M. Chandler. 1998. Insertion sequences. Microbiol. Mol. Biol. Rev. 62: 725–774.
Machordom, A. and E. Macpherson. 2004. Rapid radiation and cryptic speciation in squat lobsters of the genus *Munida* (Crustacea, Decapoda) and related genera in the South West Pacific: Molecular and morphological evidence. Mol. Phylogenet. Evol. 33(2): 259–279.
Maeda, N. and O. Smithies. 1986. The evolution of multigene families: Human haptoglobin genes. Ann. Rev. Genet. 20: 81–108.
Magni, G.E. 1969. Spontaneous mutations. *In*: Proc. 12th Int. Congr. Genet. Tokyo 3: 247–259.
Malecote, G. 1973. Isolation by distance, genetic structure of populations. pp. 72 –75. *In*: N.E. Morton (ed.). Genetic Structure of Population. Univ. of Hawaii, Honolulu.
Manchenko, G.P. 2002. Handbook of Detection of Enzymes on Electrophoretic Gels. CRC Press LLC.
Mau, B. and M.A. Newton. 1997. Phylogenetic inference for binary data on dendograms using Markov chain Monte Carlo. Journal of Computational and Graphical Statistics 6: 122–131.
Mau, B., M.A. Newton and B. Larget. 1999. Bayesian phylogenetic inference via Markov chain Monte Carlo methods. Biometrics 55: 234–249.
Margoliash, E. and E.L. Smith. 1965. Structural and functional aspects of cytochrome c in relation to evolution. pp. 221–242. *In*: V. Bryson and H.J. Vogel (eds.). Evolving Genes and Proteins. Academic Press, N.Y.

Marshall, D.R. and A.H.D. Brown. 1975. The charge-state model of protein polymorphism in natural populations. J. Mol. Evol. 6: 149–163.
Marr, J.C. 1957. The problem of defining and recognizing subpopulations of fishes. U.S. Fish and Wildlife Service, Spec. Sci. ReP., Fish 208: 1–6.
Martinez-Navarro, E.M., J. Galian and J. Serrano. 2005. Phylogeny and molecular evolution of the Tribe Harpalini (Coleoptera, Carabidae) inferred from mitochondrial cytochrome-oxidase I. Mol. Phylogenet. Evol. 35: 127–146.
Mather, K. and B.J. Harrison. 1949. The manifold effect of selection. Heredity 3: 1–52, 131–162.
Maxson, L.R. 1984. Molecular probes of phylogeny and biogeography in toads of the widespread genus *Bufo*. Mol. Biol. Evol. 1: 345–356.
Maxson, L.R. and A.C. Wilson. 1974. Convergent morphological evolution detected by studying proteins of tree frog in the *Hyla eximia* species group. Science 185: 66–68.
Maxson, L.R. and A.C. Wilson. 1975. Albumin evolution and organismal evolution in tree frogs (Hylidae). Syst. Zool. 24: 1–15.
Maxson, L.R., V.M. Sarich and A.C. Wilson. 1975. Continental drift and the use of albumin as an evolutionary clock. Nature 255: 397–400.
Mayr, E. 1947. Systematics and Species Origin from the Viewpoint of Systematicist. Foreign Literature, Moscow, 502 p.
Mayr, E. 1968. Zoological Species and Evolution. Mir Publ., Moscow, 398 p.
Mayr, E. 1982. Process of speciation in animals. pp. 1–20. *In*: C. Barigozzi (ed.). Mechanisms of Speciation. Alan R. Liss, N.Y.
Mettler, L. and T. Gregg. 1972. Genetics of population and evolution. Moscow: Mir Publ., 323 p. (Translated from: Population genetics and evolution/Lawrence E. Mettler, Thomas G. Gregg. Prentice-Hall, Englewood Cliffs, New Jersey, 1969).
McCune, A.R. and N.R. Lovejoy. 1998. The relative rate of sympatric and allopatric speciation in fishes: Tests using DNA sequence divergence between sister species and among clades. pp. 172–185. *In*: D.J. Howard and S.H. Berlocher (eds.). Endless Forms: Species and speciation. Oxford University Press, Oxford – N.Y.
McLaughlin, P.J. and M.O. Dayhoff. 1970. Eukaryotes versus prokaryotes: An estimate of evolutionary distance. Science 168: 1469–1470.
McLaughlin, P.J. and M.O. Dayhoff. 1972. Evolution of species and proteins: A time scale. *In*: M.O. Dayhoff (ed.). Atlas of Protein Sequence and Structure. National Biomedical Research Foundation, Silver Springs, Md. 5: 47–66.
McLellan, T. 1984. Molecular charge and electrophoretic mobility in cetacean hemoglobins of known sequence. Biochem. Genet. 22: 181–200.
McLellan, T., F.-L. Ames and K. Nikaido. 1983. Genetic variation in proteins: Comparison of one-dimensional and two-dimensional electrophoresis. Genetics 104: 381–390.
MEGA 4 (MEGA 5) Tamura, K., D. Peterson, N. Peterson, G. Stecher, M. Nei and S. Kumar, MEGA5. 2011. Molecular Evolutionary Genetics Analysis using Maximum Likelihood, Evolutionary Distance, and Maximum Parsimony Methods. Mol. Biol. Evol. 28: 2731–2739. http://www.megasoftware.net/.
Mina, M.V. 1986. Microevolution of Fish: Evolutionary Aspects of Phyletic Diversity. Nauka Publ., Moscow, 207 p.
Mitton, J.B. and B.A. Piers. 1980. The distribution of individual heterozygosity in natural populations. Genetics 95(4): 1043–1054.
Mitton, J.B. and M.C. Grant. 1984. Associations among protein heterozygosity, growth rate, and developmental homeostasis. Ann. Rev. Ecol. Syst. 15: 479–499.
Mitton, J.B. and W.M. Lewis. 1989. Relationships between genetic variability and life-history features of bony fishes. Evolution 43(8): 1712–1723.
Miya, M., H. Takeshima, H. Endo, N.B. Ishiguro, J.G. Inoue, T. Mukai, T.P. Satoh, M. Yamaguchi, A. Kawaguchi, K. Mabuchi, S.M. Shirai and N. Nishida. 2003. Major patterns of higher teleostean phylogenies: a new perspective based on 100 complete mitochondrial DNA sequences. Mol. Phyl. Evol. 1: 121–38.
Moore, W.S. 1977. An evolution of narrow hybrid zone in vertebrates. Quarterly Rev. Biol. 52: 263–277.
Moriwaki, K., K. Hirai, C. Kohigashi, N. Miyashita, A. Kryukov, L. Frisman and Y. Yamaguchi. 2001. Geographical distribution of a recombinant beta-hemoglobin haplotype P and possible migration of wild mouse populations in Asia. Evolution, genetics, ecology and biodiversity: International

conference. *In*: Alexei P. Kryukov and Yuri Ph. Kartavtsev (eds.). Vladivostok—Vostok Marine Biological Station. September 24–30, 2001: Abstracts. Vladivostok: Institute of Marine Biology, 37 p.

Morton, N.T., J.F. Crow and H.J. Muller. 1956. An estimate of the mutational damage in man from data on consanguineous marriages. Proc. Natl. Acad. Sci. US 42: 855–863.

Mukai, T. and C.C. Cockerham. 1977. Spontaneous mutation rates at enzyme loci in *Drosophila melanogaster*. Proc. Nat. Acad. Sci. U.S.A. 74: 2514–2517.

Müller, H.J. 1959. Advances in radiation mutagenesis through studies on Drosophila. Progress in Nuclear Energy. Pergamon Press, N.Y. 6(2): 146–160.

Nagylaki, T. 1984. The evolution of multigene families under interchromosomal gene conversion. Genetics 106: 529–548.

Neave, F. 1958. The origin and speciation of *Oncorhynchus*. Trans. Roy Soc. Canada. 52: 25–49.

Needleman, S.B. and C.D. Wunsch. 1970. A general method applicable to the search for similarities in the amino acid sequence of two proteins. J. Mol. Biol. 48(3): 443–453. www.ncbi.nlm.nih.gov/pubmed/5420325.

Neel, J.V. 1949. The inheritance of sickle cell anemia. Science 110: 164–166.

Neel, J.V. 1973. "Private" genetic variants and the frequency of mutation among South American Indians. Proc. Nat. Acad. Sci. U.S.A. 70: 3311–3315.

Neel, J.V. and E.D. 1978. Rothman Indirect estimates of mutation rate in tribal American Indians. Ibid 75: 5585–5588.

Neel, J.V., H.W. Mohrenweiser and M.H. Meisler. 1980. Rate of spontaneous mutation at human loci encoding protein structure. Proc. Nat. Acad. Sci. U.S.A. 77: 6037–6041.

Neel, J.V., C. Satoh, K. Goriki, M. Fujita, N. Takahashi, J.-T. Asakawa and R. Hazama. 1986. The rate with which spontaneous mutation alerts the electrophoretic mobility of polypeptides. Proc. Nat. Acad. Sci. U.S.A. 83: 389–393.

Neff, N.A. and G.R. Smith. 1979. Multivariate analysis of hybrid fishes. Syst. Zool. 28: 176–196.

Nei, M. 1972. Genetic distances between populations. Amer. Nat. 106(949): 283–292.

Nei, M. 1973 Analysis of gene diversity in subdivided populations. Proc. Nat. Acad. Sci. U.S.A. 70: 3321–3323.

Nei, M. 1975. Molecular Population Genetics and Evolution. North Holland Publ. Co., Amer Elsevier Publ. Co., Inc., Amsterdam–Oxford–N.Y., 288 p.

Nei, M. 1977a. Estimation of mutation rate from rare protein variants. Amer. J. Hum. Genet. 29: 225–232.

Nei, M. 1977b. F-statistics and analysis of gene diversity in subdivided populations. Ann. Hum. Genet. L. 41: 225–233.

Nei, M. 1978. Estimation of average heterozygosity and genetic distance from a small number of individuals. Genetics 89: 583–590.

Nei, M. 1983. Genetic polymorphism and the role of mutations in evolution. pp. 165–190. *In*: M. Nei and R. Koehn (eds.). Evolution of Genes and Proteins. Sinauer Assoc., Sunderland, Massachusetts.

Nei, M. 1985. Human evolution at molecular level. pp. 41–64. *In*: T. Ohta and K. Aoki (eds.). Population Genetics and Molecular Evolution. Japan Sci. Soc. Japan Science Society Press, Tokyo.

Nei, M. 1987. Molecular Evolutionary Genetics. Columbia Univ. Press, N.Y., 512 p.

Nei, M. and R. Chakraborty. 1973. Genetic distance and electrophoretic identity of proteins between taxa. J. Mol. Evol. 2: 323–328.

Nei, M. and F. Tajima. 1981. DNA polymorphism detectable by restriction endonucleases. Genetics 97: 145–163.

Nei, M. and R. Koehn (eds.). 1983. Evolution of Genes and Proteins. Sinauer Assoc., Sunderland, Mass.

Nei, M. and A.K. Roychoudhury. 1982. Genetic relationship and evolution of human races. Evol. Biol. 14: 1–59.

Nei, M. and F. Tajima. 1985. Evolutionary change of restriction cleavage sites and phylogenetic inference for man and apes. Mol. Biol. Evol. 2: 189–205.

Nei, M. and S. Kumar. 2000. Molecular Evolution and Phylogenetics. Oxford Univ. Press, N.Y., 333 p.

Nei, M., F. Tajima and Y. Tateno. 1983. Accuracy of estimated phylogenetic trees from molecular data. II. Gene frequency data. J. Mol. Evol. 19: 153–170.

Nei, M., J.C. Stephens and N. Saitou. 1985. Methods for computing the standard errors of branching points in an evolutionary tree and their application to molecular data from humans and apes. Mol. Biol. Evol. 2: 66–85.

Neifah, A.A. and E.R. Lozovskaya. 1984. Genes and Development of Organism. Nauka Publ., Moscow, 188 p.

Nelson, J.S. 1966. Hybridization between two cyprinid fishes, *Hybopes plumbea* and *Rhinichthys cataractae*, in Alberta. Can. J. Zool. 44: 963–968.
Nelson, J.S. 1973. Occurrence of hybrids between longnose sucker (*Catastomus catastomus*) and white sucker (*C. commersoni*) in upper Canananaskis Resorvoir, Alberta. J. Fish. Res. Board. Can. 30: 557–560.
Nevo, E. and H. Cleve. 1978. Genetic differentiation during speciation. Nature 275: 125.
Nevo, E., A. Beiles and R. Ben-Shlomo. 1984. The evolutionary significance of genetic diversity: ecological, demographic and life history correlates. *In*: G.S. Mani (ed.). Evolutionary Dynamics of Genetic Diversity. Lecture Notes in Biomathematics 53: 4–213.
Needleman, S.B. and C.D. Wunsch. 1970. A general method applicable to the search for similarities in the amino acid sequence of two proteins. J. Mol. Biol. 48(3): 443–453.
Novick, A. and L. Szilard. 1950. Experiments with hemostat on spontaneous mutation of bacteria. Proc. Nat. Acad. Sci. U.S.A. 36: 708–719.
Ohnishi, S., M. Kawanishi and T.K. Watanabe. 1983. Biochemical phylogenies of Drosophila: Protein differences detected by two-dimensional electrophoresis. Genetica 61: 55–63.
Ohno, S. 1970. Evolution by Gene Duplication. Springer-Verlag, New York, 160 p.
Ohno, S. 1972. So much "junk" DNA in our genome. Brookbaven Symp. Biol. 23: 366–370.
Ohno, S. 1973. Genetic Mechanisms of Progressive Evolution. Mir Publ., Moscow, 227 p.
Ohno, S. 1981. Original domain for the serum albumin family arose from repeated sequences. Proc. Nat. Acad. Sci. U.S.A. 78: 7657–7661.
Ohno, S. 1984. Repeats of base oligomers as the primordial coding sequences of the primeval earth and their vestiges in modern genes. J. Mol. Evol. 20: 313–321.
Ommani, F. 1975. Fish. Mir Publ., Moscow, 192 p.
Ohta, T. 1971. Associative overdominance caused by linked detrimental mutations. Genetical Research 18: 277–286.
Ohta, T. 1982. Linkage disequilibrium due to random genetic drift in finite subdivided population. Proc. Nat. Acad. Sci. U.S.A. 79: 1940–1944.
Ohta, T. and J.H. Gillespie. 1996. Development of neutral and nearly neutral theories. Theor. Popul. Biol. 49(2): 128–142.
On-Line Biology Book. 2002. Text: M.J. Farabee. http://gened.emc.maricopa.edu/bio/ BIO181/BIOBK/ BioBookTOC.html.
Page, R.D.M. 1996. Treeview: An application to display phylogenetic trees on personal computers. Comparative and Applied Bioscience 12: 357–358.
Pauling, L., H. Itano, S.J. Singer and I.C. Well. 1949. Sickle cell anemia, a molecular disease. Science 110: 543–548.
Paterson, H.E.H. 1978. More evidence against speciation by reinforcement. South African J. Sci. 74: 369–371.
Paterson, H.E.H. 1985. The recognition concept of species. pp. 21–29. *In*: E.S. Vrba (ed.). Species and Speciation. Transvaal Museum Monograph, Pretoria.
Peacock, K. and D. Boulter. 1975. Use of amino acid sequence data in phylogeny and evaluation of methods using computer simulation. J. Mol. Biol. 95: 513–527.
Penney, R.W. and M.J. Hart. 1999. Distribution, genetic structure, and morphometry of *Mytilus edulis* and *M. trossulus* within a mixed species zone. J. Shellfish Res. 18(2): 367–374.
Pereira, V. 2004. Gene history repeats itself. Heredity 93: 3–4.
Phillipp, D.P., W.F. Childers and G.S. Whitt. 1983. A biochemical genetic evaluation of the northern and Florida subspecies of largemouth bass. Transactions of Amer. Fisheries Soc. 112: 1–20.
Phillips, R.B. and K.D. Zajicek. 1982. Q band chromosomal banding polymorphisms in lake trout (*Salvelinus namaycush*). Genetics 101: 227–234.
Phillips, R.B., K.D. Zajicek and F.M. Utter. 1985. Q band chromosome polymorphism in Chinook salmon (*Oncorhynchus tshawytscha*). Copeia: 273–278.
Pilbeam, D. 1984. The descent of hominids and hominids. Sci. Amer. 250(3): 84–96.
Plotkin, J.B., J. Dushoff and H.B. Fraser. 2004. Detecting selection using a single genome sequence of *M. tuberculosis* and *P. falciparum.* Nature 428(6986): 942–945.
POPULUS 1.4. Alstad D. Department of Ecology, Evolution & Behavior University of Minnesota 318 Church St. SE Minneapolis, MN, USA 55455–0302/Tel: 612–625–0488. E-mail: DNA@UMNACVX.bitnet_.
Posada, D. and K.A. Grandal. 1998. MODELTEST: Testing the model of DNA substitution. Bioinformatics 14: 817–818.

Post, T.J. and T. Uzzell. 1981. The relationship of *Rana sylvatica* and the monophyly of the *Rana boylei* group. Syst. Zool. 30: 170–180.
Powell, J.R. 1983. Interspecific cytoplasmic gene flow in the absence of nuclear gene flow: Evidence from *Drosophila*. Proc. Nat. Acad. Sci. U.S.A. 80: 492–495.
Powers, D.A. 1987. A multidisciplinary approach to the study of genetic variation in species, new directions in physiological ecology. pp. 102–134. *In*: M.L. Feder and A.F. Bennet (eds.). Cambridge University Press, New York.
Prager, E.M. and A.C. Wilson. 1971. The dependence of immunological cross-reactivity upon sequence resemblance among lysozymes. J. Biol. Chem. 246: 5978–5989.
Prager, E.M., A.H. Brush, R.A. Nolan and A.C. Wilson. 1974. Slow evolution of transferrin and albumin in birds according to micro-complement fixation analysis. J. Mol. Evol. 3: 243–262.
Pritham, E.J. 2009. Transposable elements and factors influencing their success in eukaryotes. Journal of Heredity 100(5): 648–655.
Purves, W.K., G.H. Orians, H. Heller and C. Raig. 1994. Life: The Science of Biology, 4th Edition. Sinauer Associates Inc. Sunderland, Mass.
Radding, C.M. 1982. Strand transfer in homologous genetic recombination. Ann. Rev. Genet. 16: 405–437.
Rao, S.R. 1980. Cluster analysis in application to the study of human races mixture. pp. 148–167. *In*: J. Van Raisen (ed.). Classification and cluster. Mir Publ., Moscow.
Ratner, V.A. 1998. Chronics of great discovery: Ideas and faces. Rus. J. Nature 4: 68–79 and 11: 18–28.
Ramshaw, J.A., J.A. Coyne and R.C. Lewontin. 1979. The sensitivity of gel electrophoresis as a detector of genetic variation. Genetics 93: 1019–1037.
Rannala, B. and Z. Yang. 1996. Probability distribution of molecular evolutionary trees: a new method of phylogenetic inference. J. Mol. Evol. 43: 304–311.
Rand, D.M. and L.M. Kann. 1998. Mutation and selection at silent and replacement sites in the evolution of animal mitochondrial DNA. Genetics 102-103: 393–407.
Rasmussen, D.I. 1964. Blood group polymorphism and inbreeding in natural populations of the deer mouse, *Peromyscus maniculatus*. Evolution 18: 219–229.
Rawson, P.D., K.L. Joyner, K. Meetze and T.J. Hilbish. 1996. Evidence for intragenic recombination within a novel genetic marker that distinguishes mussels in the *Mytilus edulis* species complex. Heredity 77: 599–607.
Rawson, P.D. and T.J. Hilbish. 1998. Asymmetric introgression of mitochondrial DNA among European populations of Blue mussels (*Mytilus* spp.). Evolution 52(1): 100–108.
Raymond, M. and F. Rousset. 1995. GENEPOP (version 1.2): population genetics software for exact tests and ecumenicism. J. Heredity 86: 248–249.
Remington, C.L. 1968. Suture-zones of hybrid interaction between recently joined biotas. *In*: T. Dobhansky, M.K. Hecht and W.C. Steere (eds.). Evolutionary Biology. Appleton–Century Crofts, N.Y. 2: 321–428.
Roberts, D.F. 1955. The dynamics of racial intermixture in the Americans Negro: Some anthropological considerations. Amer. J. Human Gen. 10: 117–144.
Robertson, H.M. and D.J. Lampe. 1995. Recent horizontal transfer of a mariner transposable element among and between Diptera and Neuroptera. Mol. Biol. Evol. 12: 850–862.
Robinson, D. and L. Foulds. 1981. Comparison of phylogenetic trees. Math. Biosciences 53: 131–147.
Ronquist, F. and J.P. Huelsenbeck. 2003. Mr. BAYES 3: Bayesian phylogenetic inference under mixed models. Bioinformatics 19: 1572–1574.
Rodriguez-Trelles, F., R. Tarrio and F. Ayala. 2003. Convergent neofuntionalization by positive Darwinian selection after ancient recurrent duplication for xantine dehydrogenase gene. Proc. Natl. Acad. Sci. U.S.A. 100: 13413–13417.
Rogers, J.S. 1972. Measures of genetic similarity and genetic distance. Univ. Texas Publ. 7213: 145–153.
Ross, M.R. and T.M. Cavender. 1981. Morphological analysis of four experimental intergeneric cyprinid hybrid crosses. Copeia: 377–387.
Rukhlov, F.N. and O.S. Lubaeva. 1980. Results of tagging of pink salmon *Oncorhynchus gorbuscha* (Walb.) fry at the ranching farms in the Skhalin District. Questions of Ichtiology (Voprosy Ichtiologii) 20(1): 134–144.
Rutaisire, J., A.J. Boot, C. Masembe, S. Nyakaana and V. Muwanika. 2004. Evolution of *Labeo victorianus* Predates the Pleistocene Desiccation of Lake Victoria: Evidence from Mitochondrial DNA Sequence Variation. South African J. Sci. 100(11-12): 607–608.
Rychkov, Y.G. 1973. System of ancient isolates of human in North Asia in the light of the problem of stability and evolution of the populations. Questions of Anthropology (Voprosy Anttropologii) 44: 3–22.

Rychkov, Y.G. and V.A. Sheremetieva. 1976. Genetics of circumpolar populations of Eurasia in the connection with the problem of human adaptation. *In:* Resources of Biosphere. Nauka Publ., Leningrad 3: 10–41.
Ryman, N.F., F.W. Allendorf and G. Stahl. 1979. Reproductive isolation with little genetic divergence in sympatric populations of brown trout (*Salmo trutta*). Genetics 92: 247–262.
Saitou, N. and M. Nei. 1987. The neighbor-joining method: a new method for reconstructing phylogenetic trees. Mol. Biol. Evol. 4: 1406–1425.
Saldanha, P.H. 1957. Gene flow from white into Negro population in Brazil. Amer. J. Hum. Genet. 9: 299–309.
Salmenkova, E.A. 1973. Genetics of isozymes of fish. Advances of Modern Biol. 75(2): 217–235.
Salmenkova, E.A. 1989. Main results and goals of population genetic research of salmon fish. pp. 7–29. *In:* V.S. Kirpichnikov (ed.). Genetics in aquaculture. Nauka Publ., Leningrad.
Sanders, B. and J. Wright. 1962. Immunogenetic studies in two trout species of the genus *Salmo*. Annals of the New York Acad. Sci. 97: 116–130.
Sanderson, M.J. and H.B. Shaffer. 2002. Troubleshooting molecular phylogenetic analyses. Annual Rev. Ecol. Syst. 33: 49–72.
Sarich, V.M. and A.C. Wilson. 1966. Qauntative immunochemistry and the evolution of primate albumins: Micro-complement fixation. Science 154: 1563–1566.
Sarich, V.M. and A.C. Wilson. 1967. Immunological time scale for hominid evolution. Science 158: 120–1203.
Sasaki, T., Y.P. Kartavtsev, T. Uematsu, V. Sviridov and N. Hanzawa. 2007. Phylogenetic independence of Far Eastern Leuciscinae (Pisces: Cyprinidae) inferred from mitochondrial DNA analysis. Gene and Genetic Systems 82: 329–340.
Schneider, S., D. Roessli and L. Excoffier. 2000. Arlequin Ver. 2.000: A Software for population genetic data analysis. Univ. of Geneva, Geneva.
Schopf, J.M. and M.R. Walter. 1983. Archean microfossils: New evidence of ancient microbes. pp. 60–99. *In*: M.H. Smith and T. Toule (eds.). Earth's Earliest Biosphere: Its Origin and Evolution. University of Georgia Press, Athens.
Schopf, J.M., J.M. Hayes and M.R. Walter. 1983. Evolution of Earth's earliest ecosystems: Recent progress and unsolved problems. pp. 361–384. *In*: J.W. Schopf (ed.). Earth's Earliest Biosphere: Its Origin and Evolution. Princeston University Press, Princeston, N.J.
Schwartz, F.J. 1972. World literature on to fish hybrids, with an analysis by family, species and hybrid. Publications of the Gulf Coast Res. Lab. Amuseum. 3: 1–328.
Schwartz, F.J. 1981. World literature on to fish hybrids, with an analysis by family, species and hybrid. NOAA Techn. Report NMFS SSRF–750, Supplement 1. U.S. DeP. Commerce, 507 p.
Setzer, P.Y. 1970. An analysis of natural hybrid swarm by means of chromosome morphology. Transactions of Amer. Fisheries Soc. 99: 139–146.
Shah, D.M. and C.H. Langley. 1979. Inter- and intraspecific variation in restriction maps of *Drosophila* mitochondrial DNAs. Nature 271: 696–699.
Shaw, C.R. 1965. Electrophoretic variation in enzymes. Science 149: 936–943.
Shaw, C.R. and R. Prasad. 1970. Starch gel electrophoresis of enzymes. A compilation of recipes. Biochem. Genet. 4: 292–520.
Shull, G.H. 1914. Duplicate genes for capsule in *Bursa bursapastoris*. Z. Ind. Abst. Vererb. 12: 97–149.
Sibley, C.G. and J.E. Ahlquist. 1984. The phylogeny of hominid primates as indicated by DNA–DNA hybridization. J. Mol. Evol. 20: 2–15.
Simmons, M.P. and M. Miya. 2004. Efficiently resolving the basal clades of a phylogenetic tree using Bayesian and parsimony approaches: a case study using mitogenomic data from 100 higher teleost fishes. Mol. Phyl. Evol. 31: 351–362.
Simon, R.C. and R.E. Noble. 1968. Hybridization in *Oncorhynchus* (Salmonidae). I. Viability and inheritance in artificial crosses of chum and pink salmon. Transactions of Amer. Fisheries Soc. 97: 109–118.
Simpson, G.G. 1961. Principles of Animal Taxonomy. The Species and Lower Categories. Columbia Univ. Press, N.Y.
Singh, S.M. 1982. Enzyme heterozygosity associated with growth at different development stages in oysters. Can. J. Genet. Cytol. 24(4): 451–458.
Singh, S.M. and E. Zouros. 1978. Genetic variation associated with growth rate in the American oyster (*Crassostrea virginica*). Evolution 32(2): 342–353.

Singh, R.S., R.C. Lewontin and A.A. Felton. 1976. Genetic heterogeneity within electrophoretic "alleles" of xantine dehydrogenases in *Drosophila pseudoobscura*. Genetics 84: 609–629.

Sites, J.W. and J.C. Marshall. 2004. Operational Criteria for Delimiting Species. Annual Rev. Ecol. Evol. Syst. 35: 199–227.

Skibinski, D.O.F., J.A. Beardmore and T.F. Cross. 1983. Aspects of the population genetics of *Mytilus* (Mytilidae: Mollusca) in the British Isles. Biol. J. Linn. Soc. 19: 173–183.

Skurihina, L.A., Y.P. Kartavtsev, M.V. Pan'kova and A.Yu. Chichvarkhin. 2001. Study of two species of mussels, *Mytilus trossulus* and *Mytilus galloprovincialis* (Bivalvia, Mytilidae) and their hybrids in Peter the Great Bay of the Sea of Japan by PCR-markers. Russian J. Genetics 37(12): 1448–1451 (English Translation).

Slightom, J.L., A.E. Blechl and O. Smithies. 1980. Human fetal Gγ- and Aγ-globin genes: Complete nucleotide sequence suggest that DNA can be exchanged between these duplicated genes. Cell 21: 627–638.

Smith, K. 1921. Racial investigations. 6. Statistical investigations on inheritance in *Zoarces viviparus* L. *In*: Compte Rendue Traveau Lab. Carlsberg, Copenhagen 14(11): 1–60.

Smith, T.F. and M.S. Waterman. 1981. Identification of common molecular subsequences. J. Mol. Biol. 47(1): 195–197. www.ncbi.nlm.nih.gov/pubmed/7265238.

Smouse, P.E. 1986. The fitness consequences of multiple-locus heterozygosity under the multiplicative over dominance and inbreeding depression models. Evolution 40: 946–957.

Smouse, P.E. and C. Chakraborty. 1986. The use of restriction fragment length polymorphism in paternity analysis. Amer. J. Hum. Genet. 38: 918–939.

Sneath, P.H. and R.R. Sokal. 1973. Numerical Taxonomy. Freeman, San Francisco.

Sokal, R.R. and C.D. Michener. 1958. A statistical method for evaluating systematic relationships. University of Kansas Sci. Bull. 28: 1409–1438.

Soloman, D.J. and A.R. Child. 1978. Identification of juvenile natural hybrids between Atlantic salmon (*Salmo salar* L.) and trout (*S. trutta* L.). J. Fish Biol. 12: 499–501.

Sourdis, J. and C. Krimbas. 1987. Accuracy of phylogenetic trees estimated from DNA sequence data. Mol. Biol. Evol. 4: 159–166.

Spolsky, C. and T. Uzzell. 1984. Natural interspecies transfer of mitochondrial DNA in amphibians. Proc. Nat. Acad. Sci. U.S.A. 81: 5802–5805.

Statistica for Windows. 1994. Users Guide. StatSoft Inc., Tulsa OK, East 14th Street, 1064 pp.

Stebbins, G.L. 1950. Variation and Evolution in Plants. Columbia University Press, New York.

Stein, J.P., J.F. Catterall, P. Cristo, A.R. Means and B.W. O'Malley. 1980. Ovomucoid intervining sequences specify functional domains and generate protein polymorphism. Cell 21: 681–687.

Stent, G.S. 1971. Molecular genetics. An introductory narrative. W.H. Freeman & Co. San Francisco (Cited from Russian edition: Stent G.S. Molecular genetics. Mir Publ., Moscow 1974).

Stepien, C.A. and J.E. Faber. 1998. Population genetic structure phylogeography and spawning phylopatry in Walleye (*Stizostedion vitreum*) from mitochondrial DNA control region sequences. Mol. Ecol. 7: 1757–1769.

Stephens, J.C. and M. Nei. 1985. Phylogenetic analysis of polymorphic DNA sequences at the Adh locus in *Drosophila melanogaster* and its sibling species. J. Mol. Evol. 22: 289–300.

Strunnikov, V.A. 1986. Genetic basis of heterosis and combinational ability in silkworm. Russian J. Genetics 22(5): 666–677.

Suzuki, H., M. Nunome, K. Moriwaki, H. Yonekawa, K. Tsuchiya and A.P. Kryukov. 2007. A thirty-years project of the genetic survey on the wild mice. Modern Achievements in Population, Evolutionary, and Ecological Genetics: International Symposium, Vladivostok–Vostok Marine Biological Station, September 9–14, 2007: Program & Abstracts.–Vladivostok,–44 p.–Engl: 38.

Suzuki, H, M.G. Filippucci, G.N. Chelomina, J. Sato, K. Serizawa and E. Nevo. 2008. A biogeographic view of *Apodemus* in Asia and Europe inferred from nuclear and mitochondrial gene sequences. Biochemical Genetics 46(5-6): 329–346.

Swofford, D.L. and R.B. Selander. 1981. Biosys–1: A Fortran program for the comprehensive analysis of electrophoretic data in population genetics and systematics. J. Hered. 72(4): 281–283.

Swofford, D.L., G.J. Olsen, P.J. Waddel and D.M. Hillis. 1996. Phylogenetic inference. pp. 407–514. *In*: D.M. Hillis, C. Moritz and B. Mable (eds.). Molecular Systematics. Sinauer. Assoc. Inc., Sunderland, Massachusetts.

Swofford, D.L. 2000. PAUP*: Phylogenetic Analysis Using Parsimony and Other Methods (Software), Sunderland: Sinauer Ass.

Sytnikov, A.V., Y.P. Kartavtsev and S.M. Nikiforov. 1997. Evaluation of the effective size in a population of shrimp *Pandalus kessleri* by temporal method and via direct counting of parents during the spawning season. Rus. J. Genetics 33(10): 1354–1361 (In Russian, Translated in English).

Szulkin, M., M. Bierne and P. David. 2010. Heterozygosity-fitness correlations: A time for reappraisal. Evolution 64: 1202–1217.

Tabachnik, W.J. and J.R. 1978. Powell Genetic structure of East African domestic population of *Aedes aegypti*. Nature 272: 535–537.

Tajima, F. 1983. Evolutionary relationships of DNA sequences in finite populations. Genetics 105: 437–460.

Takahata, N. and M. Slatkin. 1984. Mitochondrial gene flow. Proc. Nat. Acad. Sci. U.S.A. 81: 1764–1767.

Takehana, Y., N. Nagai, M. Matsuda, K. Tsuchiya and M. Sakaizumi. 2003. Geographic Variation and Diversity of the Cytochrome *b* Gene in Japanese Wild Populations of Medaka, *Oryzias latipes*. Zool. Sci. 20(10): 1279–1291.

Tamura, K., D. Peterson, N. Peterson, G. Stecher, M. Nei and S. Kumar. 2011. MEGA5: Molecular evolutionary genetics analysis using maximum likelihood, evolutionary distance, and maximum parsimony methods. Molecular Biology and Evolution 28: 2731–2739.

Tateno, Y., M. Nei and F. Tajima. 1982. Accuracy of estimated phylogenetic trees from molecular data. I. Distantly related species. J. Mol. Evol. 18: 387–404.

Templeton, A.R. 1981. Mechanisms of speciation—population genetic approach. Ann. Rev. Ecol. Syst. 12: 23–48.

Templeton, A.R. 1983. Phylogenetic inference from restriction endonuclease cleaverage site maps with particular reference to the evolution of humans and apes. Evolution 37: 221–244.

Templeton, A.R. 1998. Species and speciation: Geography, population structure, ecology, and gene trees. pp. 32–43. *In*: D.J. Howard and S.H. Berlocher (eds.). Endless Forms: Species and Speciation. Oxford University Press, N.Y.–Oxford.

Templeton, A.R. 2001. Using phylogeographic analyses of gene trees to test species status and processes. Molecular Ecology 10(3): 779–791.

Templeton, A.R. 2004. Statistical phylogeography: methods of evaluating and minimizing inference errors. Molecular Ecology 13(4): 789.

Tennessen, J.A. 2008. Positive selection drives a correlation between non-synonymous/synonymous divergence and functional divergence. Bioinformatics 24(12): 1421–1425.

Thompson, E.A. 1973. The Icelandic admixture problems. Ann. Hum. Genet. 37: 69–80.

Thompson, J.D., D.G. Higgins and T.J. Gibson. 1994. CLUSTAL W: improving the sensitivity of progressive multiple sequence alignment through sequence weighting, position-specific gap penalties and weight matrix choice. Nucleic Acids Res. 22: 4673–4680.

Thorpe, J.P. 1982. The molecular clock hypothesis: Biochemical evolution, genetic differentiation, and systematics. Ann. Rev. Ecol. Syst. 13: 139–168.

Timofeev-Resovsky, A.V., N.N. Vorontsov and A.V. Yablokov. 1977. A Short Essay on the Theory of Evolution. Nauka Publ., Moscow, 301 p.

Tinkle, D.W. 1965. Population structure and effective size of a lizard populations. Evolution 19: 569–573.

Tipping, M.E. Bayesian Inference: An Introduction to Principles and Practice in Machine Learning. Available from: http://www.miketipping.com/papers.htm.

Topal, M.D. and J.R. Fresco. 1976. Complementary base paring and the origin of substitution mutations. Nature 263: 285–289.

Turelli, M. 1984. Heretable genetic variation via mutation-selection balance: A population genetic perspective. Theor. Poul. Biol. 25: 138–193.

Turelli, M. and L. Ginzburg. 1983. Should individual fitness increase with heterozygosity? Genetics 104: 191–209.

Utter, F. and P. Aebersold. 1987. Population genetics and fishery management. pp. 21–45. *In*: N. Ryman and F. Utter (eds.). Population Genetics and Fishery Management. Washington University Press, Seattle, London.

Utter, F., P. Aebersold and G. Winans. 1987. Interpreting genetic variation detected by electrophoresis. pp. 21–46. *In*: N. Ryman and F. Utter (eds.). Population Genetics and Fishery Management. Washington University Press, Seattle–London.

Uzzell, T. and K.W. Corbin. 1971. Fitting discrete probability distribution to evolutionary events. Science 172: 1089–1096.

Van Valen, L. 1976. Ecological species, multispecies, and oaks. Taxon 25: 233–239.

Van Wagner, C.E. and A.J. Baker. 1990. Association between mitochondrial DNA and morphological evolution in Canada geese. J. Mol. Evol. 31: 373–382.

Vasiliev, V.P. 1985. Evolutionary Karyology of Fish. Nauka Publ., Moscow, 209 p.

Venter, J.C., M.D. Adams, E.W. Mayers, M.D. Adams, E.W. Myers, P.W. Li, R.J. Mural, G. Granger et al. 2001. The sequence of the human genome. Science 291: 1304–1351.

Vawter, A.T., R. Rosenblatt and G.C. Gorman. 1980. Genetic divergence among fishes of the Eastern Pacific and the Caribbean: Support for the molecular clock. Evolution 34: 705–711.

Vernon, E.H. 1957. Morphometric comparison of three races of kokanee (*Oncorhynchus nerka*) within a large British Columbia Lake. J. Fish. Res. Bd. Can. 14: 573–598.

Voelker, R.A., H.E. Schaffer and T. Makai. 1980. Spontaneous allozyme mutations in *Drosophila melanogaster*: Rate of occurrence and nature of mutants. Genetics 94: 961–968.

Vogel, F. 1972. Non-randomness of base replacement in point mutation. J. Mol. Evol. 1: 334–367.

Vogel, F. and A.G. Motulsky. 1986. Human Genetics, 2nd ed. Springer, Berlin, 807 p.

Verspoor, E., N.H.C. Fraser and A.F. Youngson. 1991. Protein polymorphism in Atlantic salmon within a Scottish river: evidence for selection and estimates of gene flow between tributaries. Aquaculture 98: 217–230.

Vorontsov, N.N. 1980. Synthetic theory of evolution, its sources, main postulates, and unsolved problems. D.I. Mendeleev All Union J. Chemical Soc. 25(3): 295–314.

Vorontsov, N.N. 1989. The problem of species and speciation. International Studies in the Philosophy of Science 3(2): 173–189.

Wagner, A. and N. Chaux. 2008. Distant horizontal gene transfer is rare for multiple families of procariotic insertion sequences. Molecular Genetics and Genomics 280: 397–408.

Wahlund, S. 1928. Zusamensetzung von populationen und correlation-sercheinungen vom standpunkt der vererbungslehre aus betrachtet. Hereditas 11: 65–106.

Walsh, M.M. and D.R. Lowe. 1985. Filamentous microfossils from the 3,500 MY old overwacht group, Barberton Mountain Land, South Africa. Nature 314: 530–532.

Waples, R.S. 1991. Heterozygosity and life-history variation in bony fishes: An alternative view. Evolution 45(5): 1275–1280.

Ward, R.D. 1990. Biochemical genetic variation in genus *Littorina* (Prosobranchia: Mollusca). Hydrobiologia 193: 53–69.

Ward, R.D., D.O.F. Skibinski and M. Woodwark. 1992. Protein heterozygosity, protein structure, and taxonomic differentiation. *In*: M.K. Hecht et al. (eds.). Evolutionary Biology. Plenum Press, N.Y. 26: 73–159.

Ward, R.D., T.S. Zemlak, B.H. Innes, P.A. Last and P.D.N. Hebert. 2005. DNA barcoding Australia fish species. Philos. Trans. R. Soc. Lond. B. Biol. Sci. 360: 1847–1857.

Wasinger, V.C., S.J. Cordwell, A. Cerpa-Poljak, J.X. Yan, A.A. Gooley, M.R. Wilkins, M.W. Duncan, R. Harris, K.W. Williams and I. Humphery-Smith. 1995. Progress with gene-product mapping of the Mollicutes: *Mycoplasma genitalium*. Electrophoresis 16: 1090–1094.

Watterson, G.A. 1975. On the number of segregating sites in genetical models without recombination. Theor. Popul. Biol. 7: 256–276.

Weir, B.S. 1979. Inferences about linkage disequilibrium. Biometrics 35: 235–254.

Weiss, G.H. and M. Kimura. 1965. A mathematical analysis of the stepping stone model of genetic correlation. J. Appl. Probab. 2: 129–149.

Whitlock, M. 1993. Lack of correlation between heterozygosity and fitness in forked fungus beetles. Heredity 70: 574–581.

Whitmore, D.H. 1983. Introgressive hybridization of smallmouth bass (*Micropterus dolomieui*) and Gaudelupe bass. Copeia: 672–679.

Whittaker, R.H. 1969. New concepts of kingdoms or organisms. Science 163: 150–160.

Wiens, J.J. and T.A. Penkrot. 2002. Delimiting species using DNA and morphological variation and discordant species limits in spiny lizards (Sceloporus). Systematic Biology 51(1): 69–91.

Wiley, E.O. 1981. Phylogenetics. The Theory and Practice of Phylogenetic Systematics. John Wiley and Sons, N.Y.

Williams, S.T., N. Knowlton, L.A. Weight and J.A. Jara. 2001. Evidence for three major clades within snapping shrimps genus Alpheus inferred from nuclear and mitochondrial sequence data. Mol. Phylogenet. Evol. 20(3): 375–389.

Wilson, A.C. 1976. Gene regulation in evolution. pp. 225–234. *In*: F.J. Ayala (ed.). Molecular Evolution. Sinauer Assoc., Inc., Sunderland, Massachusetts.

Wilson, A.C., S.S. Carlson and T.J. White. 1977. Biochemical evolution. Ann. Rev. Biochem. 46: 573–639.

Wilton, A.N., C.C. Laurie-Ahlberg, T.H. Emigh and J.W. Curtsinger. 1982. Naturally occurring enzyme activity variation in *Drosophila melanogaster*. II. Relationship among enzymes. Genetics 102: 207–221.

Wolfe, K. and D. Shields. 1997. Molecular evidence for an ancient duplication of the entire yeast genome. Nature 387: 708–713.

Woodruff, D.S. 1973. Natural hybridization and hybrid zones. Syst. Zool. 22: 213–218.

Workman, P.L. and J.D. Niswander. 1970. Population studies on the Southwestern indian tribes. II. Local genetic differentiation in the Papago. Amer. J. Hum. Genet. 1: 24–29.

Workman, P.L., H. Harpending, J.M. Lalouel, C. Lynch, J.D. Niswander and R. Singleton. 1973. Population studies on the Southwestern indian tribes. VI. Papago population structure: A comparison of genetic and migration analysis. pp. 166–194. *In*: N.E. Morton (ed.). Genetic Structure of Populations. University of Hawaii Press Haw. Univ. Press, Honolulu.

Wright, J.E., K. Johnson, A. Hollister and B. May. 1983. Meiotic models to explain classical linkage, pseudolinkage, and chromosome pairing in tetraploid derivative salmonid genomes. Isozymes: Curr. ToP. Biol. Med. Res. 10: 239–260.

Wright, S. 1921. Systems of mating. Genetics 6: 111–178.

Wright, S. 1931. Evolution in Mendelian populations. Genetics 16: 97–159.

Wright, S. 1935. The analysis of variance and correlations between relatives with respect to deviations from an optimum. J. Genet. 30: 243–256.

Wright, S. 1938. Size of populations and breeding structure in relation to evolution. Science 87: 430–431.

Wright, S. 1943. Isolation by distance. Genetics 28: 114–138.

Wright, S. 1951. The genetical structure of populations. Ann. Eugenics 15: 323–354.

Wright, S. 1965. The interpretation of population structure by F-statistics with special regard to systems of mating. Evolution 19: 395–420.

Wright, S. 1968. Evolution and the Genetics of Population. Vol. 1. Genetic and Biometric Foundations. Univ. Chicago Press, Chicago.

Wright, S. 1969. Evolution and the Genetics of Population. Vol. 2. The Theory of Gene Frequencies. Univ. Chicago Press, Chicago, 511 p.

Wright, S. 1978. Variability Within and Among Populations. Evolution and the Genetics of Populations. Vol. 4. University of Chicago Press, Chicago – London.

Wu, C.-I. and H. Hollocher. 1998. Subtle is nature: The genetics of species differentiation and speciation. pp. 339–351. *In*: D.J. Howard and S.H. Berlocher (eds.). Endless Forms: Species and Speciation. Oxford University Press, N.Y.–Oxford.

Wu, W., T.R. Schmidt, M. Goodman and L. Grossman. 2000. Molecular evolution of cytochrome c oxidase subunit 1 in primates: Is there coevolution between mitochondrial and nuclear genomes? Mol. Phylogenet. Evol. 17(2): 294–304.

Wyles, J.S., J.G. Kunkel and A.C. Wilson. 1983. Birds, behavior, and anatomical evolution. Proc. Nat. Acad. Sci. U.S.A. 80: 4394–4397.

Yablokov, A.V. 1987. Population biology. Highest Education Publ. (*Vistshaya shkola*), Moscow, 303 p.

Yager, L.N., J.F. Kaumeyer and E.S. Weinberg. 1984. Evolving sea urchin histone genes—nucleotide polymorphisms in the H4 gene and spacers of *Strongylocentritus purpuratus*. J. Mol. Evol. 20: 215–226.

Yang, S.Y. and J.L. Patton. 1981. Genetic variability and differentiation in the Galapagos finches. The Auk 98: 230–242.

Yanulov, K.P. 1962. On the gatherings of sea bream (*Sebastes mentella* Travin) in Labrador-Newfoundland area. Soviet Fishery Research in North-West Part of Atlantic Ocean. Moscow. Rus. J. Fishery (Rybnoe Hoziastvo) pp. 285–296.

Yonekawa, H., K. Moriwaki, O. Gotoh, J.-I. Hayashi, J. Watanabe, N. Miyashita, M.L. Petras and Y. Tagashira. 1981. Evolutionary relationships among five subspecies of *Mus musculus* based on restriction enzyme cleavage patterns of mitochondrial DNA. Genetics 98: 801–816.

Yonekawa, H., K. Tsuda, K. Tsuchia, L.V. Yakimenko, K.V. Korobitsyna, G.N. Chelomina, L.N. Spiridonova, L.V. Frisman, A.P. Kryukov and K. Moriwaki. 2000. Genetic diversity, geographic distribution and evolutionary relationships of *Mus musculus* subspecies based on polymorphism of mitochondrial DNA. pp. 90–108. *In*: A.P. Kryukov and L.V. Yakimenko (eds.). Problems of Evolution. Dalnauka, Vladivostok.

Yunis, J.J. and O. Prakash. 1982. The origin of man: A chromosomal pictorial legacy. Science 215: 1525–1530.

Zhimulev, A.V. 2002. General and Molecular Genetics. Novosibirsk Univ. Publ., Novosibirsk, 459 p.
Zhivotovsky, L.A. 1984. Integration of Polygenic Systems in Population. Nauka Publ., Moscow, 183 p.
Zhivotovsky, L.A. 1991. Population Biometry. Nauka Publ., Moscow, 271 p.
Zhivotovsky, L.A., K.I. Afanasiev and G.A. Rubtsova. 1987. Selective processes at allozyme loci in pink salmon *Oncorhynchus gorbuscha* (Walbaum). Rus J. Genetics (Moscow) 23(10): 1876–1883.
Zouros, E. 1979. Mutation rates, population sizes, and amount of electrophoretic variation of enzyme loci in natural populations. Genetics 92: 623–646.
Zouros, E. 1987. On the relation between heterozygosity and heterosis: An evaluation of the evidence from marine mollusks. Isozymes: Current Topics in Biol. Med. Res. 15: 255–270.
Zouros, E. and D.W. Folts. 1984. Minimal selection requirements for the correlation between heterozygosity and growth, and for the deficiency of heterozygotes, in oyster populations. Dev. Genet. 4: 393–405.
Zouros, E. and D.W. Folts. 1987. The use of allelic isozyme variation for the study of heterosis. *In*: M.C. Ruttazi, J.S. Scandalios and G.S. Whitt (eds.). Isozymes: Current Topics in Biol. Med. Res. Alan Liss, N.Y. 13: 1–59.
Zouros, E. and A.L. Mallet. 1989. Genetic explanations of the growth/heterozygosity correlation in marine mollusks. pp. 317–323. *In*: J.S. Ryland and P.A. Tyler (eds.). Reproduction Genetics and Distribution of Marine Organisms. Olsen & Olsen, Fredensborg.
Zouros, E., S.M. Singh and H.E. Miles. 1980. Growth rate in oysters: an overdominant phenotype and its possible explanations. Evolution 34(5): 856–867.
Zuckerkandl, E. and L. Pauling. 1962. Molecular disease, evolution, and genetic heterogeneity. pp. 189–225. *In*: M. Kasha and B. Pullman (eds.). Horizons in Biochemistry. Academic Press, N.Y.
Zuckerkandl, E. and L. Pauling. 1965. Evolutionary divergence and convergence in proteins. pp. 97–166. *In*: V. Bryson and H.J. Vogel (eds.). Evolving Genes and Proteins. Academic Press, N.Y.

SUBJECT INDEX

R

S

T